IFMBE Proceedings 89

The IFMBE Proceedings Book Series is an official publication of *the International Federation for Medical and Biological Engineering* (IFMBE). The series gathers the proceedings of various international conferences, which are either organized or endorsed by the Federation. Books published in this series report on cutting-edge findings and provide an informative survey on the most challenging topics and advances in the fields of medicine, biology, clinical engineering, and biophysics.

The series aims at disseminating high quality scientific information, encouraging both basic and applied research, and promoting world-wide collaboration between researchers and practitioners in the field of Medical and Biological Engineering.

Topics include, but are not limited to:

- Diagnostic Imaging, Image Processing, Biomedical Signal Processing
- Modeling and Simulation, Biomechanics
- Biomaterials, Cellular and Tissue Engineering
- Information and Communication in Medicine, Telemedicine and e-Health
- Instrumentation and Clinical Engineering
- Surgery, Minimal Invasive Interventions, Endoscopy and Image Guided Therapy
- Audiology, Ophthalmology, Emergency and Dental Medicine Applications
- Radiology, Radiation Oncology and Biological Effects of Radiation

IFMBE proceedings are indexed by SCOPUS, EI Compendex, Japanese Science and Technology Agency (JST), SCImago. They are also submitted for consideration by WoS.

Proposals can be submitted by contacting the Springer responsible editor shown on the series webpage (see "Contacts"), or by getting in touch with the series editor Ratko Magjarevic.

Yuri Dekhtyar · Inga Saknite
Editors

19th Nordic-Baltic Conference on Biomedical Engineering and Medical Physics

Proceedings of NBC 2023, June 12–14, 2023, Liepaja, Latvia

Springer

Editors
Yuri Dekhtyar
Riga Technical University
Riga, Latvia

Inga Saknite
University of Latvia
Riga, Latvia

ISSN 1680-0737 ISSN 1433-9277 (electronic)
IFMBE Proceedings
ISBN 978-3-031-37131-8 ISBN 978-3-031-37132-5 (eBook)
https://doi.org/10.1007/978-3-031-37132-5

This Springer imprint is published by the registered company Springer Nature Switzerland AG
The registered company address is: Gewerbestrasse 11, 6330 Cham, Switzerland

Foreword

Looking again into the entire future of biomedical engineering.

Biomedical engineering and medical physics are both highly research and innovation directed fields and require interdisciplinary and multidisciplinary approach in tackling the challenges in medicine and health care. In the last decades, research and innovation in medical technology were the number 1 in patent applications in the European Patent Office, and the expectation is continuation of the growth at the same rapid pace in the following years. Technological progress in medical technology is considered as one of the main factors in boosting EU economy.

However, when the World Health Organization (WHO) declared a Public Health Emergency of International Concern on 30 January 2020, and characterized the outbreak as a pandemic in March, the new situation caught the European healthcare system completely unprepared. Biomedical and clinical engineers were focused and intensively involved in finding solutions to stop and prevent the spread of the pandemic, from innovative designs to allocation of necessary resources to combat the disease. Based on the experience gained during the pandemics, at the Fourth meeting of the European Parliament Interest Group held on 21 March 2023, and supported by the European Alliance of Medical and Biological Engineering and Science (EAMBES) and the International Federation for Medical and Biological Engineering (IFMBE), issues of "Pandemic Management and Preparedness – Telemedicine and the Role of Innovative Technologies in Securing a Safer Future" were presented and plans for comprehensive solutions proposed.

During the pandemics, professionals had to learn how to extensively use communication technologies not only in health care, but also for mutual communication. Many editions of international events were designed in full online capacity or in a hybrid mode, and though such a *modus operandi* showed as useful, we are very happy to welcome all authors and participants of this 19th Nordic-Baltic Conference on Biomedical Engineering and Medical Physics organized in Liepaja, Latvia, from 12 to 14 June 2023, again, mainly as a face-to-face scientific and networking event co-sponsored by the International Federation for Medical and Biological Engineering.

Biomedical engineers always have and will continue to play a leading role in research, development, and innovation of new health technologies and in improving the quality of life of all people in this only world we have.

On behalf of the IFMBE, with best wishes for a successful meeting.

Ratko Magjarević
President, IFMBE

Preface

The 19th Nordic-Baltic Conference on Biomedical Engineering and Medical Physics (NBC 2023) is the third time that an IFMBE event comes to Latvia. The 14th NBC conference was the first one held in Latvia back in 2008. In 2013, the IFBME International Symposium on Biomedical Engineering and Medical Physics was dedicated to the 150th anniversary of Riga Technical University. Both events were successful in presenting achievements and new knowledge gained in a wide range of technologies for applications in medicine.

The NBC 2023 was organized in Liepaja, the largest city in the Kurzeme region and the third-largest city in Latvia. It has grown from a provincial village to a regional industrial hub with deep-rooted historical and cultural traditions. Liepaja is a symbol of the rapid scientific developments in the region. In the last decade, the annual number of scientific publications by researchers based in Liepaja has grown four times, compared to the growth in the fields of biomedical engineering and medical physics worldwide.

The NBC 2023 brought together a diverse group of attendees including researchers of various career stages, industry partners, medical professionals, and biologists. The conference covered a wide range of topics: biomechanics, rehabilitation, materials, micro-, nano-, and molecular technologies, personalized medicine, diagnostics, therapy, medical physics, education, training, safety, and quality.

In addition to scientific content, NBC 2023 attendees participated in workshops to advance their professional skills including on how to excel in writing scientific publications. A special attention was given to student participants. As a satellite event of NBC 2023, the 6th Summer School Nonlinear Life brought together and promoted scientific discussion amongst students and recent graduates. Importantly, a panel discussion was dedicated to the role of women in biomedical sciences and engineering.

We thank all NBC 2023 sponsors who helped promote this event, as well as all attendees for their scientific contributions. We believe that the ground-breaking, multidisciplinary publications collected in this special issue further advance the fields of biomedical engineering and medical physics.

Yuri Dekhtyar
Inga Saknite

Organization

The Organizational Committee

Martins Piksis (Chairman)	Latvian Medical Engineering and Physics Society
Katrina Caikovska	Latvian Medical Engineering and Physics Society
Yuri Dekhtyar	Latvian Medical Engineering and Physics Society
Arturs Grigorjevs	Latvian Medical Engineering and Physics Society
Aleksejs Katasevs	Latvian Medical Engineering and Physics Society
Maksims Polakovs	Latvian Medical Engineering and Physics Society
Vineta Vanaga	Latvian Medical Engineering and Physics Society
Eriks Badamsins	Riga Technical University
Talis Juhna	Riga Technical University
Inga Saknite	University of Latvia
Janis Spigulis	University of Latvia
Gundega Tomele	University of Liepāja

The Scientific Committee

Adliene Diana, Lithuania
Ajcevic Milos, Italy
Aldskogius Hakan, Sweden
Antoniac Iulian, Romania
Aune Ragnhild Elizabeth, Norway
Balazsi Csaba, Hungary
Byrne Hugh J., Ireland
Cabrini Marina, Italy
Caruna Carmel, Malta
Chougule Arun, India
Couteau Christophe, France
Damilakis John, Greece
De Paolis Lucio Tommaso, Italy
Dekhtyar Yuri (Chairman), Latvia
D'Elios Mario Milco, Italy
Dimitrova Todorka, Bulgaria
Erdem Arzum, Turkey
Fridolin Ivo, Estonia

Gonzalez Joaquin Roca, Spain
Griškevičius Julius, Lithuania
Gross Kārlis-Agris, Latvia
Helgason Thordur, Iceland
Hofmann Ulrich, Germany
Jamsa Timo, Finland
Jobbagy Akos, Hungary
Katashev Alexei, Latvia
Krumina Gunta, Latvia
Koutinas Michalis, Cyprus
Kozlova Elena, Sweden
Kuang Yu, USA
Kubisz Leszek, Poland
Ładyżyński Piotr, Poland
Ločs Janis, Latvia
Magjarevic Ratko, Croatia
Malmivuo Jaakko, Finland
Marozas Vaidotas, Lithuania
Martin Martinez Jose Miguel, Spain
Meigas Kalju, Estonia
Mihailescu Ion N., Romania
Milano Franco, Italy
Moshi Geso, Australia
Nazarenko Sergei, Estonia
Pilt Kristjan, Estonia
Popa Catalin, Romania
Proskurins Jevgenijs, Latvia
Rotomskis Richardas, Lithuania
Rozenfeld Anatoly B, Australia
Saknite Inga (Secretary), Latvia
Spigulis Janis, Latvia
Stan Sergiu-Dan, Romania
Tabakov Slavik, UK
Tamulevičius Tomas, Lithuania
Tamulevičius Sigitas, Lithuania
Trajanovic Miroslav, Serbia
Vaseashta Ashok, USA

IFMBE

LMIFB

LIEPĀJAS UNIVERSITĀTE

1862
RIGA TECHNICAL UNIVERSITY

UNIVERSITY OF LATVIA

European Alliance for EAMBES Medical and Biological Engineering & Science

EUT+ EUROPEAN UNIVERSITY OF TECHNOLOGY

Contents

The Impact of Engineering Enabling Technologies on the Further Development of Personalized Orthopedics

Miroslav Trajanovic(✉), Nikola Vitkovic, Nikola Korunovic, Dragan Misic, and Jovan Arandjelovic

Faculty of Mechanical Engineering, University of Nis, Aleksandra Medvedeva 14, 18000 Nis, Serbia

miroslav.trajanovic@gmail.com

Abstract. Personalized orthopedics became possible in pre-clinical and clinical practice primarily through the application of engineering solutions in the prevention, diagnosis, and treatment of orthopedic patients. In order to predict the further development of personalized orthopedics, it is necessary to look at which enabling engineering technologies can be applied and where their impact is expected. This paper identifies and analyzes those technologies.

Keywords: Personalized Orthopedics · Enabling Technologies · Engineering

1 Introduction

It has been known for a long time that a personalized approach in medical practice provides the best benefits to patients, but this was not possible due to the lack of sufficiently detailed data on the patient's condition and characteristics, equipment, and techniques for personalized approach. However, the development of many engineering technologies in the last fifty years has resulted in innovative solutions that have also been applied in orthopedics. Thus, step by step, the conditions for a personalized approach in orthopedic practice were created. This tendency has not stopped but is developing faster and faster.

The aim of this study is to identify, based on bibliometric analysis, the main engineering enabling technologies that contribute to the development of personalized orthopedics, to point out their importance, and to briefly present those that have the greatest impact.

2 Bibliometric Analysis

2.1 The Importance of Engineering Technologies for the Development of Orthopedics

There are indications that engineering technologies have significantly influenced the development of orthopedics in the last fifty years. Also, it is considered that they especially enabled a personalized approach in orthopedics [1]. However, it is not clear how much impact engineering technologies have had on orthopedics and which engineering disciplines have had the greatest impact on the development of orthopedics.

Y. Dekhtyar and I. Saknite (Eds.): NBC 2023, IFMBE Proceedings 89, pp. 1–8, 2023.
https://doi.org/10.1007/978-3-031-37132-5_1

In order to get an answer to this question, a simple bibliographic research was done using Core Collection database of the Clarivate Analytics Web of Science (WoS) platform [2]. Bearing in mind that various terms related to orthopedics are used in the scientific literature, and with the intention of being comprehensive, the following search query was used: Orthopedics OR Orthopedic OR Orthopaedics OR Orthopaedic OR Bones. The "All fields" option was used for the search, which includes title, abstract, author keywords, Keywords Plus (terms automatically generated from the titles of cited articles) and any other metadata available for that article. The search result showed that there are 1,035,427 publications in the Core Collection that contain at least one of the search terms. The search covered the period from 1996 to 2023.

Search terms were found in a total of 254 Web of Science categories. Table 1 shows the first 10 categories sorted in descending order by the number of articles in which the search terms were found. Other categories are not shown to save space. It is obvious that the largest number, a total of 159,817 articles, is in the Orthopedics category, which is 15.435% of total 1,035,427 publication.

Table 1. Top ten Web of Science categories with the largest number of articles that contain the searched terms.

Web of Science Categories	Number of articles	% of 1,035,427
Orthopedics	159817	15.435
Surgery	102888	9.937
Endocrinology Metabolism	77551	7.490
Oncology	74956	7.239
Hematology	69577	6.720
Dentistry Oral Surgery Medicine	57025	5.507
Engineering Biomedical	50034	4.832
Cell Biology	48409	4.675
Medicine Research Experimental	45249	4.370
Immunology	43951	4.245

Among the 254 Web of Science categories in which the searched terms were found, there are also 36 engineering categories. In the Table 2 only first 15 engineering categories are sorted and presented in descending order by the number of articles in which the search terms were found.

The total number of articles from engineering categories is 271,720 or 17.191%. Of course, some of these articles are not necessarily related exclusively to orthopedics, but also to other disciplines, such as surgery, but obviously the influence of engineering disciplines on medicine, and therefore on orthopedics, is huge.

Table 2. Top fifteen engineering Web of Science categories with the largest number of articles that contain the searched terms.

Web of Science Categories	Number of articles	% of 1,035,427
Engineering Biomedical	50034	4.832
Radiology Nuclear Medicine Medical Imaging	40881	3.948
Materials Science Biomaterials	34123	3.296
Materials Science Multidisciplinary	20622	1.992
Cell Tissue Engineering	19292	1.863
Chemistry Multidisciplinary	12953	1.251
Physics Applied	9868	0.953
Nanoscience Nanotechnology	7543	0.728
Chemistry Physical	6857	0.662
Polymer Science	5903	0.57
Engineering Electrical Electronic	5095	0.492
Materials Science Ceramics	4854	0.469
Metallurgy Metallurgical Engineering	4587	0.443
Physics Condensed Matter	4225	0.408
Engineering Mechanical	3013	0.291

2.2 The Most Influential Enabling Engineering Technologies in the Field of Orthopedics

In order to recognize the most important engineering technologies that contribute to the development of personalized orthopedics, another bibliometric analysis was performed on the WoS platform. This time the following query was used (Orthopedics OR Orthopedic OR Orthopedics OR Orthopedic OR Bones) AND Personalized (Topic). The "Topic" option was used for the search, which includes title, abstract, author keywords, and Keywords Plus.

As a result of the query, 2855 papers containing the requested terms were found. With further refinement, only papers from 2021 to 2023 were selected, and all papers that do not belong to pure orthopedic disciplines were rejected. Thus, the scope of research was reduced to 983 published papers. A statistical analysis of keywords defined by the author was performed for the selected papers. The results of this analysis are presented in Table 3.

Although the analysis done is indicative, it is not perfect. The problem lies in the fact that the granularity of the keywords that the authors mention is different, and different keywords are assigned to the same things. For example, the main keyword in works dealing with the application of titanium alloys for the manufacture of implants is found in the forms: biomaterials, titanium alloy, Ti 6Al 4V, or Ti6Al4V ELI (Grade 23). The same is valid for the case of 3D printing and Additive manufacturing.

Table 3. Frequency of top ten keywords related to engineering enabling technologies in orthopedics.

Engineering technology	Frequency
Additive manufacturing	26
3D Printing	15
3D bioprinting	8
Finite element analysis	8
Computed tomography	6
Machine learning	6
Computer aided design	6
Tissue engineering	5
Artificial intelligence	7
Biomaterials	7

3 Enabling Engineering Technologies in Orthopedics

3.1 Additive Manufacturing

Additive manufacturing technologies, often referred to as 3D printing, has one of the biggest impacts on the development of personalized orthopedics. The possibility of these technologies to produce efficiently objects of very complex shapes has enabled the production of personalized implants, scaffolds, prostheses, orthoses [3] and surgical tools [4]. Also, these technologies can be used to create models of a patient's bones or joints for planning surgical procedures and thus optimize surgical outcomes (Fig. 1).

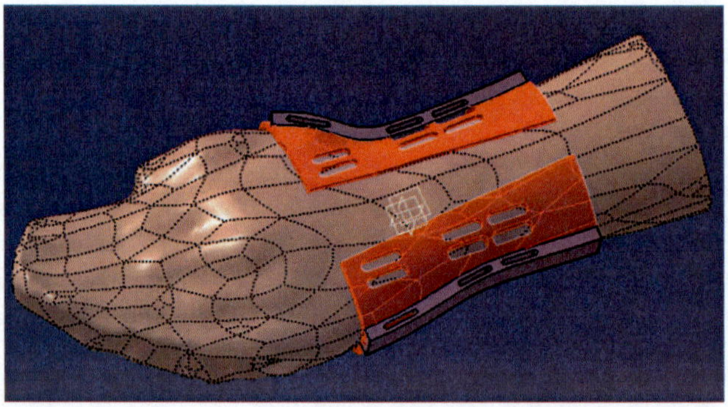

Fig. 1. Personalized orthosis made of Ninja Flex with FDM additive manufacturing technology [3].

Additive manufacturing is a mature technology that, together with new materials, can not only achieve any shape, but also enables the creation of porous structures that, in the case of scaffolds, allow better innervation and a larger surface area for cells to adhere and proliferate, which promotes new tissue formation [5]. To further promote the growth of bone tissue within porous implants and scaffolds, surface coatings are used.

One of the new additive manufacturing technologies that is being worked on in many laboratories is bioprinting [6]. It uses 3D printing techniques to create living tissue constructs by layering and depositing cells, growth factors, and biomaterials in a controlled and precise manner. Most of bioprinters are using extrusion as a method for building layers, but other methods, such as stereolithography, inkjet or laser-asisted method. Most of the research involving bioprinters is directed towards the production of bone tissue and cartilage.

Another promising technology is 4D printing, which is additive manufacturing with the added capability of creating objects that can change their shape or properties over time, in response to external stimuli such as temperature, moisture, pressure, or light. This is achieved by using advanced materials, such as shape-memory alloys, smart polymers, or hydrogels, that can be programmed to respond to specific environmental conditions. The use of 4D printing to create implants with predetermined shape and size allows for a precise fit at defect sites. During the post-printing stage, the implant may undergoes functional transformation and mimics biological features, which promotes tissue remodeling and maturation [7].

Additive manufacturing is particularly important for the further development of regenerative medicine. Stem cell therapy and platelet-rich plasma therapy may become more widespread in orthopedics. This could allow for the repair of damaged bone tissue and the regeneration of lost cartilage.

3.2 Biomaterials

Biomaterials are biocompatible materials that are used in medical devices, implants, and tissue engineering to replace or repair damaged or diseased tissues. The most commonly used metal materials are stainless steel (316 L), titanium alloy (Ti6Al4V), and cobalt-chromium alloys (F75 CoCr). They are mainly used for the production of implants, fixation devices, scaffolds and auxiliary instruments. These materials are strong, durable, and resistant to corrosion, making them ideal for use in load-bearing applications such as hip and knee replacements [8].

Ceramics such as alumina (Al2O3), zirconia (ZrO2), and calcium phosphate are also used in orthopedic implants. These materials are biocompatible, meaning they are well-tolerated by the body, and can be shaped and sized to fit individual patients.

Polymers such as polyethylene, polyurethane, Teflon and silicone are commonly used in orthopedic implants, particularly in joint replacements. These materials are lightweight, flexible, and can be easily molded to fit a variety of shapes and sizes.

The emergence of new advanced materials has enabled the production of scaffolds with improved properties. Currently, work is being done on the production and improvement of composite scaffolds that can be metal, ceramic or polymer based. Additive manufacturing is mainly used for the production of composite scaffolds [9].

While autografts are currently considered the best option for orthopedic tissue reconstruction, there is a limit to the amount of tissue that can be harvested without causing harm to the donor site. Tissue engineering approaches, which involve using decellularized bone, cartilage, skeletal muscle, tendon, and ligament from allogeneic or xenogeneic sources, have emerged as a promising alternative treatment option. The extracellular matrix, which acts as a natural scaffold for cell attachment, proliferation, and differentiation, plays a vital role in these approaches. Decellularization of in vitro cell-derived matrices can also enable the generation of autologous constructs using tissue-specific or progenitor cells. While decellularized bone tissue is widely used in orthopedic applications, the potential of using decellularized cartilage, skeletal muscle, tendon, and ligament cell-derived matrices is only now beginning to be explored for potential clinical translation in the orthopedic setting.

3.3 Computer Aided Design

Computer aided design (CAD) has many different applications in personalized orthopedics, so it has significantly contributed to its development. The most common application of CAD in orthopedics is for the design of personalized implants, prosthetics, scaffolds, and surgical tools. In order to achieve anatomical personalization, the patient's CT image is usually the starting point.

Second important application of CAD is in surgical planning. Planning surgeries in advance, allows surgeons to better visualize the procedure and optimize the approach. This can lead to more accurate and efficient surgeries with reduced risk of complications. Similar approach can be used for a rehabilitation planning.

3D geometrical models of bones, implants and other orthopedic devices are necessary input for other engineering technologies such as finite element method analysis, additive manufacturing, computer aided manufacturing, and artificial reality technologies, which include virtual reality (VR), augmented reality (AR), and mixed reality (MR).

In orthopedic practice, it is often the case that the orthopedist does not have a CT scan of the entire bone at his disposal, so planning the operation is difficult. The reason for this can be a major trauma, bone cancer or osteoporosis [12]. In such cases, reverse engineering technology is applied, the result of which is a 3D geometric model of the original bone. The method is based on artificial intelligence and a large number of samples of recordings of healthy bones of the same type. The accuracy of this method is very good and meets the requirements for the production of personalized implants, prostheses and scaffolds.

3.4 Finite Element Method

The finite element method (FEM) has numerous applications in orthopedics. One of its main uses is in the prediction of behavior of orthopedic implants, scaffolds and prosthetics, such as joint replacements, spinal implants, and bone plates. FEM can simulate the biomechanical behavior of the implant within the body, predicting how it will perform over time and allowing for optimization of its design. Also, FEM is used for multiscale computational modeling of bone tissue, which contributes not only to a better understanding of bone behavior, but also represents the basis for creating quality models for simulating the interaction between bone and implants [10].

FEM makes a great contribution to personalized orthopedics in the optimization of the mechanical characteristics and shape of scaffolds and implants. This is achieved through a design study process during which, by changing the parameters and material of the implant or scaffold and successive FEM analyses, the optimal construction solution is reached [11] (Fig. 2).

Fig. 2. Three instances of bone segment-scaffold assembly, in which struts angle was modified to take representative values of 32°, 52° and 72° respectively [11].

4 Conclusion

The strong progress of personalized orthopedics in the last thirty years has been achieved primarily through the application of various engineering technologies. Additive manufacturing, biomaterials, finite element method, medical imaging, virtual reality, robotics and artificial intelligence had the greatest impact.

All listed engineering technologies are in constant development. Thanks to them, an even faster development of personalized orthopedics is expected. We can expect growing use of additive manufacturing, not only for production of orthopedic devices but also in regenerative medicine. Regenerative medicine techniques, such as tissue engineering and stem cell therapies, may play a larger role in orthopedics in the future. These techniques may enable the regeneration of damaged or diseased tissues, reducing the need for traditional implants and devices. All this will be supported by the use of existing and further development of new biomaterials.

Also, we can expect increased use of artificial intelligence particularly in the areas of diagnosis and treatment planning. AI algorithms may be able to analyze patient data and medical images to identify patterns and predict outcomes, enabling more precise and individualized care.

As personalized medicine continues to expand and patient involvement increases, patients are anticipated to have a greater impact on their own healthcare. Wearable technologies like smartwatches, smartphones, smart clothing, smart glasses, and other sensors are expected to play a significant role in this development.

Acknowledgment. This research was financially supported by the ERASMUS+ project "Collaborative e-platform for innovation and educational enhancement in medical engineering" – CALLME. Project Reference: 2022-1-RO01-KA220-HED-000087703.

References

1. Canciglieri, O.J., Trajanovic, M.F. (eds.): Personalized Orthopedics - Contributions and Applications of Biomedical Engineering. Springer Nature, Berlin (2002)
2. Web of Science. https://www.webofscience.com/wos/woscc/basic-search. Accessed 6 Apr 2023
3. Aranđelović, J., Korunović, N., Stamenković, B., Arsić, M., Trajanović, M.: Design methodology of a personalised wrist orthosis for fractures and rehabilitation. In: Zdravković, M., Trajanović, M., Konjović, Z. (Eds.) ICIST 2021 Proceedings, pp. 154–157 (2021)
4. Wixted, C. M., Peterson, J. R., Kadakia, R. J., Adams, S. B.: Three-dimensional printing in orthopaedic surgery: current applications and future developments. JAAOS: Glob. Res. Rev. **5**(4), e20.00230–11 (2021)
5. Vu, A.A., Burke, D., Bandyopadhyay, A., Bose, S.: Effects of surface area and topography on 3D printed tricalcium phosphate scaffolds for bone grafting applications. Addit. Manuf. **39**, 101870 (2021)
6. Santoni, S., Gugliandolo, S.G., Sponchioni, M., Moscatelli, D., Colosimo, B.M.: 3D bioprinting: current status and trends—a guide to the literature and industrial practice. Bio-Des. Manuf. **5**(1), 14–42 (2021). https://doi.org/10.1007/s42242-021-00165-0
7. Wan, Z., Zhang, P., Liu, Y., Lv, L., Zhou, Y.: Four-dimensional bioprinting: current developments and applications in bone tissue engineering. Acta Biomater. **101**, 26–42 (2020)
8. Szczęsny, G., Kopec, M., Politis, D.J., Kowalewski, Z.L., Łazarski, A., Szolc, T.: A review on biomaterials for orthopaedic surgery and traumatology: from past to present. Materials **15**, 3622 (2022)
9. Chen Y., Li W., Zhang C., Wu Z., Liu J.: Recent developments of biomaterials for additive manufacturing of bone scaffolds. Adv. Healthcare Mater. **9**(23), 2000724 (2020)
10. Podshivalov, L., Fischer, A., Bar-Yoseph, P.Z.: On the road to personalized medicine: multiscale computational modeling of bone tissue. Arch. Comput. Methods Eng. **21**(4), 399–479 (2014). https://doi.org/10.1007/s11831-014-9120-1
11. Stojkovic, M., Korunovic, N., Trajanovic, M., Milovanovic, J., Trifunovic, M., Vitkovic, M.: Design study of anatomically shaped lattice scaffolds for the bone tissue recovery. In: Papadrakakis, M., Kojic, M., Tuncer I. (eds.) SEECCM III, 3rd South-East European Conference on Computational Mechanics-an ECCOMAS and IACM Special Interest Conference, pp 381–393. NTUA, Athens (2013)
12. Vitkovic, N., Radovic, L., Trajanovic, M., Manic, M.: 3D point cloud model of human bio form created by the application of geometric morphometrics and method of anatomical features. In: Human Tibia Example, FILOMAT, vol. 33, no. 4, pp. 1217–1225 (2019)

The Development of Objective and Quantitative Eye-Tracking-Based Method for the Diagnostics of Oculomotor Dysfunctions

Gunta Krumina(✉), Ilze Ceple, Viktorija Goliskina, Evita Kassaliete, Tomass Ruza, Evita Serpa, Aiga Svede, and Liva Volberga

Faculty of Physics, Mathematics and Optometry, Department of Optometry and Vision Science, University of Latvia, Jelgavas str. 1, 1004 Riga, Latvia
Gunta.Krumina@lu.lv

Abstract. Vision specialists, neurologists, and psychologists commonly rely on manual techniques to evaluate oculomotor system performance: (1) direct observation tests such as the NSUCO oculomotor test and (2) visual-verbal formats like Pierce, King-Devick, and DEM tests. Unfortunately, these tests mainly cover a rough and subjective evaluation of oculomotor performance and are limited by observer's experience or measurement obtaining. Our study aims to address these limitations by developing objective and quantitative approach to assess eye movement performance in children and adults using eye-tracking technology. We have developed a method for detailed eye movement analysis in clinical practice, which evaluates (1) saccades (reflexive, voluntary, and anti-saccades), (2) smooth pursuit eye movements (horizontal, vertical, and circular), (3) fixation stability, and (4) reading eye movement parameters, including saccades, regressions, number of fixations, fixation duration, and reading speed. Our method was tested on 378 children aged 6–13 years using *Tobii Pro Fusion* eye-tracker. By applying our newly developed quantitative methodology, we were able to identify the benefits and limitations of our approach. We believe our method will be particularly useful for vision specialists, and we aim to continue refining and improving it in future studies.

Keywords: Children · Eye Movement Recording · Fixations · Reading Eye Movements · Saccades · Smooth Pursuit

1 Introduction

Vision is a complex process that involves a precise cooperation between the eyes perceiving the visual information and central nervous system analysing the perceived information. Eye movements are the fastest and most frequent motions made by the human body, and their control system, known as oculomotor system [1], involves complex and sophisticated neural networks. Eye movements are essential in many aspects of everyday life, including reading, academic achievement, sports, and many other activities.

Y. Dekhtyar and I. Saknite (Eds.): NBC 2023, IFMBE Proceedings 89, pp. 9–17, 2023.
https://doi.org/10.1007/978-3-031-37132-5_2

In reading, efficient eye movement performance is crucial for proper tracking of the text and comprehension of the material. The ability to make quick and accurate eye movements, such as saccades and fixations, is essential for fluent reading [1]. Children who struggle with oculomotor development may experience difficulties with reading and comprehension, which can impact their academic achievement [2]. In addition to reading, eye movements are also important in sports, particularly those that require tracking of moving objects such as baseball, tennis, or soccer. Accurate and efficient eye movements are necessary for tracking the ball or opponent, predicting their movements, and making quick decisions [3]. Eye movements also play a role in other activities such as driving, navigating through crowded environments, and even social interactions. A precise eye movement performance allows individuals to scan their surroundings, identify important information, and respond appropriately. Overall, the importance of eye movements in everyday life cannot be overstated. Efficient eye movements are essential for a wide range of activities and can impact academic, athletic, and social success [4].

There are several methods applied in clinical practice to assess oculomotor performance. The NSUCO test is a widely used clinical tool that assesses oculomotor system's ability, accuracy, head movement, and body movement, including tasks for scoring saccades and pursuits. It is used to diagnose a range of conditions related to oculomotor dysfunction [5]. The visual-verbal Pierce test assesses the time to execute a task by involving saccades and is used to detect specific reading disorders that may be related to oculomotor system [6]. The King-Devick test is similar to the Pierce test, a rapid number naming test that assesses saccadic eye movements and is commonly applied in screening for concussion in sports medicine [7]. However, both of these tests have demonstrated a poor reliability [8]. The DEM test is design to control for rapid automaticity naming (RAN) by incorporating a vertical subtest that differentiates between poor saccadic performance and poor RAN ability. It can help in identifying developmental delays and disabilities that may affect academic performance [9]. The authors [10, 11] concluded that clinicians should be cautious about using the DEM test in isolation as a method of diagnostics or monitoring the effectiveness of saccadic dysfunction treatment. Although each of these methods offers a distinct approach to evaluating oculomotor performance and can provide valuable insights for diagnosing and treating conditions related to eye movement dysfunction, they all rely heavily on subjective measurements and the examiner's experience.

There are several new technologies being used to examine and assess eye movements. Recent advances in eye-tracking technology, using video-oculographic (VOG) recordings with high-resolution cameras, allow for precise measurement and quantification of various eye movement parameters. This information can be used to diagnose different eye movement disorders, assess reading difficulties, and evaluate the effectiveness of treatments [12]. Virtual reality is also being used to create realistic environments that can test eye movements in a more naturalistic way. This technology is particularly useful for assessing eye movements during activities like driving, sports, and other real-world situations [13]. Electrooculography (EOG) measures the electrical signals generated by the eye muscles during eye movements and can also be applied to diagnose eye movement disorders and evaluate treatment effectiveness [14]. Functional magnetic resonance imaging (fMRI) is a brain imaging technique that can be used to study brain activity during eye movement tasks. This technology can provide insights into the neural mechanisms underlying eye movements and can be used to diagnose and monitor eye movement disorders [15]. Overall, these technologies provide clinicians with more detailed and accurate information about eye movements, enabling better diagnosis and treatment of eye movement disorders.

However, each of these methods evaluates a specific stage of eye movement programming and is not always practical in clinical settings due to factors such as reliance on the practitioner's experience, high costs, and inconvenience. To address these limitations, our study drew upon the practical experience of optometrists who regularly conduct vision assessments and encounter eye movement-related problems. Specifically, we aimed to develop an objective and quantitative approach for assessing eye movement performance in both children and adults using eye-tracking technology.

2 Developed Method

During the initial phases of our methodological development for assessing eye movements, our primary focus was on examining oculomotor dysfunctions in children. Our aim was to incorporate our previously established screening method for visual function and perception in school-aged children [16, 17] with the next major hurdle in our developmental phase. We guided the development of our visual stimuli and methodology using extant scientific literature on eye movement assessment tools, methods, norms, and established correlations between oculomotor dysfunctions and reading dysfunctions. Our methodology comprises four distinct components: the assessment of (1) saccades, (2) smooth pursuit, (3) fixation stability, and (4) reading eye movements. The data was gathered in a school setting. All children were positioned at a distance of 65 cm and their chin and forehead were fixed, except during the DEM and reading tasks, where only the forehead was stabilized due to the need for the child to read aloud. The chin rest was not utilized during these tasks to prevent potential head movements. The screen resolution was 1920 × 1020 (Dell P2419, 23.8″). Before conducting the saccades, smooth pursuit, and fixation stability tasks, we calibrated the eye movements. We used *Tobii Pro Fusion* eye-tracker to record eye movements, applying two recording frequencies: 120 Hz for saccades and smooth pursuit tasks, and 250 Hz for fixation stability and reading tasks.

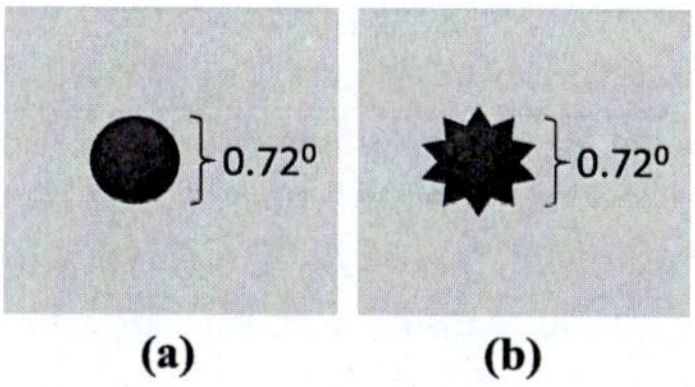

Fig. 1. The stimuli for saccades tasks: **(a)** stimuli for reflexive and voluntary saccades measurements and **(b)** stimuli for anti-saccades measurements.

Stimuli

Saccades were measured in three different tasks: reflexive, voluntary, and anti-saccade performance. In all cases, the stimulus size was 0.7° (see Fig. 1). To avoid confusion in the task instructions, a star was chosen as the visual stimulus in the anti-saccade task (see Fig. 1b). The stimuli for the saccade tasks were presented 14.0° horizontally from the centre to the right or left direction, and 11.31° vertically from the centre to the up and down directions. The background colour was grey (RGB: 180; 180; 180). *Reflexive saccades* were measured by randomly displaying a dot in one of the four directions (right, left, up, or down) for 5 s. Before each stimulus presentation, a fixation cross appeared at the centre of the screen. The subject's task was to fixate on the cross and then quickly and accurately fixate on the stimulus dot. The evaluated parameters included latency, velocity, accuracy, and amplitude. *Voluntary saccades* were measured using the same dot stimulus. In the first set, two dots were presented on the screen in a horizontal position, and in the second set, two dots were presented in a vertical position. The subject was instructed to perform jumps (saccades) back and forth between the two dots as fast and precise as possible for 15 s. The evaluated parameters included accuracy, velocity, and the number of cycles completed (i.e., how many times the subject jumped back and forth between the two dots). *Anti-saccades* were measured by presenting a fixation cross at the centre of the screen before each stimulus presentation. The subject was instructed to change their viewing direction as quickly and accurately as possible along with the onset of the stimulus, in the opposite direction to the direction of the stimulus demonstration. For example, if the stimulus was demonstrated to the right, the subject had to change their gaze at the same amplitude but towards the left. The stimuli were presented randomly, covering all four directions. The evaluated parameters in the anti-saccade task included the pro-saccades error (i.e., the percentage of times the subject moved their eyes in the wrong direction), anti-saccades value (i.e., the percentage of times the subject correctly moved their eyes in the opposite direction of the stimulus demonstration), latency, velocity, and accuracy (i.e., hypo/hyper performance).

Smooth pursuit eye movements were assessed in three different ways. In all tasks, the size of the dot stimulus was similar to that used in the saccade tasks, with a size of 0.7°, and the location of the dot was 10.8° from the computer screen (see Fig. 2). The speed of the stimulus was constant in all tasks at 6°/sec. The first task involved horizontal smooth pursuit, where the stimulus appeared on the right side of the screen after the fixation cross disappeared. The subject was instructed to fixate on the cross initially and then track the moving dot as precisely as possible. To avoid predictable initiation of smooth pursuit eye movements, the initial fixation was demonstrated a time interval. The dot performed three sets of back-and-forth motion from one side of the screen to the other. In the vertical smooth pursuit task, the same stimuli, instructions, and presentation of stimuli were applied, except for the direction, which was from bottom to top (back and forth). In the circular smooth pursuit task, the same stimuli, instructions, and presentation of stimuli were used, with the first three circular movements performed in an anti-clockwise direction and the next three in a clockwise direction. The evaluated parameters of smooth pursuit tasks included latency, accuracy, movement symmetry, and the number of catch-up saccades if there was no smooth pursuit movement.

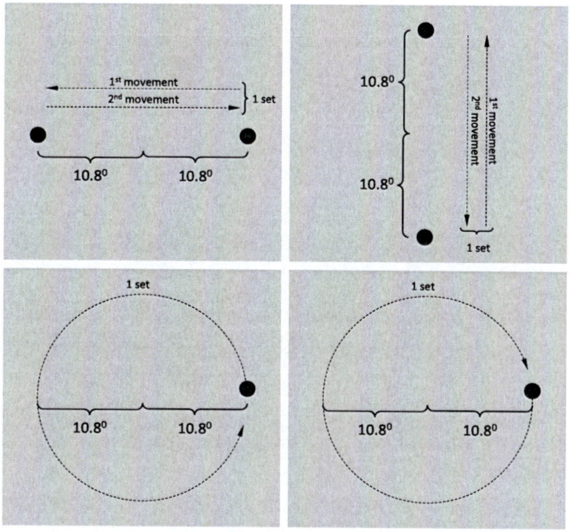

Fig. 2. The location and movement of stimulus in the measurements of horizontal (top left figure), vertical (top right figure), and circular (bottom figures) smooth pursuit

(a) **(b)**

Fig. 3. The stimuli for measurements of fixation stability: **(a)** cross with dot and **(b)** a bee.

Fixation Stability. We employed two types of stimuli (see Fig. 3), as determined from a comprehensive literature review. In order to obtain a more stable fixation, we applied the cross with and incorporated dot stimulus [18] at a size of 0.6°. To enhance engagement of younger participants, we also utilized an attention-grabbing stimulus in the form of a bee (sized at 1.76° both horizontally and vertically). The fixation point was initially displayed at the screen centre for a duration of 3 s, following which the stimulus – either a cross with a dot or the bee. The fixation stability stimulus was then presented for a period of 10 s, during which the subject was instructed to maintain fixation as closely as possible to the centre of the stimulus or the bee's eyes. Both stimuli were presented three times. The evaluated parameters for fixation stability included accuracy and micro-saccade amplitude.

Reading Eye Movements. We employed two approaches: simulation of reading and real text reading (see Fig. 4). For the simulation of reading, we employed the DEM (Developmental Eye Movement) test which has been shown to be effective for reading in previous studies [9] and is suitable for comparison with our data. We used Part C of the DEM test and normalized the data for Latvian school-aged children, considering the

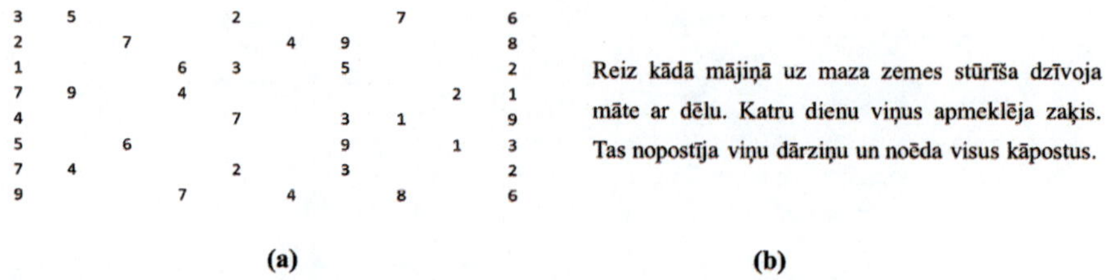

Fig. 4. The stimuli for measurements of reading eye movements parameters: **(a)** the DEM test Part C numbers, and **(b)** reading text in Latvian, e.g., for 1st grade child.

language specificity [19]. In addition to the DEM test, we prepared age-appropriate four text samples for real text reading. The width of the DEM reduced Part C (8 rows out of 16) and reading tests were 14.9° wide. The height of the DEM test was 6.2° while the height of the reading samples varied for different age groups (1st, 2nd, 3rd, and 4th to 6th grades) to accommodate appropriate letter sizes for each age group. During the tests, the subject was instructed to call out the numbers on the DEM test and read the texts aloud as quickly as possible. To avoid head movements during calling and reading, the subject's chin was not fixed.

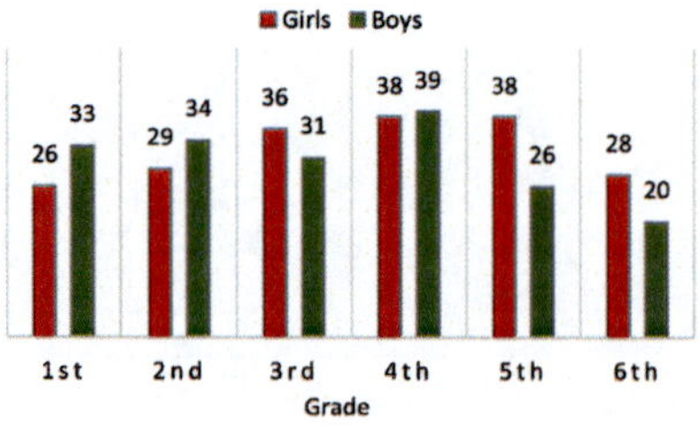

Fig. 5. Distribution of school-aged children by grade and gender who participated in the method testing. by grade and gender.

Pros and Cons of Our Method

The objective of this study was to evaluate the feasibility of measuring eye movement parameters in school-aged children using a specific method. Specifically, the study aimed to assess the practicality of conducting such measurements in a school environment, to gather data for the purpose of developing norms, and to identify factors that should be taken into account during data collection, as well as issues where the method could be improved. Our oculomotor evaluation method involves measuring four groups of eye movements – saccades, smooth pursuit, fixation stability, and reading eye movements – to provide a thorough assessment of oculomotor system. The entire process, including repetitions for eye movement calibration, takes about 20 min for each child. We evaluated 378 children (see Fig. 5) in first to sixth grade. The average age of children was 9.3 years (SD = 1.7). Reading performance and eye movement parameters were evaluated in children with parental consent, based on specific visual criteria. These criteria included a minimum near visual acuity of 0.8–1.0 on the decimal system, either with or without correction if needed. Children with strabismus were excluded from the study.

The selection of participants was carried out by school speech therapists using a random sampling method, and reading skills were assessed at baseline.

There are currently analysing the data and developing age norms for each eye movement parameter. As there are no existing age-appropriate norms of Latvian language-specific information for these measurements, we are creating our own. Additionally, we evaluated the reading performance of all participants using Acadience Reading test with the help of school speech specialists. For a selected group of 80 children (40 with reading dysfunction (low Acadience Reading score) and 40 without reading dysfunction, selected randomly), we conducted a comprehensive eye examination, which included eye refraction, binocular and eye accommodation functions, and the Acadience Reading test adapted to Latvian, also administered by school speech specialists. Our objective was to examine the relationship between oculomotor system parameters, reading parameters, and visual disorders.

We have identified several strengths and limitations of our developed method which we aim to address in future research. One of the major strengths is the potential use of this method outside of specialist practice, such as in school environments, since recently different mobile and small eye-tracking devices have been introduced in the market, like *Tobii Pro Fusion*. The objective assessment of various eye movement parameters eliminates any observer bias, and the method's versatility allows for the choice and application of different stimuli for wider applications beyond oculomotor system screening. However, the current challenge lies in the analysis of all parameters and correlations, which requires further development of specially designed algorithms to handle the large amount of data recorded. Furthermore, we aim to collect more data to develop an AI model that can improve oculomotor dysfunction diagnosis.

In conclusion, developing an oculomotor dysfunction method based on eye-tracking technology, with high sensitivity and negative predicting value for detecting oculomotor dysfunctions would be of considerable benefit.

3 Discussion and Conclusion

The early experimental eye movement research has shown that when evaluating oculomotor performance clinicians primarily rely on direct observation of patients' eye movements, possibly due to the ease of qualitatively assessing salient characteristics of various abnormalities, such as strabismus, nystagmus, and inaccurate tracking, using unaided vision. However, recent advances in eye tracking technology have enabled objective eye movement recording, leading to more detailed spatial and temporal resolution of oculomotor behaviour, precise quantitative determinations of eye movement characteristics, and generation of permanent records [20, 21]. Currently, high-resolution and high-precision devices have become available and are being investigated for their potential in evaluating ocular follow responses and diagnosing binocular vision [23, 24]. The objective of our study was to develop a methodology for assessing the oculomotor system that could be applied to various devices. However, it is important to note that this task presents a considerable challenge, given that each device employs its own unique data processing algorithm.

The implementation of the novel method for testing visual functions in optometric practice will assist parents, teachers, speech specialists, and psychologists in comprehending the aetiology of diverse reading and learning difficulties. In the context of such dysfunctions, aa precise diagnosis is essential to determine the most suitable vision therapy.

With our advanced knowledge in optometry and visual processes, we can establish correlations between general visual functions, such as binocular vision, stereovision, and eye accommodation, and specific eye movement parameters. By evaluating visual functions in children with and without reading and learning difficulties, using mobile eye trackers, we aim to enhance the role of optometrists in the vision care system by introducing a new product for the detection of eye movement disorders. Additionally, we aim to improve the general understanding of oculomotor dysfunctions as specific eye movement disorders and develop a clear management system for each type of detected eye movement disorder. Specifically, we propose the implementation of a specific vision therapy that could effectively alleviate the problem. For instance, using saccadic eye movement training to improve the amplitude of saccadic eye movements through targeted actions, rather than passive stimulation.

The assessment of oculomotor system is imperative not only for children with reading disabilities [22] but also for adults who have suffered various head traumas [25], unclear neurodegenerative diagnoses [26], or have disorders that can affect visual system functions. Our method for diagnosing oculomotor dysfunctions will be instrumental in developing a new branch of vision care services since vision therapy, as opposed to vision correction, is the primary therapy for treating eye movement disorders.

Acknowledgements. We would like to express our gratitude to schools participating in the study (Marupe State Gymnasium, Marupe Elementary School, Riga Cultures Secondary School, and Kuldiga Center Secondary School), as well as speech therapists J.Hanzovska, L.Meiersone, M.Vorza, I.Petuhova, S.Depa, S.Jirgensone. The study is supported by the Latvian Council of Science (project No lzp-2021/1–0219, and SIA Mikrotikls and University of Latvia Foundation (project No 2260) and two anonymous reviewers for their helpful comments.

References

1. Rayner, K.: Eye movements in reading and information processing: 20 years of research. Psychol. Bull. **124**(3), 372–422 (1998)
2. Alvarez-Peregrina, C., Sánchez-Tena, M.Á., Andreu-Vázquez, C., Villa-Collar, C.: Visual health and academic performance in school-aged children. Int. J. Environ. Res. Public Health **17**(7), 1–8 (2020). Art. no.2346
3. Barbieri, F.A., Rodrigues, S.T.: Editorial: the role of eye movements in sports and active living. Front. Sports Active Living **2**, 1–2 (2020). Art.no. 603206
4. Land, M.F.: Eye movements and the control of actions in everyday life. Prog. Retin. Eye Res. **25**, 296–324 (2006)
5. Maples, W.C., Atchley, J., Ficklin, T.: Northeastern State University college of optometry's oculomotor norms. J. Behav. Optom. **3**(6), 143–150 (1992)
6. Pierce, J.: Pierce saccade test. Bloomington, Cook (1972)

7. Galetta, K.M., Liu, M., Leong, D.F., Ventura, R.E., Galetta, S.L., Balcer, L.J.: The King-Devick test of rapid number naming for concussion detection: meta-analysis and systematic review of the literature. Concussion **1**(2), 1–15 (2015). Art.no. CNC8
8. Oride, M.K., Marutani, J.K., Rouse, M.W., DeLand, P.N.: Reliability study of the Pierce and King-Devick saccade tests. Am. J. Optom. Physiol. Opt. **63**(6), 419–424 (1986)
9. Facchin, A.: Spotlight on the developmental eye movement (DEM) test. Clin. Optom. **13**, 73–81 (2021)
10. Rouse, M.W., Nestor, E.M., Parot, C.J., Deland, P.N.: A reevaluation of the developmental eye movement (DEM) test's repeatability. Optom. Vis. Sci. **81**(12), 934–938 (2004)
11. Orlansky, G., et al.: Reliability of the developmental eye movement test. Optom. Vis. Sci. **88**(12), 1507–1519 (2011)
12. Eide, M.G., et al.: Eye tracking complementing manual vision screening for detecting oculomotor dysfunction. In: 2019 E-Health and Bioengineering Conference (EHB), Iasi, Romania, pp. 1–5 (2019)
13. Adhanom, I.B., MacNeilage, P., Folmer, E.: Eye tracking in virtual reality: a broad review of applications and challenges. In: Virtual Reality, pp. 1–25 (2023). Accessed 18 Jan 2023
14. López, A., Ferrero, F., Postolache, O.: An affordable method for evaluation of ataxic disorders based on electrooculography. Sensors **19**(17), 1–16 (2019). Art.no. 3756
15. Sweeney, J.A., Luna, B., Keedy, S.K., McDowell, J.E., Clementz, B.A.: fMRI studies of eye movement control: investigating the interaction of cognitive and sensorimotor brain systems. Neuroimage **36**(Suppl 2), T54–T60 (2007)
16. Krūmiņa, G., et al.: Skolas vecuma bērnu redzes skrīnings un tuvuma redzes funkciju novērtēšanas metode (In English: School-aged children vision screening and method for assessing near visual functions). Latvijas Universitāte, Rīga, p. 153 (2013)
17. Krūmiņa, G., et al.: Skolas vecuma bērnu redzes uztveres novērtēšanas metode (In English: The assessment of vision perception in school-aged children). Latvijas Universitāte, Rīga, p. 103 (2013)
18. Thaler, L., Schütz, A.C., Goodale, M.A., Gegenfurtner, K.R.: What is the best fixation target? the effect of target shape on stability of fixational eye movements. Vision. Res. **76**, 31–42 (2013)
19. Serdjukova, J., Ekimane, L., Valeinis, J., Skilters, J., Krumina, G.: How strong and weak readers perform on the developmental eye movement test (DEM): norms for Latvian school-aged children. Read. Writ. **30**(2), 233–252 (2016). https://doi.org/10.1007/s11145-016-9671-7
20. Brunyé, T.T., Drew, T., Weaver, D.L., Elmore, J.G.: A review of eye tracking for understanding and improving diagnostic interpretation. Cognitive Research: Principles and Implications 4(1), 1–16 (2019). Art. No 7
21. Zammarchi, g., Conversano, C.: Application of eye tracking technology in medicine: a bibliometric review. Vision **5**, 1–14 (2021). Art. no. 56
22. Bilbao, C., Piñero, D.P.: Diagnosis of oculomotor anomalies in children with learning disorders. Clin. Exp. Optom. **103**(5), 597–609 (2020)
23. Quaia, C., FitzGibbon, E.J., Optican, L.M., Cumming, B.G.: Binocular summation for reflexive eye movements: a potential diagnostic tool for stereodeficiencies. Invest. Ophthalmol. Vision Sci. **59**(15), 5816–5822 (2018)
24. Miladinović, A., Quaia, C., Ajčević, M., Diplotti, L., Cumming, B.G., Pensiero, S., Accardo, A.: Ocular-following responses in school-age children. PLoS One **17**(11), 1–15 (2022). Art. no. 0277443
25. Sussman, E.S., Ho, A.L., Pendharkar, A.V., Gjajar, J.: Clinical evaluation of concussion: the evolving role of oculomotor assessments. J. Neurosurg. **40**(4), 1–7 (2016). Art. no. E7
26. Antoniades, C.A., Kennard, C.: Ocular motor abnormalities in neurodegenerative disorders. Eye **29**(2), 200–207 (2015)

A Novel Algorithm for the Compensation of Hemoglobin Interference on Bilirubin Measurement Applied to a Two-Wavelengths Reflectance Photometer

Lorenzo Zucchini(✉), Miloš Ajčević, and Agostino Accardo

Department of Engineering and Architecture, University of Trieste, Trieste, Italy
lorenzo.zucchini@phd.units.it

Abstract. Unconjugated Bilirubin (UCB) concentration in plasma can be higher in newborns than in adults; while a controlled and temporary increase of UCB concentration in plasma is physiological, a prolonged or excessive increase can lead to severe neurological damage or death. In high-resource settings several methods for bilirubin measurement have been in use for decades, including spectrophotometers that provide relatively accurate results and interference compensation of other analytes, such as hemoglobin. In Low- to Middle Income Countries (LMICs) however, the presence of expensive and complex laboratory instrumentation is not obvious. In this study a novel algorithm for the compensation of hemoglobin on bilirubin measurement was proposed and evaluated on a simple, inexpensive two-wavelengths reflectance photometer. Two bilirubin levels with five different hemoglobin concentrations were prepared; a linear function was used to approximate the relation between reflectance at 570 nm and hemoglobin concentration. The relative errors were reduced from 94% to 15.4% for the worst case. Further investigations are required to optimize the algorithm, including the use of different wavelengths and wider ranges of concentration.

Keywords: Bilirubin · Neonatal Hyperbilirubinemia · Hemoglobin · Reflectance Photometry · LMICs

1 Introduction

Bilirubin is the major product of erythrocytes breakdown; under normal conditions unconjugated bilirubin (UCB) is released in plasma where bounds to albumin, until it is excreted in bile as conjugate bilirubin [1, 2]. In neonates, due to several factors, UCB can form in higher concentrations: more than half of newborns experience a temporary and controlled increase of UCB known as hyperbilirubinemia during their first days of life, which is a physiological condition [3, 4]. However, roughly one in ten newborns develops clinically significative hyperbilirubinemia: due to the neurotoxicity of bilirubin, this condition can be severe and lead to kernicterus or death [2–4]. In these cases, frequent monitoring of serum bilirubin is needed to establish a proper treatment and reduce chances of neurological damage [5, 6].

Y. Dekhtyar and I. Saknite (Eds.): NBC 2023, IFMBE Proceedings 89, pp. 18–24, 2023.
https://doi.org/10.1007/978-3-031-37132-5_3

Several guidelines for the management of neonatal hyperbilirubinemia were established [7–9]; however, in Low- to Middle Income Countries (LMICs), the lack of infrastructures, technologies, and accessibility hinder their effective use [10–12].

Direct spectrophotometry for the measurement of bilirubin has been in use for decades and several variants, both for laboratory and point-of-care applications, have been proposed [13–19]. The main limitation of this method is the interference from other pigments, mainly hemoglobin, which can be present in plasma samples due to hemolysis or bad blood samples handling or collection. Interferent analytes compensation is normally accomplished through the use of reagents [19] or the use of lamps combined with optical filtering systems capable of generating up to 512 different wavelengths [20]. These aspects increase complexity, dimensions, cost and reduce usability of direct photometry-based devices limiting their applicability in low-resource settings such as LMICs [21].

The aim of this work is to propose and test a novel algorithm for hemoglobin interference compensation on a simple, portable, inexpensive two-wavelengths reflectance photometer for the measurement of total serum bilirubin applicable to LMIC scenarios.

2 Materials and Methods

2.1 The Measuring Setup

A two-wavelengths reflectance photometer was used for the measurement of bilirubin and hemoglobin samples (Fig. 1). Two discrete LEDs irradiated the sample at 465nm and 570 nm, respectively (KPHHS-1005QBC-D-V and KPT-1608CGCK, Kingbright, Taiwan); a photodiode (TEMD6200FX01, Vishay, USA) was used to measure reflected light intensity at each wavelength; these components, as well as the sample, were enclosed in a dark chamber blocking any external light. Other electronic components necessary for the acquisition and digitalization of the signal were also included.

The 35 μL plasma samples were positioned on lateral flow test strips, composed of a fiberglass filter coupled with a nitrocellulose membrane laid on a plastic support; since no light was transmitted through the sample, the fraction of light not absorbed by the sample was reflected towards the photodiode. In this way, the photometer acquired the intensity of reflected light.

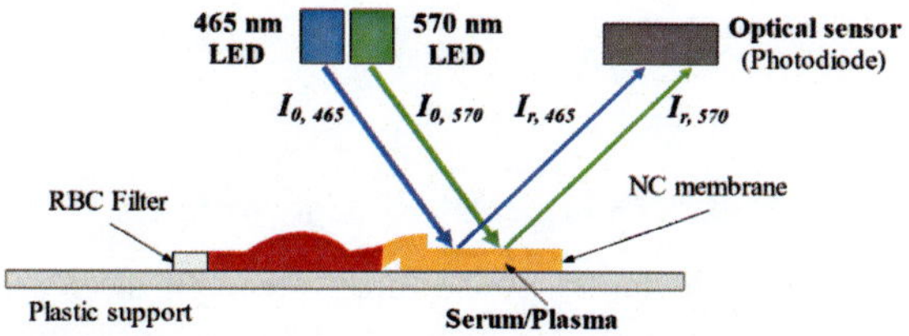

Fig. 1. A drawing of the discrete LED light sources, the irradiated sample, and the photodiode placement. All these elements were enclosed in the darkened optical chamber and controlled by other electronics.

2.2 Test Samples

A panel containing two different levels of bilirubin concentration and five different levels of hemoglobin was prepared.

Commercial purified bilirubin (Bilirubin powder ≥ 98% – Code B4126, Sigma-Aldrich) was dissolved in DMSO to obtain a mother bilirubin solution. Meanwhile, 10 mL of blood was drowned from one adult volunteer in a tube containing K3 EDTA anticoagulant and, after centrifugation, two aliquots of plasma were transferred to separated test tubes. The bilirubin mother solution was added in different proportions to each of the test tubes to obtain 5.0 mg/dL and 15.0 mg/dL bilirubin concentrations, respectively. To obtain a mother hemoglobin solution the same blood sample was lysed and centrifuged again to separate lysed cells fragments; the high hemoglobin concentration plasma was collected and used to prepare five levels of hemoglobin concentration for each one of the test tubes.

The final concentration of hemoglobin for each sample was verified using conventional methods. The final panel of samples is shown in Table 1; each sample was tested in replicates of three.

Table 1. The panel of samples for this experiment. Three samples were prepared for each of the two different bilirubin levels in combination with five different hemoglobin concentrations.

Sample	Expected Bilirubin [mg/dL]	Hemoglobin [g/dL]
B_1_H_1	5.0	0.21, 0.23, 0.21
B_1_H_2		0.32, 0.30, 0.30
B_1_H_3		0.46, 0.46, 0.48
B_1_H_4		0.62, 0.57, 0.59
B_1_H_5		0.68, 0.68, 0.71
B_2_H_1	15.0	0.16, 0.16, 0.16
B_2_H_2		0.27, 0.30, 0.27
B_2_H_3		0.39, 0.39, 0.39
B_2_H_4		0.50, 0.52, 0.50
B_2_H_5		0.64, 0.64, 0.68

2.3 Hemoglobin Interference Compensation Algorithm

The reflectance photometer device response to different bilirubin concentrations was characterized by testing non-hemolyzed samples with different bilirubin concentration in a sufficiently wide range for this experiment: specifically, the 465 nm wavelength was used as it is relatively close to the bilirubin absorption peak. A function was used to approximate the relation between 465 nm reflectance and bilirubin concentration. This reflectance-bilirubin calibration curve was used to estimate bilirubin concentration.

At 465 nm both bilirubin and hemoglobin show significative absorption, while at 570 nm only hemoglobin shows significative absorption and bilirubin response is close to zero. For this reason, a given hemoglobin concentration causes a specific decrease in reflectance at each wavelength (ΔR_{570} and ΔR_{465}, respectively) with respect to a non-hemolyzed sample. To subtract the influence of hemoglobin from reflectance at 465 nm, thus, to measure bilirubin without hemoglobin interference, a relation between ΔR_{570} and ΔR_{465} had to be defined.

Reflectance data for non-hemolyzed samples (B_1_H_1 and B_2_H_1 in Table 1) was defined as reference for 570 and 465 nm wavelengths and indicated as $R_{570,\ ref}$ and $R_{465,\ ref}$, respectively. For each of the other tests, ΔR_{570} and ΔR_{465} were calculated as reported in Eq. 1 and Eq. 2, where $R_{570,\ [Hb]}$ and $R_{465,\ [Hb]}$ are the reflectance values for a certain hemoglobin concentration at each wavelength.

$$\Delta R_{570} = R_{570,ref} - R_{570,[Hb]} \tag{1}$$

$$\Delta R_{465} = R_{465,ref} - R_{465,[Hb]} \tag{2}$$

Subsequently, ΔR_{570} and ΔR_{465} were correlated using a linear function. In such way, a certain difference between $R_{570,\ [Hb]}$ and $R_{570,\ ref}$ could have been associated with a difference between $R_{465,\ [Hb]}$ and $R_{465,\ ref}$, calculating the difference in reflectance at 465 nm caused by hemoglobin. The ΔR_{465} itself was then added to the resulting reflectance and bilirubin concentration was calculated.

3 Results

Reflectance at 465 nm was highly correlated with hemoglobin, leading to major overestimation errors at both levels (Table 2).

A significative correlation was found between ΔR_{465} and ΔR_{570} ($r = 0.85$, $p < 0.02$); a linear equation was used to approximate this relation (Fig. 2).

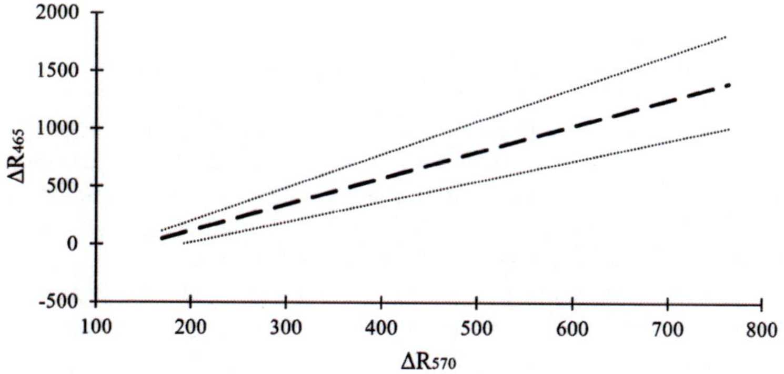

Fig. 2. The linear equation used to approximate the relation between ΔR_{465} and ΔR_{570} (dashed line). The upper and lower dotted lines represent the interpolation of lower and higher bilirubin level samples, respectively.

In Table 2 compensated bilirubin data is reported before and after the application of the proposed algorithm. Without any compensation the relative errors were higher than 20% for almost every hemoglobin concentration apart from the non-hemolyzed samples, and as high as 94% in the worst case. The algorithm effectively reduced relative errors and resulted in a 15.4% error in the worst case.

Table 2. Results of the bilirubin measures without compensation and with compensation algorithm. Relative percentage errors are reported ($\varepsilon_{\%}$).

		No compensation		After compensation	
Sample	Expected [mg/dL]	Measured [mg/dL]	ε%	Measured [mg/dL]	$\varepsilon_{\%}$
B_1_H_1	5.0	5.3	3.3	5.0	3.4
B_1_H_2		6.4	28.7	5.4	4.6
B_1_H_3		7.7	54.7	5.7	9.4
B_1_H_4		8.4	68.0	5.5	6.1
B_1_H_5		9.7	94.0	5.7	11.0
B_2_H_1	15.0	15.9	1.8	14.6	4.6
B_2_H_2		17.0	13.3	14.1	7.6
B_2_H_3		18.3	22.0	13.4	12.3
B_2_H_4		19.9	32.7	12.9	15.4
B_2_H_5		22.8	52.2	13.1	13.9

The discrepancy highlighted in Fig. 2 between the two tested bilirubin levels suggests that bilirubin absorption at 570 nm is not negligible, resulting in an increased dispersion of data points; to confirm this aspect, all reflectance data at 570 nm from each bilirubin level was averaged, and a slight correlation was observed (r = 0.38).

4 Discussion

In this study, a method for hemoglobin influence compensation on bilirubin measurement was proposed and tested on a simple two-wavelengths reflectance spectrometer; the instrument used two discrete LEDs emitting light at 465 nm and 570 nm, respectively, and a photodiode measuring reflected light intensity. The sample size was 35 μL.

The testing of a panel containing two levels of bilirubin and five hemoglobin concentrations resulted in extensive overestimation errors using 465 nm light without any compensation, since the absorption spectra of bilirubin and hemoglobin overlap significantly at such wavelength. The choice of adding a 570 nm light was based on the hypothesis that at this wavelength bilirubin shows negligible absorption, thus reflectance signal is only attributable to hemoglobin concentration. The results indicated a relation, albeit limited, between reflectance at 570 nm and bilirubin concentration: this resulted in different data distributions for the two levels, hence a not negligible deviation from the ideal case of having a calibration law depending only on hemoglobin concentration. Although this limitation, the algorithm led to a significant reduction of overestimation errors at both levels of bilirubin: the use of a linear law allowed a reduction of the relative errors from 94% to 15.4% in the worst case.

In a previous work, a simple two-wavelengths spectrophotometer with hemoglobin compensation capabilities was proposed [18]: the authors reported a coefficient of determination r^2 of 0.89 on bilirubin determination in serum, however no data was reported concerning hemoglobin influence. In another study, a three-wavelengths handheld reader was used to test plasma samples with bilirubin concentrations ranging from 0.0 to 36.3 mg/dL [22]: the authors speculated that measured spectra showed little to no change in absorption due to hemoglobin, although no data regarding hemoglobin concentrations was disclosed. Lastly, a wearable device for real-time detection of jaundice capable of measuring bilirubin, SpO_2, and heart rate was recently proposed [23], resulting in good agreement with commercial transcutaneous bilirubinometers.

Future effort will be aimed at optimizing the wavelength used for hemoglobin detection, thus at finding a wavelength for which bilirubin absorption is negligible and hemoglobin absorption is adequate for this application. In addition, a larger panel of bilirubin and hemoglobin concentrations will be needed to confirm the applicability of this approach to the screening for neonatal hyperbilirubinemia in low-resource settings.

5 Conclusions

An algorithm for the compensation of hemoglobin interference on bilirubin measurement was proposed and tested. Despite the need of additional investigation and optimizations, the results obtained in this study were promising and suggested the possibility of achieving further improvements.

References

1. Fevery, J.: Bilirubin in clinical practice: a review: Bilirubin in clinical practice. Liver Int. **28**(5), 592–605 (2008). https://doi.org/10.1111/j.1478-3231.2008.01716.x
2. Olusanya, B.O., Kaplan, M., Hansen, T.W.R.: Neonatal hyperbilirubinaemia: a global perspective. Lancet Child Adolesc. Health **2**(8), 610–620 (2018). https://doi.org/10.1016/S2352-4642(18)30139-1
3. Greco, C., et al.: Neonatal jaundice in low- and middle-income countries: lessons and future directions from the 2015 Don Ostrow Trieste Yellow retreat. Neonatology **110**(3), 172–180 (2016). https://doi.org/10.1159/000445708
4. Mitra, S., Rennie, J.: Neonatal jaundice: aetiology, diagnosis and treatment. Br. J. Hosp. Med. **78**(12), 699–704 (2017). https://doi.org/10.12968/hmed.2017.78.12.699
5. Bhutani, V.K., Johnson, L., Sivieri, E.M.: Predictive ability of a predischarge hour-specific serum bilirubin for subsequent significant hyperbilirubinemia in healthy term and near-term newborns. Pediatrics **103**(1), 6–14 (1999). https://doi.org/10.1542/peds.103.1.6
6. Subcommittee on Hyperbilirubinemia: Management of hyperbilirubinemia in the newborn infant 35 or more weeks of gestation. Pediatrics **114**(1), 297–316 (2004). https://doi.org/10.1542/peds.114.1.297

7. Kemper, A.R., et al.: Clinical practice guideline revision: management of hyperbilirubinemia in the newborn infant 35 or more weeks of gestation. Pediatrics **150**(3), e2022058859 (2022). https://doi.org/10.1542/peds.2022-058859
8. Amos, R.C., Jacob, H., Leith, W.: Jaundice in newborn babies under 28 days: NICE guideline 2016 (CG98). Archiv. Dis. Childhood – Educ. Pract. **102**(4), 207–209 (2017). https://doi.org/10.1136/archdischild-2016-311556
9. Sgro, M., Kandasamy, S., Shah, V., Ofner, M., Campbell, D.: Severe neonatal hyperbilirubinemia decreased after the 2007 Canadian guidelines. J. Pediatr. **171**, 43–47 (2016). https://doi.org/10.1016/j.jpeds.2015.12.067
10. Iskander, I., Gamaleldin, R.: Acute Bilirubin encephalopathy: some lessons learned. Semin. Perinatol. **45**(1), 151353 (2021). https://doi.org/10.1016/j.semperi.2020.151353
11. Erdeve, O.: Management of neonatal jaundice in low-income and middle-income countries. BMJ Paediatr. Open **4**(1), e000845 (2020). https://doi.org/10.1136/bmjpo-2020-000845
12. Bhardwaj, U., Kohli, V., Thukral, A.: Management of hyperbilirubinemia in newborn infants 35 or more weeks of gestation: American Academy of Pediatrics, 2022. Indian Pediatr. **60**(1), 63–66 (2023). https://doi.org/10.1007/s13312-023-2697-4
13. Kelly, A., McKenna, J.P., McLelland, A., Percy, R.A., Spooner, R.J.: A bichromatic method for total bilirubin with a Centrifichem 400. Clin. Chem. **25**(8), 1482–1484 (1979). https://doi.org/10.1093/clinchem/25.8.1482
14. Thaler, M., Luppa, P.B., Schlebusch, H.: Bilirubin Measurement – an Updated Survey1. LaboratoriumsMedizin **32**(1) (2008). https://doi.org/10.1515/JLM.2008.005et
15. Rolinski, B., et al.: Evaluation of total bilirubin determination in neonatal whole-blood samples by multiwavelength photometry on the Roche OMNI S point-of-care analyzer. Point of Care **4**(1), 3 (2005). https://doi.org/10.1097/01.poc.0000157097.59514.62
16. Pierre Ndabakuranye, J., et al.: 70 years of bilirubin sensing: towards the point-of-care bilirubin monitoring in cirrhosis and hyperbilirubinemia. Sens. Diagnost. **1**(5), 932–954 (2022). https://doi.org/10.1039/D2SD00033D
17. Meites, S., Hogg, C.K.: Direct spectrophotometry of total serum bilirubin in the newborn. Clin. Chem. **6**(5), 421–428 (1960). https://doi.org/10.1093/clinchem/6.5.421
18. Kudavelly, S., Keswarpu, P., Balakrishnan, S.: A simple and accurate method for estimating bilirubin from blood. In:2011 IEEE International Instrumentation and Measurement Technology Conference, pp. 1–4. IEEE: Hangzhou (2011). https://doi.org/10.1109/IMTC.2011.5944035.
19. Garber, C.C.: Jendrassik-Grof analysis for total and direct bilirubin in serum with a centrifugal analyzer. Clin. Chem. **27**(8), 1410–1416 (1981). https://doi.org/10.1093/clinchem/27.8.1410
20. Hallemann, H., Putz, K., Schweiger, G., Spitzer, S., Strohmeier, M.: Technical aspects of bilirubin determination in whole blood. Point of Care: J. Near-Patient Test. Technol. **4**(1), 9–10 (2005). https://doi.org/10.1097/01.poc.0000157101.51422.72
21. Hulzebos, C.V., et al.: Diagnostic methods for neonatal hyperbilirubinemia: benefits, limitations, requirements, and novel developments. Pediatr. Res. **90**(2), 277–283 (2021). https://doi.org/10.1038/s41390-021-01546-y
22. Keahey, P.A., et al.: Point-of-care device to diagnose and monitor neonatal jaundice in low-resource settings. Proc. Natl. Acad. Sci. **114**(51), E10965–E10971 (2017). https://doi.org/10.1073/pnas.1714020114
23. Inamori, G., et al.: Neonatal wearable device for colorimetry-based real-time detection of jaundice with simultaneous sensing of vitals. Sci. Adv. **7**(10), eabe3793 (2021). https://doi.org/10.1126/sciadv.abe3793

Method for Muscle Fatigue Detection Using Inertial Sensors

Beāte Banga[1](✉), Alexei Katashev[2], and Modris Greitāns[1]

[1] Institute of Electronics and Computer Science, 14 Dzerbenes St., Riga 1006, Latvia
beate.banga@edi.lv

[2] Riga Technical University, Institute of Biomedical Engineering and Nanotechnologies, 6B Kipsalas St., Riga 1048, Latvia

Abstract. Muscle fatigue is a common symptom that many people experience and is associated with difficulties in voluntary movement, which can lead to injuries. Currently, surface electromyography (sEMG) is considered the gold standard for muscle fatigue estimation, but its accuracy can be impacted by various factors. Therefore, new methods, such as the use of inertial sensors (IMU), are being introduced. This study aimed to explore the relationship between muscle fatigue and biomechanical parameters using inertial sensors and sEMG as a validation tool. Four participants performed an elbow flexion exercise, and the data from IMU sensor nodes and sEMG were collected. The results showed that there were correlations between the electrical activity of m. biceps brachii and rotation angles of the forearm and upper arm. Additionally, an increase in motion amplitude deviation was found to be a potential indicator of muscle fatigue. These findings suggest that inertial sensors can be used as an alternative to sEMG for detecting muscle fatigue, which has potential implications for injury prevention and rehabilitation. However, further research with a larger sample size is needed to validate these findings.

Keywords: Muscle fatigue · Inertial sensors · IMU · Surface electromyographs · sEMG · Motion capture

1 Introduction

Muscle fatigue is a common non-specific symptom experienced by many individuals. It is often defined as an overwhelming feeling of tiredness, lack of energy and exhaustion. Muscle fatigue comes in different forms and this paper will analyze acute muscle fatigue caused by intense muscle use.

Acute muscle fatigue impairs voluntary movements, but rest restores performance. Its impact on athlete performance and injury risk has led to increased attention in sports. Although the phenomenon has been observed for a while, studies on changes in human biomechanics due to acute muscle fatigue are recent [1].

sEMG is the current "gold standard" for acute muscle fatigue but can be inaccurate due to various reasons (noise from heart rate, respiration, etc.). New methods, such as IMUs, are being sought to detect acute muscle fatigue. This work investigates the

Y. Dekhtyar and I. Saknite (Eds.): NBC 2023, IFMBE Proceedings 89, pp. 25–32, 2023.
https://doi.org/10.1007/978-3-031-37132-5_4

relationship between fatigued and non-fatigued muscle activity and movement changes using IMUs, with sEMG used for validation.

IMUs measure the orientation of an object in space and these sensors can be placed on the body to record the absolute orientation of body parts in space.

This article investigates the relationship between muscle fatigue and changes in joint angles, as it has been shown in the literature that muscle fatigue has an effect on biomechanical parameters [2].

The work has been developed at the Electronics and Computer Science Institute (ECSI), in collaboration with the Biomedical Engineering and Nanotechnology Institute of Riga Technical University (RTU BINI), using an IMU system developed at ECSI and the existing sEMG at RTU BINI. Four volunteer participants took part in the study.

2 Methods

As it has been shown in the literature [3], with increase of muscle fatigue, a decrease in coordination ability is observed, this study investigates whether the IMU can record these changes. IMUs are a new generation of wearable devices with emerging applications in human movement recording. The sEMG is used as a validation tool, which is considered the "gold standard" for recording muscle fatigue, and in the practical part the increase in the amplitude of the electrical potential of the muscle is recorded to determine whether muscle fatigue has occurred, as this is one of the indicators of muscle fatigue [1].

2.1 Exercise

For the practical part of the exercise, the elbow flexion exercise is chosen. Measurements of the electrical activity of the m. biceps brachii and measurements of the angles of movement execution are performed in parallel while the individual performs a repetitive elbow flexion exercise. The exercise is performed without stopping for one minute, during which the individual is required to follow as closely as possible all the rules of a technically correct exercise (see Fig. 1).

Four volunteer participants took part in the practical part of the study and were briefed before the start of the practical part and the ethical aspects of the study were agreed.

2.2 Systems Used

For the measurements, the IMU system developed at the Institute of Electronics and Computer Science and the CleveMed BioRadio 150 physiological parameter recording system at the Institute of Biomedical Engineering and Nanotechnology of Riga Technical University as sEMG are used (see Fig. 2).

IMU System. This study used a wearable IMU system. The IMU system used for data acquisition consisted of three connected sensor nodes and a data acquisition device mounted on the body. Data from the data acquisition device was sent to a computer using Bluetooth technology.

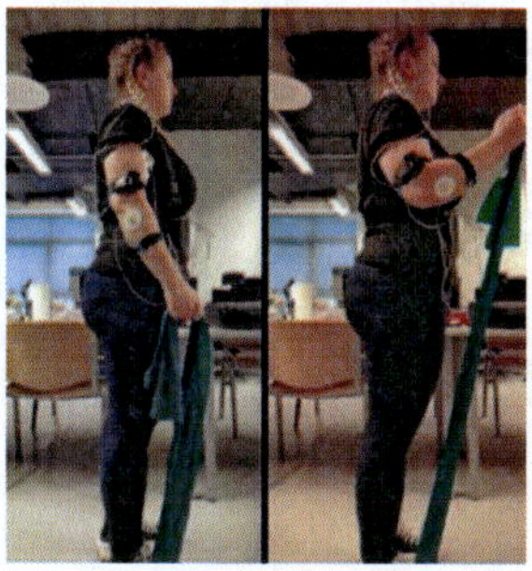

Fig. 1. Elbow flexion exercise. Volunteer participant is doing elbow flexion exercise using resistance band.

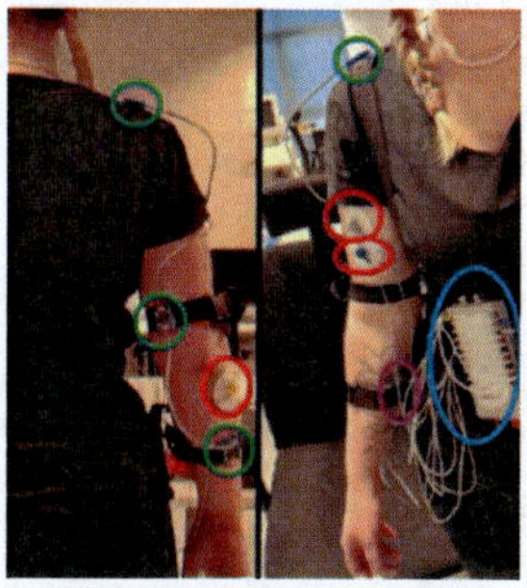

Fig. 2. IMU and sEMG placement. Green circles show the position of the IMU sensor nodes (fixed on the forearm, upper arm and shoulder but for the data analysis only data from the forearm and upper arm nodes is collected), purple circle shows IMU system's Master board and power supply unit, red circles show sEMG electrodes (placed on biceps breachii muscle) and blue circles shows sEMG signal recording system.

In the IMU system used, each sensor node consists of three types of sensors: accelerometer, magnetometer and gyroscope. Each sensor has three measurement axes (x,y and z) [4].

From sensor nodes data was sent in quaternion form. Quaternions do not directly represent the rotation of an object, so it is necessary to convert quaternions into a more easily understandable representation of rotation. This work uses the mathematical representation of Euler angles.

In the mathematical representation of Euler angles, the rotation of an object is denoted as rotation about three orthogonal axes, denoted as - Roll (rotation about x-axis), Pitch (rotation about y-axis) and Yaw (rotation about z-axis). The angles of rotation about the axes are conventionally denoted as follows: rotation about the x-axis by Φ, rotation about the y-axis by θ, rotation about the z-axis by ψ. In formulas (1), (2) and (3) these angles are calculated from quaternions q_0 (scalar part), and q_1, q_2, q_3 (vector part).

$$\Phi = \arctan \frac{2(q_0 q_1 + q_2 q_3)}{1 - 2\left(q_1^2 + q_2^2\right)}. \tag{1}$$

$$\theta = \arcsin(2(q_0q_2 - q_3q_1)) \tag{2}$$

$$\psi = \arctan\frac{2(q_0q_3 + q_1q_2}{1 - 2(q_2^2 + q_3^2)} \tag{3}$$

sEMG System. Electromyography, or EMG, is a technique for recording the electrical potentials generated by skeletal muscle cells [5].

With sEMG, muscle fatigue can be observed as: changes in the mean absolute value of the signal, changes in the amplitude and duration of the muscle action potential and a general shift to lower frequencies [6]. In this work, the method of comparing changes in muscle action potential amplitude.

Only one electrical potential readout channel was used as the electrodes were mounted only on the biceps muscle.

As the electric potential is in both positive and negative directions, the first data processing step is to obtain absolute values of the electric potential. A rectification method using a moving average was used to estimate the shape of the electric potential. For each repetition obtained, the average value of one hundred measurements was calculated. The mean value, absolute deviation and relative deviation of the peak value were calculated.

3 Results

3.1 IMU System

Using formulas (1), (2) and (3), Euler angles were calculated for the upper arm are shown in the graph in Fig. 3. Although the upper arm must remain stationary during the exercise, it can be observed that there are oscillations in the upper arm data.

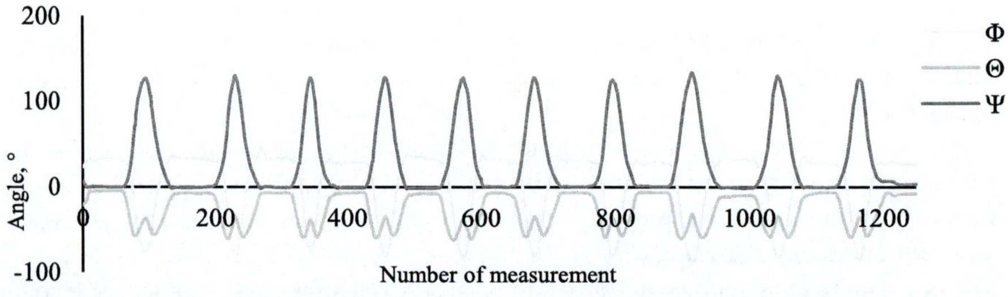

Fig. 3. Upper arm Euler angles. The angle ψ is the forward or backward rotation of the upper arm, the angle Φ shows the rotation about its axis and the angle θ shows the crouching and retraction away from the trunk.

Euler angles were also calculated for the lower arm and are shown in the graph in Fig. 4.

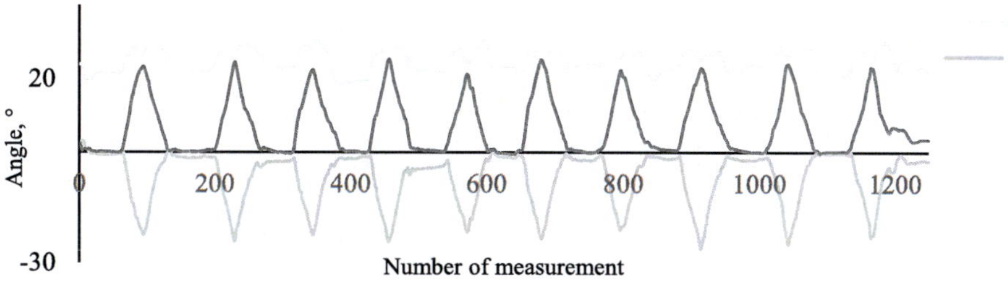

Fig. 4. Lower arm Euler angles. The angle ψ is the forward or backward rotation of the lower arm, the angle Φ shows the rotation about its axis and the angle θ shows the crouching and retraction away from the trunk.

3.2 sEMG

To assess the mean electrical potential shape of a non-tired muscle, a forearm flexion exercise was performed 10 times. Noises exceeding the electrical potential that the muscle is capable of generating were observed (peak values above 5 mV in the graph), that is why the data was then filtered so that data between 30 and 400 Hz is left, and the absolute values are calculated using the rectification method described above. The data are plotted on a graph (see Fig. 5).

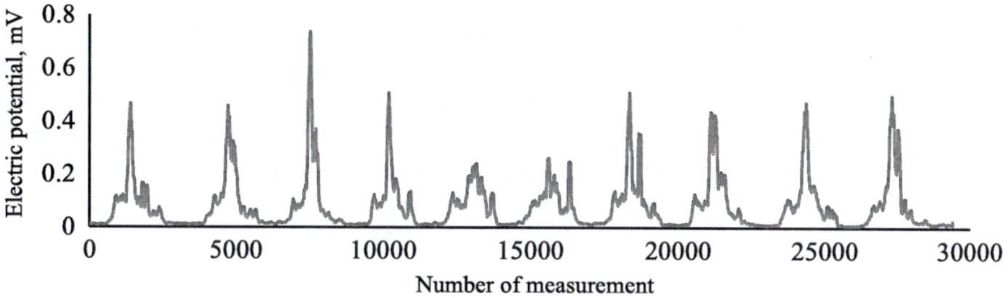

Fig. 5. sEMG data from biceps brachii.

3.3 sEMG and IMU

To compare sEMG and IMU data, they were plotted on the same graph (see Fig. 6). On this graph it can be observed that during the elbow flexion movement, changes in angles appear in the IMU data and changes in the electrical potential are also observed in the sEMG data.

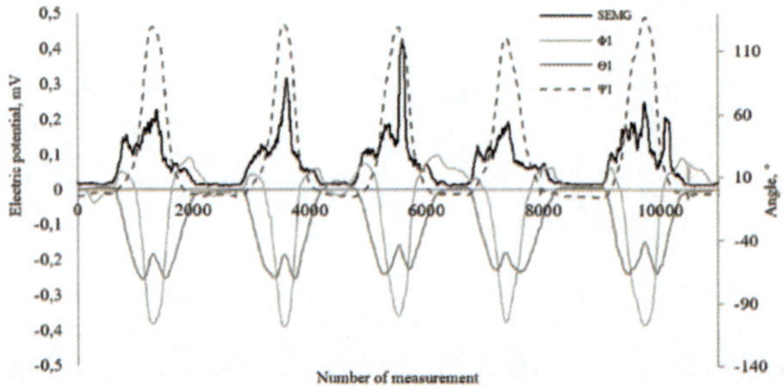

Fig. 6. sEMG and IMU (ψ1, Φ1, θ1) data for five elbow flexion repetitions.

4 Discussion

To determine muscle fatigue, the electrical potential and Euler angles were analyzed and compared for the first five repetitions of the exercise and the last five repetitions for both the upper arm and the forearm sensors.

It is found that the average peak value of the electrical potential of the first five elbow flexions is less than the average peak value of the electrical potential of the last five times, while the deviation in the first case is about 13% lower. This means that by the end of the exercise, the individual's coordination had weakened, making it more difficult to maintain a consistent technique and possibly other muscle groups began to compensate for the biceps muscle at times. As the electrodes were only placed on the biceps muscle, it is not possible to judge which other muscle groups participated in this exercise.

For each participant the mean value of the maximum Euler angles and biceps brachii electrical potential from the peak values as well as their standard deviation and repeatability of the exercise for the first five and last five elbow flexions were calculated. The results for the first five elbow flexors for all participants are shown in Table 1 and for the last five elbow flexors in Table 2. The first column of the tables shows the participant's serial number, the second column shows the calculated values (*aa* – mean value, SD - standard deviation and *arep* - exercise repeatability). Numbers that are highlighted using "bold" font are maximum values when comparing values in both tables. In this case, the "-" sign only indicates the direction of execution of the movement, so the absolute values of the measurements are compared.

Comparing the results between all participants, it is found that in the results obtained from sEMG, in almost all cases the values of peak amplitude and standard deviation of the measurements at the end of the last exercise were higher than at the beginning, while the repeatability of the exercise during the first five exercise sessions in all cases with sEMG is were higher. From the IMU data, it can be concluded that the mean peak Euler angle values of the amplitudes were higher at the end than at the beginning in 71% of the cases, the standard deviation of the measurements was higher at the end than at the beginning in 88% of the cases and the exercise repeatability was higher at the beginning than at the end in 80% of the cases.

When comparing the changes in the individual angles, the most pronounced changes are observed in the forearm Euler angle Φ values and the upper arm angle Θ values, where *aa* and SD have been greater than at the beginning in 91% of the cases and *arep* has been greater at the beginning than at the end.

From these results it can be concluded that when muscle fatigue occurs, the individual has difficulty in repeating the exercise in the same way each time. This change can be recorded if it is known how the individual performs the exercise before fatigue sets in, and if the repetitiveness of the exercise is less than at the start, then it can be inferred that muscle fatigue has set in.

Table 1. Results of all participants for the first five times of performing the exercise.

Participants	Calculated values	IMU from forearm			IMU from upper arm			sEMG
		Φ, °	Θ, °	Ψ, °	Φ, °	Θ, °	Ψ, °	mV*10^{-1}
1	*aa*	−163.70	**−87.81**	169.62	22.94	−44.26	**−29.48**	**3.00**
	SD	2.65	0.40	0.97	2.23	2.31	1.65	0.86
	arep	**98.38**	**99.54**	**99.43**	**90.28**	**94.78**	**94.4**	**71.33**
2	*aa*	−105.20	−67.78	129.42	24.84	−21.90	−20.26	2.93
	SD	2.87	**2.13**	5.12	3.40	1.82	4.49	0.69
	arep	**97.27**	96.86	**96.04**	86.31	**91.69**	77.84	**76.45**
3	*aa*	−128.57	−82.28	140.63	**19.90**	−17.65	**−7.31**	2.93
	SD	4.15	0.94	3.32	**6.85**	2.04	0.73	0.65
	arep	**96.77**	**98.86**	**97.64**	**65.58**	**88.44**	**90.01**	**77.82**
4	*aa*	−127.99	−63.06	**166.94**	**18.78**	**−25.63**	−14.41	1.52
	SD	6.49	1.99	7.27	**5.73**	0.91	0.98	0.27
	arep	**94.93**	**96.84**	**95.65**	69.49	**96.45**	**93.2**	**82.24**

The aim of the work was achieved and it was shown that the developed method can be used to record changes in biomechanical parameters that occur during muscle fatigue. It was proven that when muscle fatigue (decrease in exercise repeatability) was detected by sEMG, in 83% cases IMU system also detected it.

The results show that IMU system has a potential to be used for detecting muscle fatigue has the potential to be developed in the future, because this system does not have the obstacles for precise measurement as sEMG system (noise from inner organs, movement artefacts etc.). For the results to be more accurate, future studies should involve IMU system with more sensor nodes attached to different parts of body.

Table 2. Results of all participants for the last five times of performing the exercise.

Participants	Calculated values	IMU from forearm			IMU from upper arm			sEMG
		Φ, °	Θ, °	Ψ, °	Φ, °	Θ, °	Ψ, °	mV*10^{-1}
1	*aa*	**−175.82**	−84.09	**178.69**	**28.13**	**−45.24**	18.23	2.97
	SD	**12.60**	**1.88**	**11.21**	**3.32**	**4.97**	**4.77**	**1.11**
	arep	92.83	97.76	93.73	88.2	89.01	73.83	62.63
2	*aa*	**−111.16**	**−69.49**	**137.51**	**32.38**	**−27.73**	**32.47**	**4.24**
	SD	**8.11**	1.69	**8.85**	**4.38**	**3.49**	**5.70**	**1.56**
	arep	92.7	**97.57**	93.56	**86.47**	87.41	**82.45**	63.21
3	*aa*	−129.11	−84.92	152.36	5.36	**−25.59**	7.21	**4.30**
	SD	**7.62**	**1.23**	**10.32**	2.54	**5.99**	**1.36**	**1.58**
	arep	94.1	98.55	93.23	52.61	76.59	81.14	63.26
4	*aa*	**−170.64**	**−68.64**	162.00	9.37	−13.47	**21.600**	**1.61**
	SD	**8.72**	**2.83**	**16.22**	2.73	**3.99**	**10.082**	**0.43**
	arep	94.89	95.88	89.99	**70.86**	70.38	53.32	73.29

Acknowledgements. This research was funded in part by the Latvian Council of Science, project "Smart Materials, Photonics, Technologies and Engineering Ecosystem" project No VPP-EM-FOTONIKA-2022/1-0001.

References

1. Wan, J.J., Qin, Z., Wang, P.Y., Sun, Y., Liu, X.: Muscle fatigue: general understanding and treatment. Exp. Mol. Med. **49**(10), 384 (2017)
2. Morin, J.B., Samozino, P., Millet, G.Y.: Changes in running kinematics, kinetics, and spring-mass behavior over a 24-h run. Med. Sci. Sports Exerc. **43**(5), 829–836 (2011)
3. Murillo-Escobar, J., Jaramillo Munera, Y.E., Orrego-Metaute, D.A., Delgado-Trejos, E., Cuesta Frau, D.: Muscle fatigue analysis during dynamic contractions based on biomechanical features and permutation entropy. Math. Biosci. Eng. **17**(3), 2592–2615 (2020)
4. BioRadio 150 User's Guide. https://manualzz.com/doc/25749560/bioradio-150-user-s-guide
5. Lin, Z., Xiong, Y., Dai, H., Xia, X.: An experimental performance evaluation of the orientation accuracy of four nine-axis MEMS motion sensors. In: 5th International Conference on Enterprise Systems (ES), pp. 185–189. IEEE (2017)
6. Reaz, M.B.I., Hussain, M.S., Mohd-Yasin, F.: Techniques of EMG signal analysis: detection, processing, classification and applications (Correction). Biol. Proced. Online **8**, 163 (2006)

Accurate Registration of 3D Models with Organs in Augmented Reality Through Inertial Sensors

Valerio De Luca[1](✉), Roberto Bortoletto[2], Davide Ranaldo[3], and Lucio Tommaso De Paolis[1]

[1] Department of Engineering for Innovation, University of Salento, Lecce, Italy
{valerio.deluca,lucio.depaolis}@unisalento.it
[2] 221e srl, Abano Terme, Italy
roberto.bortoletto@221e.com
[3] Casa di Cura Villa Maria, Padova, Italy
info@davideranaldo.it

Abstract. Augmented Reality (AR) can improve the accuracy of surgical procedures by displaying 3D reconstructions of organs directly on the patient's body. Moreover, it can improve the surgeon's level of attention by freeing him from the task of mentally associating information from various sources. A crucial aspect for the reliability of an AR system is the registration process, which aims to guarantee the correspondence between the 3D model and the real organ during the whole operation.

The research work introduced in this paper aims to improve the accuracy of the registration process by updating the alignment between real and virtual organs even in presence of deformations induced by the insertion of surgical instruments or small displacements that organs might undergo during surgery. By detecting even imperceptible movements, the inertial sensors allow dynamic updating of the alignment between real and virtual organs to ensure an error of a few millimetres.

The solution introduced in this study could therefore constitute a first step towards improving precision in AR-guided surgery, although more in-depth studies are needed to assess the optimal positioning of the sensors, which in the future could be done automatically by an application that suggests it in the preparatory phase of surgery.

Keywords: Augmented Reality · Mini-invasive Surgery · Registration Accuracy · Inertial Sensors

1 Introduction

Minimally invasive surgery techniques make it possible to reduce the period of hospitalisation and the risk of postoperative complications by minimising incisions on the patient's body. In this context, augmented reality can provide a kind of 'X-ray vision' of the patient's anatomy by displaying 3D reconstructions of organs directly on the patient's body [1, 2]. Moreover, the surgeon can be guided by real-time visual and audio cues that improve his concentration and avoid damage to anatomical structures [3]. This can have

Y. Dekhtyar and I. Saknite (Eds.): NBC 2023, IFMBE Proceedings 89, pp. 33–41, 2023.
https://doi.org/10.1007/978-3-031-37132-5_5

a major impact on the accuracy of surgeries and the time required to perform them, as well as improving expected results from the effective use of minimally invasive surgery.

Fundamental to the reliability of an augmented reality system is the registration process [2], which aims to ensure the correspondence between the 3D model and the real organ throughout the entire operation [3]: high accuracy is required to guarantee an error of less than 3 mm even after deformations or displacements that the organ may undergo during surgery.

The typical methods for tracking the pose and orientation of real objects in augmented reality applications can be based on markers, sensors or features recognized through computer vision techniques [4].

For applications in medicine and surgery, where a very high level of precision and accuracy is required, professional optical trackers [5, 6] are often used: they use infrared cameras to detect the position of various semi-reflective spheres arranged according to known geometries [7]. The main limitation of optical tracking is the need for a direct line of sight between the optical markers and the camera sensor, which is rigidly fixed in the operating room.

An alternative solution may be electromagnetic tracking, based on an electromagnetic field generator: it does not require such a direct line of sight, but the presence of metallic or ferromagnetic sources in the working area may cause interference in the measurement [8].

To ensure maximum accuracy, however, it is necessary to detect even the smallest deformations of anatomical structures, which can be induced either by the insertion of surgical instruments or by the patient's normal cardio-respiratory activity.

Organ-specific biomechanical models [9] can provide an accurate estimate of surface deformations, but would not be able to provide the result in real time, especially in the presence of significant or non-linear deformations [10].

To make the problem computationally tractable, sensors or markers can be placed on the patient's body. The workflow described in [11] for liver resection surgery is based on a finite element model obtained from preoperative image segmentation and driven by 3D markers tracked by infrared cameras.

This paper aims to introduce hybrid approaches based on the combined use of inertial sensors with visual markers detected by a stereo-camera with reference to orthopaedic hip and knee surgery. Moreover, an approach to integrate semi-reflective spheres detected by an optical tracker into the considered scenario will be evaluated.

2 Methods

2.1 Hybrid Tracking Architecture

For the first prototype of the application, the HTC VIVE Pro 2 system [12] (Fig. 1) was chosen, which can offer a more detailed and higher-quality visual experience than stand-alone visors. The VIVE Pro 2 headset supports a maximum overall resolution of 4896×2448 pixels at a refresh rate of 120 Hz and provides a field of view of 120°. Two handheld controllers are supplied with the device, through which the user can interact in a virtual environment. The movements of the user's head and the position and orientation of the two handheld controllers are tracked by two infrared cameras.

Although it was initially designed for virtual reality [13, 14], the VIVE headset can be used also as a video-see-through augmented reality display thanks to the integrated stereo camera and the SRWorks SDK available on the VIVE site for developers [15]. Whereas optical-see-through augmented reality superimposes additional information directly onto the scene seen by human eyes through the semi-transparent lenses of a visor, video-see-through augmented reality superimposes additional information into the video stream captured by a camera and presented to the user.

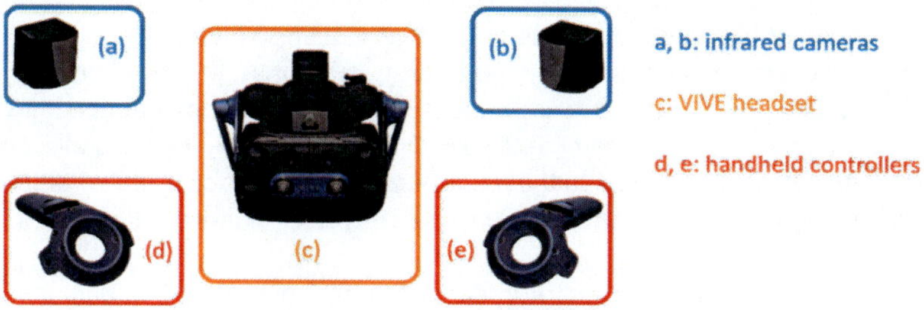

Fig. 1. HTC VIVE Pro 2 Virtual Reality system.

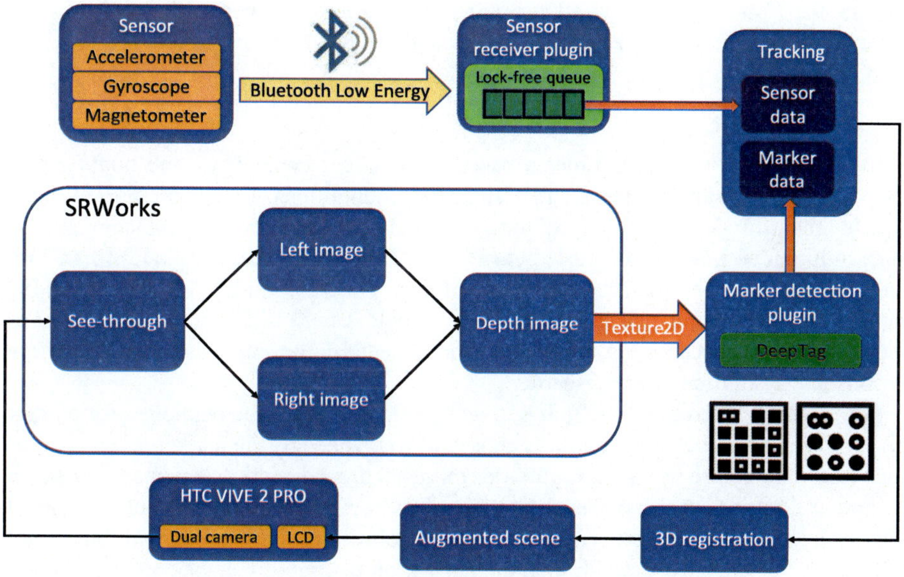

Fig. 2. System architecture.

For the development of the application, the Unity game engine [16] was chosen, which supports several desktop, mobile, console and extended reality platforms. The Unity engine is able to run C# scripts, associated with some game objects, that define a number of event functions to be executed in a predetermined order [17]. The *Update* function in particular, called once per frame, can be used to dynamically update the tracking data and more generally the scene in augmented reality.

Figure 2 shows the general architecture designed for a hybrid tracking based on inertial sensors and visual markers. The see-through module of SRWorks enables the acquisition of the real environment. The left and the right images captured by the dual camera are combined into a depth texture, which can be retrieved by means of the *GetDepthTexture* method of the *ViveSR_DualCameraImageCapture* class and passed on to a plugin for the detection of visual markers attached to inertial sensors sending accelerometer, gyroscope and magnetometer data. More details on marker detection and inertial sensors will be provided in Subsects. 2.2 and 2.3 respectively.

In order to guarantee compatibility with Unity and portability on various types of platforms, variants of a sensor receiver plugin were prepared for communication via the Bluetooth Low Energy (BLE) protocol. In particular, considering both stand-alone and PC-connectable visors in addition to the traditional desktop platform, efforts were made to ensure compatibility with x86, Android and Universal Windows Platform (UWP) platforms.

To ensure the efficient reception of updated data in Unity, lock-free data structures and atomic calls for updating variables were used in the plugin implementation. A thread running within the plugin atomically updates a variable as soon as a message containing new data is received from the sensors. At the same time, to read updated data from the sensors, the *Update* method of a Unity script calls a plugin function that accesses the contents of the same variable atomically and without blocking.

2.2 Marker Detection

A correct registration of 3D objects based on markers depends on the quality, size and spatial disposition of markers [4]. The experimental measurements reported in [4] revealed that, for Vuforia SDK [18], the root mean squared error in AR placement is of some millimetres when the distance between the marker and the device is less than 2 m, but it can reach even 17 cm for a distance of 6 m. On the contrary, ARcore SDK [19] seems to provide better placement accuracy, which is not influenced by distance. For both frameworks, accuracy is significantly influenced by lighting conditions. For these reasons, these solutions were discarded.

The DeepTag framework [20], based on deep learning, can be employed for marker design and accurate detection and was chosen to be invoked by the marker detection plugin in the architecture of Fig. 2. It provides more accurate detection and pose estimation than other libraries even in presence of severe Gaussian noise or motion blur. In typical marker detection frameworks [21–26] digital symbols are represented with black and white cells and decoded through image binarization to simplify shape analysis. ArUco [27, 28] tries to increase the intermarker distance and the number of bit transitions in the coding system. TopoTag [29] tries to improve the accuracy of pose estimation achievable through a checkerboard design by exploiting more keypoints. Instead of relying on edge detection and image binarization, DeepTag directly regresses keypoints and digital symbols from local shapes with Convolutional Neural Networks (CNN). In this way, it is able to recognize both traditional markers and new markers with customised local patterns.

2.3 Inertial Sensors

The adopted sensor model is Muse V3, produced by 221e S.r.l. [30], which combines inertial and environmental sensors together with on-board flash storage and Bluetooth Low Energy (BLE) connectivity. It runs proprietary algorithms that range from orientation estimation to embedded time-frequency analysis for predictive maintenance applications. It includes the following main components:

- Digital inertial sensors, which are gyroscope (up to ±2000 dps), accelerometer (up to ±32g), magnetometer (±50G);
- Digital 3-axis High Dynamic Range (HDR) accelerometer (up to ±400 g);
- Temperature (accuracy up to ±0.1 °C, −40…125 °C) and relative humidity (accuracy up to ±1.5%RH, 0…100%RH) sensors;
- Digital barometer (accuracy 0.5 hPa, 260–1260 hPa);
- Light and proximity sensor (0.0022–73000 lx, detection range up to 160 mm).

2.4 Use Case: Orthopaedic Hip and Knee Surgery

As a first scenario, a sawbone leg model (Fig. 3) will be considered to simulate orthopaedic hip and knee surgery. Two Muse V3 inertial sensors can be placed at the centres of mass of the femur and tibia, respectively. A visual marker printed on paper can be attached above each sensor.

Femur and tibia pose and joint angles can be estimated from marker position data and accelerometer, gyroscope and magnetometer data provided by the sensors using recursive updating algorithms based on a Kalman filter [31, 32].

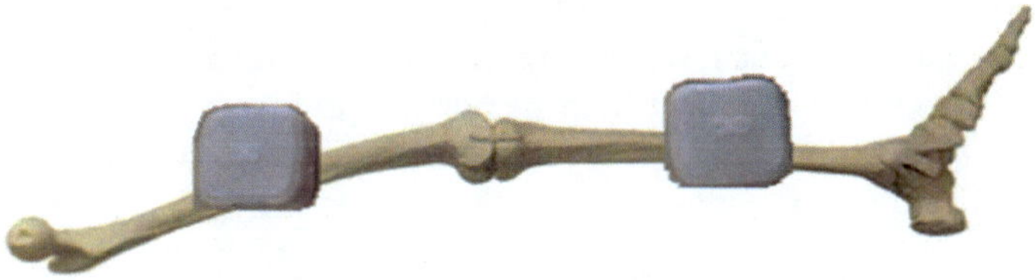

Fig. 3. A sawbone leg model with Muse V3 sensors placed on it.

2.5 The Addition of an Optical Tracker

The combined use of sensors, visual markers and optical trackers could allow both a comparative evaluation of the various combinations that can be used for tracking and the evaluation of more complex data fusion techniques that can effectively aggregate data from these three different types of technologies.

The Polaris Vicra optical tracker [5], produced by NDI Inc., exploits two infrared cameras and a position sensor to detect infrared-emitting or retroreflective spheres affixed to a tool or object (Fig. 4); using the information received from the spheres, the device (273 mm × 69 mm × 69 mm) is able to know the position and orientation of the tools

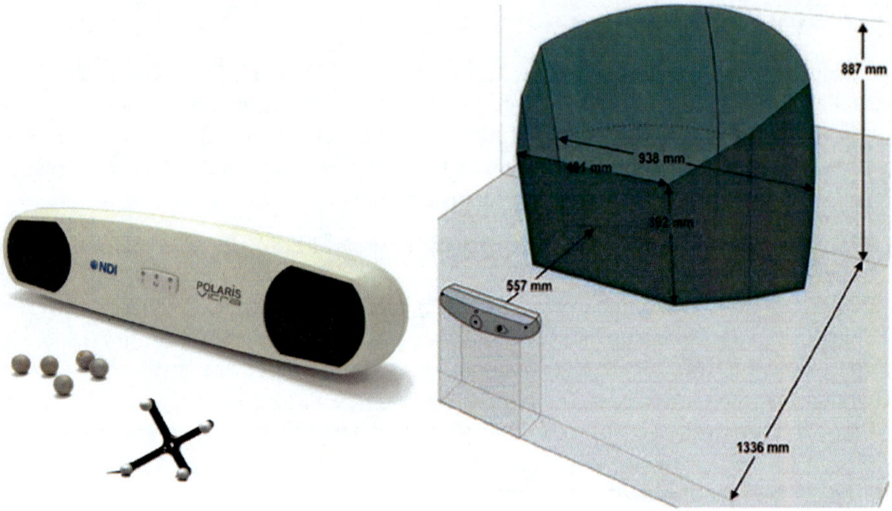

Fig. 4. Polaris Vicra tracker with reflective spheres and its measurement volume.

within a specific measurement volume, depicted in Fig. 2, with an accuracy of 0.2 mm and 0.1 tenth of a degree. Data are updated at a maximum rate of 20 Hz.

The Vicon Bonita [6] tracking system provides greater stability and robustness thanks to four infrared cameras working at a frame rate of 240 fps. It provides a wide angle of view measuring 82.7° × 66.85° and a narrow angle of view measuring 32.7° × 24.81°.

In a hybrid scenario including markers, inertial sensors and optical tracking, RUNE-Tag markers [33], which are also compatible with DeepTag, could be used: they consist of a set of high-contrast circular features spatially arranged in concentric layers and allow identification and pose estimation even in the presence of severe occlusions. The centre of such markers, which is left empty, could accommodate the semi-reflective spheres used by the optical trackers.

3 Conclusions and Future Work

In the present work, some hybrid approaches have been outlined to ensure accurate tracking in an augmented reality system for surgery, which is necessary for the correct updating of 3D model registration with respect to the patient's anatomy. The adoption of small inertial sensors equipped with Bluetooth Low Energy connectivity opens up interesting research scenarios, thanks to the possibility of affixing above them both visual markers detected by a stereo camera and semi-reflective spheres detected by more complex optical trackers.

In future steps of the experimentation, the adoption of additional stereo cameras, which have a higher resolution than that integrated in the VIVE headset, will be considered: in particular, both small pluggable cameras on the headset (such as OAK-D [34] or ZED Mini [35]) and external cameras that can be installed in a fixed, frontal position with respect to the operating scenario will be considered (such as Azure Kinect DK [36]).

Moreover, the optimal positioning of the sensors will be studied in more detail: in particular, a software tool will be developed that can detect this optimal positioning automatically and suggest it to the surgeon in a virtual reality environment that can be used for preoperative planning.

References

1. Barcali, E., Ladanza, E., Manetti, L., Francia, P., Nardi, C., Bocchi, L.: Augmented reality in surgery: a scoping review. Appl. Sci. (Switz.) **12**(14) (2022). https://doi.org/10.3390/app12146890
2. Ma, L., Huang, T., Wang, J., Liao, H.: Visualization, registration and tracking techniques for augmented reality guided surgery: a review. Phys. Med. Biol. **68**(4) (2023). https://doi.org/10.1088/1361-6560/acaf23
3. De Paolis, L.T., De Luca, V.: Augmented visualization with depth perception cues to improve the surgeon's performance in minimally invasive surgery. Med. Biol. Eng. Comput. **57**(5), 995–1013 (2018). https://doi.org/10.1007/s11517-018-1929-6
4. El Barhoumi, N., Hajji, R., Bouali, Z., Ben Brahim, Y., Kharroubi, A.: Assessment of 3D models placement methods in augmented reality. Appl. Sci. (Switz.) **12**(20) (2022). https://doi.org/10.3390/app122010620
5. NDI Polaris Vicra. https://www.ndigital.com/optical-measurement-technology/polaris-vicra. Accessed 16 Apr 2023
6. Vicon Bonita. https://biomech.hacettepe.edu.tr/manuals/BCO517_9_BonitaCams.pdf. Accessed 16 Apr 2023
7. De Paolis, L.T., Ricciardi, F.: Augmented visualisation in the treatment of the liver tumours with radiofrequency ablation. Comput. Meth. Biomech. Biomed. Eng. Imaging Vis. **6**(4), 396–404 (2018). https://doi.org/10.1080/21681163.2017.1287598
8. Sorriento, A., et al.: Optical and electromagnetic tracking systems for biomedical applications: a critical review on potentialities and limitations. IEEE Rev. Biomed. Eng. **13**, 212–232 (2020). https://doi.org/10.1109/RBME.2019.2939091
9. Chabanas, M., Alvarez, P., Rouzé, S., Castro, M., Dillenseger, J.-L., Payan, Y.: Lung deformation between preoperative CT and intraoperative CBCT for thoracoscopic surgery: a case study. In: Progress in Biomedical Optics and Imaging - Proceedings of SPIE, SPIE 2018 (2018). https://doi.org/10.1117/12.2293938
10. Mendizabal, A., Tagliabue, E., Hoellinger, T., Brunet, J.N., Nikolaev, S., Cotin, S.: Data-driven simulation for augmented surgery. Adv. Struct. Mater. **132**, 71–96 (2020). https://doi.org/10.1007/978-3-030-50464-9_5
11. Adagolodjo, Y., Golse, N., Vibert, E., De Mathelin, M., Cotin, S., Courtecuisse, H.: Marker-based registration for large deformations - application to open liver surgery. In: Proceedings of the IEEE International Conference on Robotics and Automation, pp. 4007–4012. Institute of Electrical and Electronics Engineers Inc. (2018). https://doi.org/10.1109/ICRA.2018.8462909
12. VIVE Pro 2 Headset. https://www.vive.com/us/product/vive-pro2/overview. Accessed 16 Apr 2023
13. De Paolis, L.T., De Luca, V.: The impact of the input interface in a virtual environment: the Vive controller and the Myo armband. Virtual Reality **24**(3), 483–502 (2019). https://doi.org/10.1007/s10055-019-00409-6

14. De Paolis, L.T., De Luca, V.: The effects of touchless interaction on usability and sense of presence in a virtual environment. Virtual Real. **26**(4), 1551–1571 (2022). https://doi.org/10.1007/s10055-022-00647-1
15. SRWorks SDK. https://developer-express.vive.com/resources/vive-sense/srworks-sdk/overview. Accessed 16 Apr 2023
16. Unity. https://unity.com. Accessed 16 Apr 2023
17. Unity Manual - Order of execution for event functions. https://docs.unity3d.com/Manual/ExecutionOrder.html. Accessed 16 Apr 2023
18. Vuforia. https://developer.vuforia.com. Accessed 16 Apr 2023
19. ARcore. https://developers.google.com/ar/develop. Accessed 16 Apr 2023
20. Zhang, Z., Hu, Y., Yu, G., Dai, J.: DeepTag: a general framework for fiducial marker design and detection. IEEE Trans. Pattern Anal. Mach. Intell. **45**(3), 2931–2944 (2023)
21. Olson, E.: AprilTag: a robust and flexible visual fiducial system. In: Proceedings of the IEEE International Conference on Robotics and Automation, pp. 3400–3407 (2011). https://doi.org/10.1109/ICRA.2011.5979561
22. Wang, J., Olson, E.: AprilTag 2: efficient and robust fiducial detection. In: IEEE International Conference on Intelligent Robots and Systems, pp. 4193–4198. Institute of Electrical and Electronics Engineers Inc. (2016). https://doi.org/10.1109/IROS.2016.7759617
23. Fiala, M.: ARTag, a fiducial marker system using digital techniques. In: Proceedings of the 2005 IEEE Computer Society Conference on Computer Vision and Pattern Recognition, CVPR 2005, pp. 590–596. IEEE Computer Society (2005). https://doi.org/10.1109/CVPR.2005.74
24. Fiala, M.: Designing highly reliable fiducial markers. IEEE Trans. Pattern Anal. Mach. Intell. **32**(7), 1317–1324 (2010). https://doi.org/10.1109/TPAMI.2009.146
25. Flohr, D., Fischer, J.: A lightweight ID-based extension for marker tracking systems. In: Froehlich, B., Blach, R., van Liere, R. (eds.) Eurographics Symposium on Virtual Environments, EGVE 2007 - Short Papers and Posters, pp. 59–64. The Eurographics Association (2007)
26. Romero-Ramirez, F.J., Muñoz-Salinas, R., Medina-Carnicer, R.: Speeded up detection of squared fiducial markers. Image Vis. Comput. **76**, 38–47 (2018). https://doi.org/10.1016/j.imavis.2018.05.004
27. Garrido-Jurado, S., Muñoz-Salinas, R., Madrid-Cuevas, F.J., Marín-Jiménez, M.J.: Automatic generation and detection of highly reliable fiducial markers under occlusion. Pattern Recognit. **47**(6), 2280–2292 (2014). https://doi.org/10.1016/j.patcog.2014.01.005
28. Garrido-Jurado, S., Muñoz-Salinas, R., Madrid-Cuevas, F.J., Medina-Carnicer, R.: Generation of fiducial marker dictionaries using Mixed Integer Linear Programming. Pattern Recognit. **51**, 481–491 (2016). https://doi.org/10.1016/j.patcog.2015.09.023
29. Yu, G., Hu, Y., Dai, J.: TopoTag: a robust and scalable topological fiducial marker system. IEEE Trans. Vis. Comput. Graph. (TVCG) **27**(9), 3769–3780 (2021)
30. 221e S.r.l. https://www.221e.com/. Accessed 16 Apr 2023
31. Palani, P., Panigrahi, S., Jammi, S.A., Thondiyath, A.: Real-time joint angle estimation using Mediapipe framework and inertial sensors. In: Proceedings - IEEE 22nd International Conference on Bioinformatics and Bioengineering, BIBE 2022, pp. 128–133. Institute of Electrical and Electronics Engineers Inc. (2022). https://doi.org/10.1109/BIBE55377.2022.00035
32. Li, T., Yu, H.: Upper body pose estimation using a visual-inertial sensor system with automatic sensor-to-segment calibration. IEEE Sens. J. **23**(6), 6292–6302 (2023). https://doi.org/10.1109/JSEN.2023.3241084
33. Bergamasco, F., Albarelli, A., Rodola, E., Torsello, A.: RUNE-Tag: a high accuracy fiducial marker with strong occlusion resilience. In: 2011 IEEE Conference on Computer Vision and Pattern Recognition (CVPR), pp. 113–120 (2011). https://doi.org/10.1109/CVPR.2011.5995544

34. OpenCV AI Kit: OAK-D. https://store.opencv.ai/products/oak-d. Accessed 16 Apr 2023
35. ZED Mini - Mixed-Reality Camera. https://www.stereolabs.com/zed-mini. Accessed 16 Apr 2023
36. Azure Kinect DK. https://azure.microsoft.com/it-it/products/kinect-dk. Accessed 16 Apr 2023

Dynamic Testing of Piezoresistive Fabrics for Use in Smart Wearable Personal Protective Equipment

Aleksandrs Vališevskis(✉), Uģis Briedis, and Alexander Oks

Faculty of Materials Science and Applied Chemistry, Institute of Design Technologies, Riga Technical University, Kipsalas iela 6, Riga, Latvia
Aleksandrs.Valisevskis@rtu.lv

Abstract. The aim of this paper is to study the materials that can be used for the impact detection and the force measurement in personal protective equipment. Several commercially available piezoresistive materials are investigated (EeonTex LTT-SLPA, Sefar Carbotex 03-120CF/03-160CF/03-205CF) and compared to proprietary DAid knitted piezoresistive fabric. It is demonstrated that the operating range of these materials can be extended by adding fiberglass mesh, which functions as a spatial separator and reinforcing layer, as well as by combining homogenous and heterogeneous materials in a single multilayer sensor. It is shown that EeonTex LTT-SLPA and Sefar Carbotex 03-120CF, used either separately or together, perform better than other studied materials. The derivative approximation can be used for more accurate signal analysis. The proposed solutions allow to increase the sensor operational range to a certain extent. A number of potential solutions is presented for the cases, when sensors are subjected to impacts that have substantially higher energy values.

Keywords: textile pressure sensor · personal protective equipment

1 Introduction

In the field of protective smart clothing such topics as the monitoring of the wearer's vital signs and spatial position are thoroughly studied [1]. On the other hand, solutions for the monitoring of hazardous events, such as impacts with foreign objects, have been studied to a much lesser extent. A matrix pressure sensor integrated into the personal protective equipment (PPE) can serve as a solution for monitoring of the aforementioned risk. Such sensors can be constructed in a number of ways, which differ by their principle of operation: capacitive sensors [2], piezoelectric sensors [3], triboelectric sensors [4], and piezoresistive sensors [5]. This study focuses on the piezoresistive sensors, due to their high sampling rate and constructive simplicity. The duration of the impact is 6–10 ms in case of a 180–360 g impactor travelling at 6 m/s, thus, a sampling rate of at least several kHz is required to gather enough data points [6].

In the core of such sensors there is a piezoresistive material, thus, by measuring the changes in electric resistivity it is assumed that the applied force can be assessed. The

Y. Dekhtyar and I. Saknite (Eds.): NBC 2023, IFMBE Proceedings 89, pp. 42–49, 2023.
https://doi.org/10.1007/978-3-031-37132-5_6

piezoresistive material can be chosen from a variety of different options. Some prototypes use Velostat or Linqstat films [6], which is a polymeric film impregnated with carbon black, which makes it electroconductive and imparts piezoresistive properties due to the percolation effect [7]. Such films are not breathable and are not well suited for the adaptation to anatomical 3D shapes, required for PPE. The same can be said about specially prepared piezoresistive compounds and coatings, usually made by mixing an elastomer matrix and a conductive filler, such as carbon black [8]. Taking into account the intended application in PPE and the required compatibility with the textile structure, the following textile piezoresistive materials were chosen for the current study: *EeonTex LTT-SLPA 60* kΩ conductive polymer-coated knitted fabric and *Sefar Carbotex 03-120CF/03-160CF/03-205CF* carbon/polyester woven fabrics. Besides that, the behaviour of these commercial fabrics is compared to that of the proprietary knitted textile piezoresistive material [9].

Such PPE, coupled with a monitoring system, would provide a major benefit for the safety and the health protection of the personnel operating in remote locations and/or in an environment that involves a risk of flying objects, e.g., first responders, workers in mines, quarries, roofers, construction workers etc. In our recent study we explored the electrophysical properties of the aforementioned fabrics under a slowly changing load, which can be used in textile pressure sensors [10]. However, the behaviour of these fabrics under a dynamic load, e.g., an impulse-like impact with a foreign object, is substantially different due to the structure of the materials.

2 Materials and Methods

2.1 Materials

Since the aim of this study is to select the materials, which are suitable for the use in the wearable PPE for impact detection, several textile piezoresistive materials have been chosen for testing. These materials can be easily tailored and integrated into the textile structures, such as vests or jackets.

EeonTex LTT-SLPA 60 kΩ (hereinafter called *Sample 1*), which is selected for the tests, which is a bi-directionally stretchable elastic knitted fabric containing 72% nylon and 28% spandex, and is coated with a proprietary conductive formulation, based on conductive polymers. Its area density is 163 g/m^2 and it has a thickness of 0,38 mm [11]. It is made by Eeonyx Corp., Pinole, CA 94564.

The other materials are from the *Sefar Carbotex* line made by Sefar AG, Heiden, Switzerland. These are woven fabrics, composed of polyester threads in the warp direction and unidirectional weft threads, made of a carbon-loaded composite material, which alternate with non-conductive polyester threads [12]. The structure of the Carbotex fabrics is shown in Fig. 1. Based on our previous study [10], these variants have been chosen for the experiments: Sefar Carbotex 03-120CF, 03-160CF, 03-205CF (hereinafter called, respectively, *Sample 2*, *Sample 3*, and *Sample 4*).

By combining the threads that have different dimensions and the resulting contact area, their properties can be adjusted. The detailed properties of the Sefar Carbotex fabrics are presented in Table 1.

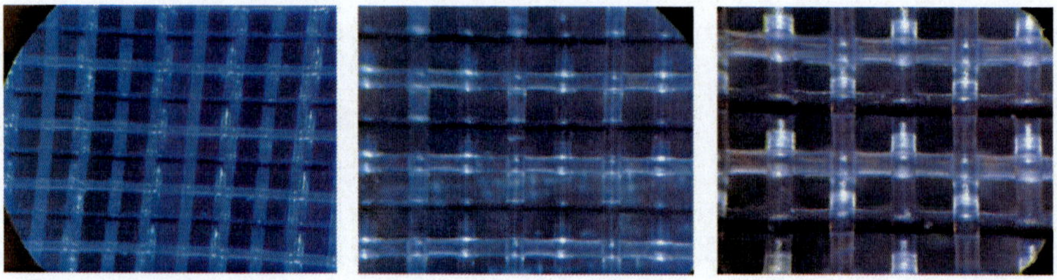

(a) Carbotex 03-120CF (b) Carbotex 03-160CF (c) Carbotex 03-205CF

Fig. 1. Structure of the Sefar Carbotex fabrics at 100x magnification.

Table 1. Properties of Sefar Carbotex fabrics

Product	Mesh opening [μm]	Open area [%]	Mesh count, warp [n/cm]	Mesh count, weft [n/cm]	Weight [g/m^2]	Thickness [μm]
03-120 CF	120	42	51	55	35	135
03-160 CF	160	44	41	46	44	148
03-205 CF	205	40	31	31	85	225

To perform the necessary measurements, the piezoresistive material is placed between strips of conductive fabric Nora-PW made by Shieldex. It is a highly conductive Ni/Cu/Ag plated nylon woven fabric with surface resistivity less than *0,03* Ω/sq., its thickness is *0,102* mm and its area density is *83* g/m^2 [13].

Since the study is focused on PPE with a potential of high-energy impacts, a number of sensors, which were tested, comprised an integrated reinforcing layer, which is formed by a fibreglass mesh with the following properties: thickness *0,04* mm, mesh size *3* × *4* mm.

2.2 Methods

In order to accomplish dynamic tests of the selected materials, an *ad hoc* drop weight tower has been constructed. The tower consists of a base with a sample holder, to which a mast is attached. The mast holds a 2 m tube with a 40 mm inner diameter, which is centred above the sample holder. Impactors are dropped through the tube and land precisely on the designated spot with their tip aligned perpendicularly to the sample. The impactors have a hemispherical tip with a 30 mm diameter. Two types of impactors are used in this set of experiments, which differ by weight: 180 g and 360 g. These impactors reach the speed of 6,2 m/s and thus, their kinetic energy is respectively 3,5 J and 7 J just before the impact. A 20 mm thick plate made from high density polyethylene foam is placed under the sample, in order to protect the structure of the rig, which also mimics the elasticity of human body. At least three impacts with each impactor weight were obtained for each sample.

An electronic data collection system based on Arduino was built for the signal acquisition and pre-processing. The system measures the resistance with a sampling rate of *3–5* kHz and has a built-in threshold for impact detection. Measuring circuit voltage is *5* V and it is based upon a voltage divider with a fixed *4,7* kΩ resistor. Due to limitations of the onboard memory, resistance is measured with *10 Ω* resolution.

Although the materials have shown a good performance during the tests with a static load (e.g., see Fig. 2), the dynamic tests have indicated that these materials have a limited dynamic range. Figure 3 shows the data obtained for EeonTex fabric: *3,5* J impacts (green line) can be distinguished from *7* J impacts (orange line): *7* J impacts have a lower resistance and a steeper gradient. However, during *7* J impacts the resistance drops to low values, and does not leave room for assessing stronger impacts. In the realm of PPE, *7* J is a modest impact, thus, in order to be able to assess energy of stronger impacts, it is necessary to increase the dynamic range of these materials by increasing their resistance (for a greater sensitivity during stronger impacts).

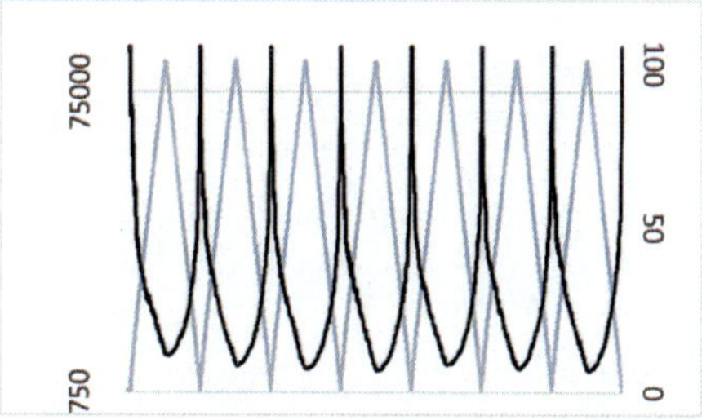

Fig. 2. Static loading of EeonTex fabric: 0–100 N compression force (grey), resistance in Ohms (black).

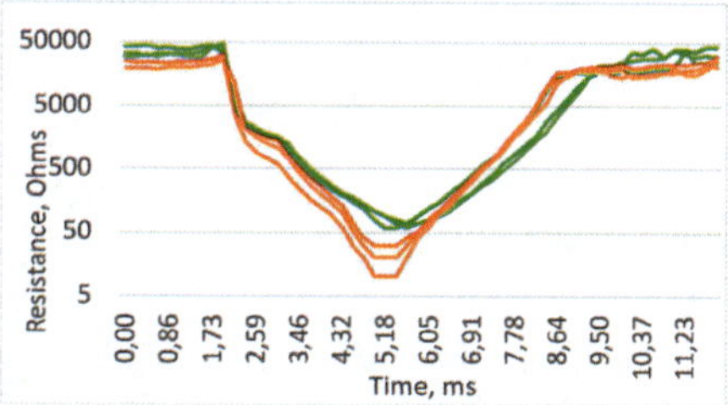

Fig. 3. Dynamic tests, EeonTex fabric: *3,5* J (green) and *7* J (orange) impactor energy.

In order to increase sensors' sensitivity to higher energy impacts, a number of solutions have been tried experimentally, namely:

a. a fiberglass mesh reinforcing layer has been introduced, which ensures a spatial separation of the sensors' layers and at the same time increases its capacity to withstand higher energy impacts;
b. double layers of materials have been tested. Materials of Carbotex family are unidirectional, with conductive threads present only in weft direction, thus, by placing two layers of such material, arranged perpendicularly to each other, one may expect to obtain new properties;
c. materials of different types have been combined in one sensor. EeonTex is a dense, knitted material and has a polymeric coating, while Carbotex is woven and contains carbon composite threads.

3 Results and Analysis

3.1 Introducing Fiberglass Mesh

In order to make these experiments, the piezoresistive layer was inserted between two layers of fiberglass mesh, which ensured spatial separation from the Shieldex Nora-PW fabric, used for signal read-out.

In Fig. 4 and 5 results for Sample 1 and Sample 2 are presented respectively.

Introduction of a fiberglass layer had a more substantial impact on Carbotex materials, as can be seen in Fig. 5 for Sample 2. Performance of Sample 1 (see Fig. 4) was less noisy, but this approach did not have a big influence on the overall resistance. Resistance of Sample 3 dropped to 50–60 Ω in case of 3,5 J impactor and 20–30 Ω for 7 J impactor. In case of Sample 4 the resistance was in the 10–20 Ω range for both impactors. Besides that, signals had a similar slope.

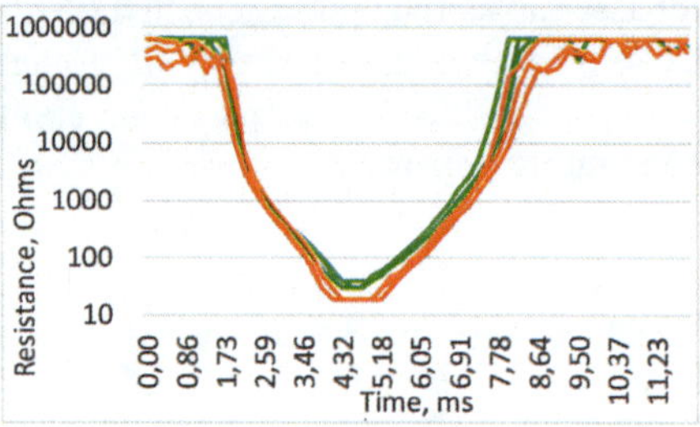

Fig. 4. Sample 1 separated by two fiberglass mesh layers: *3,5* J (green) and *7* J (orange) impactor energy.

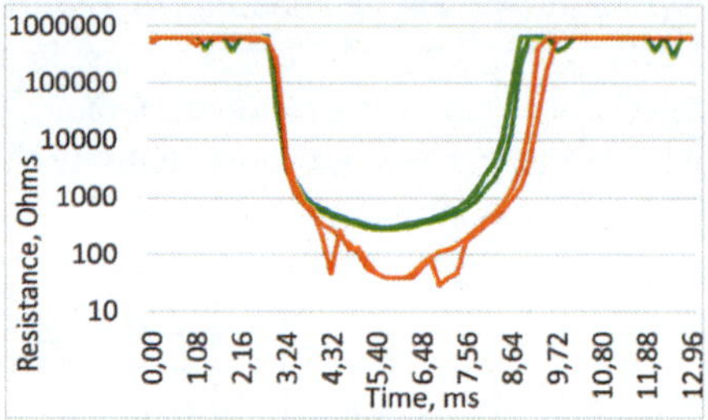

Fig. 5. Sample 2 separated by two fiberglass mesh layers: *3,5* J (green) and *7* J (orange) impactor energy.

3.2 Double Layer of Piezoresistive Material

For these tests sensors had two layers of piezoresistive materials. In case of Carbotex fabrics the two layers were arranged perpendicularly in relation to each other.

The results of the experiments that were carried out with the piezoresistive layers only were not satisfactory, because (1) in the case of EeonTex fabric the intended increase of resistance was not obtained, and (2) in the case of Carbotex materials noise was introduced, which complicated the signal analysis.

In order to overcome these drawbacks, fiberglass layers were inserted between the layers in order to ensure spatial separation and mechanical stability. In total, three layers of fiberglass mesh were used – above and below the piezoresistive layers, as well as between them.

In Fig. 6 and 7 the results for Sample 1 and Sample 2 are presented respectively.

This solution proved to be advantageous for all materials. Resistance of Sample 3 was slightly lower than that of Sample 2: 170–280 Ω in case of 3,5 J and 70–120 Ω in case of 7 J impact. Resistance of Sample 4 was even lower (50 Ω in case of 3,5 J and 20 Ω in case of 7 J impactor). As can be seen in Fig. 6 and 7, a clear separation between two energetic levels of impact is ensured.

3.3 Combining Different Types of Materials

In the last set of experiments, a sensor consisting of two layers of piezoresistive materials were tested. The Sample 1 was used in the first layer and, as the second layer, three types of Carbotex materials were tested: 03-120CF (Sample 2), 03-160CF (Sample 3), and 03-205CF (Sample 4). The direct combination of the materials did not produce the desired

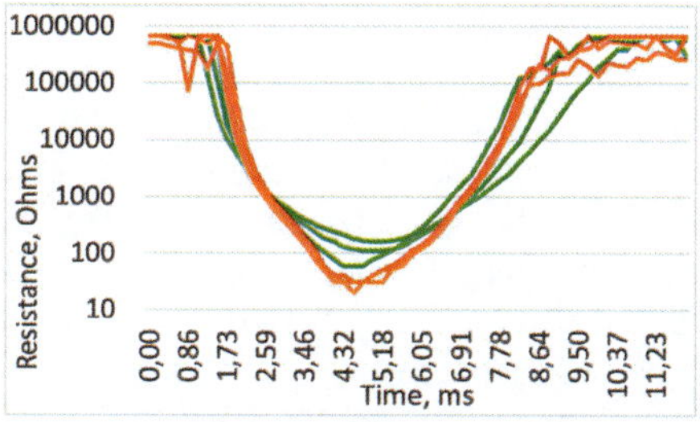

Fig. 6. Two layers of Sample 1 separated with three layers of fiberglass mesh: *3,5J* (green) and *7* J (orange) impactor energy.

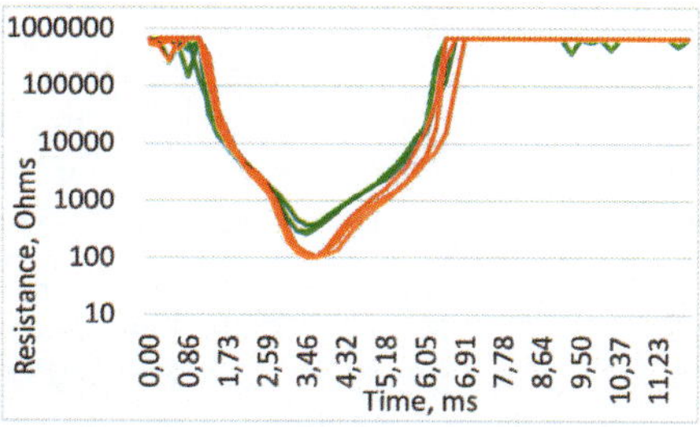

Fig. 7. Two layers of Sample 2 separated with three layers of fiberglass mesh: *3,5* J (green) and *7* J (orange) impactor energy.

effect, since the overall resistance has decreased and, even in case of the 3,5 J impactor, it dropped to 30 Ω.

Since Sample 1 and Sample 2 have demonstrated better performance and, besides that, the addition of fiberglass layers proved to be advantageous, the final tests have been performed by combining two layers of Sample 2 with Sample 1 and fiberglass meshes in such a way: mesh – Sample 2 – mesh – Sample 1 – mesh – Sample 2 – mesh.

The rationale behind adding Sample 1 fabric into the final sensor is based on its dense knitted structure. Sample 2 has low density, which, in case of higher static loads, has caused outer conductive traces to be pushed through this material, causing short-circuits in the sensors.

The results obtained with such a sensor configuration are shown in Fig. 8. Although it ensures a clear separation between 3,5 J and 7 J impactor energies, the minimum resistance was 80–110 Ω in case of lower energy and 50 Ω in case of higher energy, which leaves only some room for higher energy impacts. Hence, if the sensor is intended for use in environments with the risk of impacts that have much higher energy (an order of magnitude, which is our empirical assumption), other materials or sensing strategies should be employed, which are touched upon in the next section.

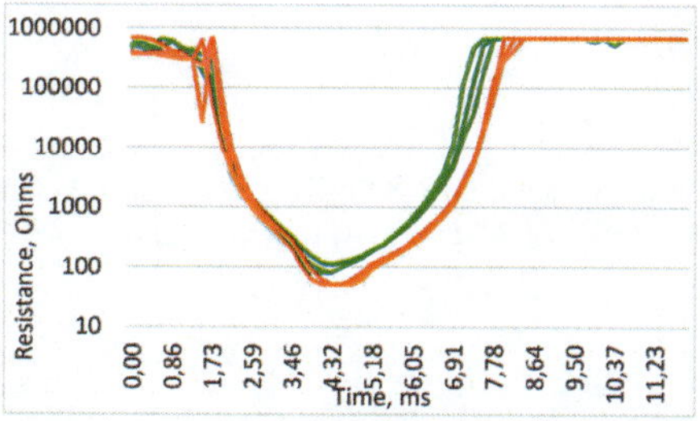

Fig. 8. Two layers of Sample 2 and one layer of Sample 1 separated with fiberglass meshes: *3,5J* (green) and *7J* (orange) impactor energy.

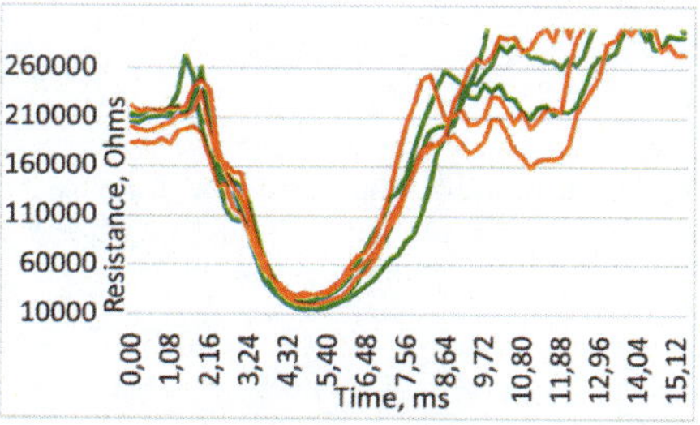

Fig. 9. Proprietary knitted fabric: *3,5J* (green) and *7J* (orange) impactor energy.

4 Discussion

As it is described in Sect. 3 and shown in Figs. 4, 5, 6, 7, and 8, the proposed solutions allow to extend sensors' range to higher loads. The use of fiberglass mesh separators has an advantageous effect on the sensors' operation. While Sample 2 has shown good results and ability to differentiate impacts that have different energy levels, in the previous tests it was prone to structural deformations, which led to short-circuits [10]. On the other hand, Sample 1 has high sensitivity and much denser structure.

The presented solutions allow to increase the range to a certain extent, but other solutions would be necessary in case of substantially higher impact energies. Some of the possible solutions can be: (a) use of embroidered traces instead of strips of conductive fabrics – this will decrease the contact surface and the resistance will increase; (b) analysing adjacent points in an array matrix sensor – by analysing multiple points one can estimate the force of the impact by taking measurements around the impact centre; (c) producing proprietary elastomeric piezoresistive materials with a lower load of conductive particles.

Much higher resistance values were obtained with a proprietary knitted piezoresistive fabric [9] (see Fig. 9), which indicates that it could be used for detecting impacts of higher energy. Although further testing is necessary by applying higher loads, because it did not show a sufficient differentiation ability with less energetic impacts.

By looking at the obtained data one can notice that different impact energies influence not only the depth of the curve, but also its slope. Thus, analytical capabilities of the sensing system can be improved by using, e.g., approximation of the first order derivative. In Fig. 10 a derivative approximation (*central finite difference*) is presented for the signal that is shown in Fig. 8.

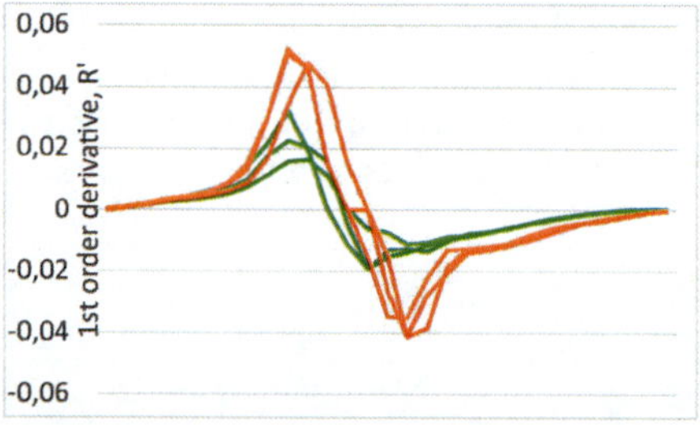

Fig. 10. Derivative approximation for a sensor with two layers of Sample 2 and one layer of Sample 1 separated with fiberglass meshes: *3,5* J (green) and *7* J (orange) impactor energy.

5 Conclusion

In this paper several commercially available piezoresistive fabrics are analysed. It is shown that, by adding reinforcing fiberglass layers, the ability to assess higher energy impacts can be improved. EeonTex LTT-SLPA (Sample 1) and Carbotex 03-120CF (Sample 2) have shown a better performance during experiments. Besides that, the operation can be improved by combining these fabrics in a single sensor. Additional data processing, e.g., approximation of the first order derivative, can be used to improve operation of the sensing system. In the case of higher impact energies other solutions should be sought, such as the use of embroidered threads, a broader 2D data analysis or developing piezoresistive material with higher resistivity. Further studies will focus on the sensor behaviour in case of impacts that have substantially higher energy values.

Acknowledgements. This research is funded by the Ministry of Defence project "Graphene-and-silica-aerogel-enhanced lightweight ballistic protection vest prototype with integrated pressure-sensitive layer for multi-zone impact detection", project No. VPP-AIPP-2021/1-0009".

References

1. Santos, G., et al.: Firefighting: challenges of smart PPE. Forests **13**, 1319 (2022)
2. Samarentsis, A.G., Makris, G., Spinthaki, S., Christodoulakis, G., Tsiknakis, M., Pantazis, A.K.: A 3D-printed capacitive smart insole for plantar pressure monitoring. Sensors. **22**(24), 9725 (2022)
3. Tahir, A.M., Chowdhury, M.E.H., Khandakar, A., et al.: A systematic approach to the design and characterization of a smart insole for detecting vertical ground reaction force (vGRF) in gait analysis. Sensors (Basel) **20**(4), 957 (2020)
4. Nguyen, T.-D., Lee, J.S.: Recent development of flexible tactile sensors and their applications. Sensors **22**(1), 50 (2022)
5. Pang, X., Zhang, Q., Shao, Y., Liu, M., Zhang, D., Zhao, Y.: A flexible pressure sensor based on magnetron sputtered MoS2. Sensors. **21**(4), 1130 (2021)
6. De Fazio, R., Perrone, E., Velázquez, R., De Vittorio, M., Visconti, P.: Development of a Self-powered piezo-resistive smart insole equipped with low-power BLE connectivity for remote gait monitoring. Sensors **21**(13), 4539 (2021)
7. Liang, J., Yang, Q.: Aggregate structure and percolation behavior in polymer/carbon black conductive composites. J. Appl. Phys. **102**, 083508:1–083508:6 (2007)
8. Georgopoulou, A., Kummerlöwe, C., Clemens, F.: Effect of the elastomer matrix on thermoplastic elastomer-based strain sensor fiber composites. Sensors **20**, 2399 (2020)
9. Semjonova, G., Vētra, J., Cauce, V., Okss, A., Kataševs, A., Eizentāls, P.: Improving the recovery of patients with subacromial pain syndrome with the DAid smart textile shirt. Sensors **20**(18) (2020). Article number 5277. ISSN 1424-8220
10. Okss, A., Vališevskis, A., Briedis, U., Baltiņa, I.: Functional fabrics for multilayered textile pressure sensors: comparison of structures and sensitivity characteristics. In: ERDev 2023, Jelgava, Latvia, 24–26 May 2023 (2023, inprint)
11. EeonTex™ Conductive Stretchable Fabric Datasheet. SparkFun Electronics, October 2014
12. Antistatic Fabrics Datasheet. Sefar AG. 06.2014
13. Shieldex Nora-PW Datasheet. Technical Textiles Inc. Rev., 06 August 2020

Studies of UV - A Inactivation of *Escherichia Coli* by 365 nm Light

Gita Revalde[1,3](✉), Anna Zajakina[2], Karina Spunde[2], Zhanna Rudevica[2], and Atis Skudra[1]

[1] Institute of Atomic Physics and Spectroscopy, University of Latvia, Jelgavas str. 3, Riga 1004, Latvia
gitar@latnet.lv
[2] Latvian Biomedical Research and Study Centre, Ratsupites str. 1, Riga 1067, Latvia
[3] Institute of Technical Physics, Riga Technical University, P. Valdena 3/7, Riga, Latvia

Abstract. In this work, we tested the effect of intense UV-A irradiation at 365 nm on *Escherichia coli* at different irradiation times. In parallel, after the exposure to UV-A light, the creation of reactive oxygen species in bacteria using cellular reactive oxygen species detection assay with 2′,7′-dichlorodihydrofluorescein diacetate was measured. The first results showed very effective inactivation (5 log10) of *Escherichia coli* after 15 min of exposure resulting in a dose of 412 J/cm^2, and subsequent increased production of cellular reactive oxygen species within 1 to 4 h after irradiation.

Keywords: UV-A · inactivation · disinfection · *E.coli* · ROS · DCFDA

1 Introduction

The last pandemic showed the necessity to develop more tools for the disinfection and inactivation of human pathogens to be better protected in the future, both in everyday life and crises. The disinfecting properties of ultraviolet UV-C (200–280 nm) light is well known, however, mainly mercury low-pressure or medium-pressure lamps with the main resonance line at 254 nm are used in experiments and equipment [1]. Only recently other UV wavelengths have begun to be studied more widely, driven mainly by the development of ultraviolet light-emitting diodes, including UV-A radiation at 365 nm. UV-A effect on the molecules is yet not fully understood, and often this radiation is used only as pretreatment, or in combination with UV-C light [2]. One of the mechanisms is the creation of reactive oxygen species (ROS), that cause cellular damage. Our research is devoted to the investigation of the influence on the inactivation of bacteria and viruses with UV light with different spectral compositions.

In this work, we tested intense UV-A radiation at 365 nm with *Escherichia coli* (*E. coli*) for different irradiation times. In parallel, the creation of ROS in bacteria after irradiation using cellular ROS detection assay was tested to understand if it has an influence.

Y. Dekhtyar and I. Saknite (Eds.): NBC 2023, IFMBE Proceedings 89, pp. 50–54, 2023.
https://doi.org/10.1007/978-3-031-37132-5_7

2 Methods

We used a UV 10 W lamp emitting light at 365 nm in the UV-A region for the experiments. The lamp spectrum and intensity were measured by the Ocean Optics HR4000+ spectrometer. The spectrometer was calibrated using a NIST-traceable reference deuterium–halogen light source DH-2000 Cal (Ocean Optics Inc.). The irradiance was measured on a plane of exposure with a cosine corrector that allows the collection of incoming light from a 180-degree angle. The set-up of the irradiation measurements is shown in Fig. 1.

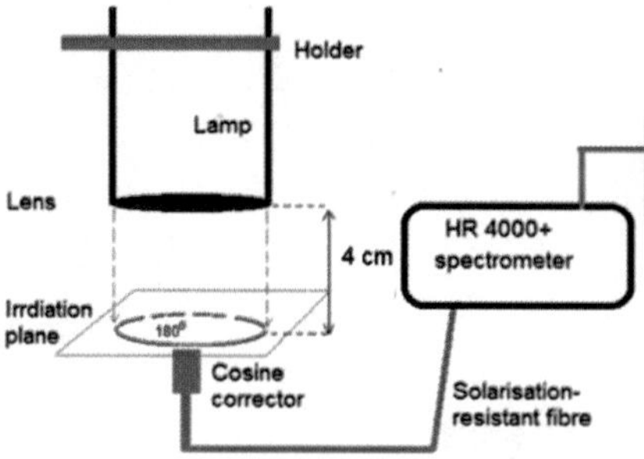

Fig. 1. The set-up for the irradiation measurements.

The intensity of the UV-A lamp was very high at ~0.46 W/cm^2. The dose D of irradiation was calculated as the intensity I multiplied by the exposure time t and is expressed in J/cm^2:

$$D = I \times t. \tag{1}$$

The tests for interaction with UV-A light were performed using *E. coli* bacteria and 2′,7′-dichlorodihydrofluorescein diacetate – DCFDA for ROS detection.

The overnight *E.coli* (ATCC 25922) cell culture was washed 3 times with phosphate-buffered saline (PBS) and diluted to obtain optical density (OD) at 600 nm OD600 = 0.1, which corresponds approximately to 1.5 × 10^8 CFU/ml (CFU – colony forming units). The culture was diluted further to obtain 1 × 10^5 CFU/ml. For the irradiation test 40 μL (4 × 10^3 CFU) of cell suspension was placed dropwise in a sterile 35 mm Petri dish and immediately irradiated for 1, 3, 5, and 15 min at room temperature.

All tests were prepared in duplicate. After irradiation 20 μl of the recovered cell suspension was transferred to the microplate with 180 μL of PBS for preparation of decimal dilutions. Diluted samples (100 μl) were plated on the Tryptic Soy Agar (Liofilchem, Italy) plates. The number of colonies (CFU) was counted after 24 h incubation at 37 °C. The conditions of test effectiveness were judged by preparing an additional sample with an untreated test culture, which was plated without UV exposure. The value of antimicrobial activity R (log10 reduction) was calculated according to the following equation:

$$R = log10(C_0/C), \tag{2}$$

where C_0 is the average CFU number of the untreated cells after incubation; C is the average CFU number of cells after irradiation.

For the analysis of ROS generation after irradiation with a UV-A lamp, bacteria were grown in the Tryptic Soy Broth medium (Liofilchem, Italy) overnight (12–16 h). The cells were washed three times with PBS before the bacterial suspension with OD600 = 0.1 was prepared (in PBS) and diluted to obtain $1{,}5 \times 10^7$ CFU/ml. The 60 μL cell drop (9×10^5 CFU) was treated in triplicate with 365 nm light for 15 min corresponding to a 412 J/cm^2 irradiation dose. Then, 50 μL of the bacterial suspension was transferred to the black clear-bottom 96-well plate and mixed with 5 μL of 250 μM H2DCFDA in PBS (Cat. No 4091-99-0; Calbiochem). The plate was placed at 37 °C and the fluorescence was measured using a fluorimeter Victor3V 1420-040 Multilabel HTS Counter (PerkinElmer, Waltham, Massachusetts, USA) at 485/535 nm immediately, after 1h, 2h, 3h, and 4h, respectively. Non-irradiated *E.coli* suspension with/without H2DCFDA reagent was measured as a control, as well as the H2DCFDA reagent diluted in PBS was used to control the ROS detection. The background fluorescence signal was subtracted from the obtained total ROS signal.

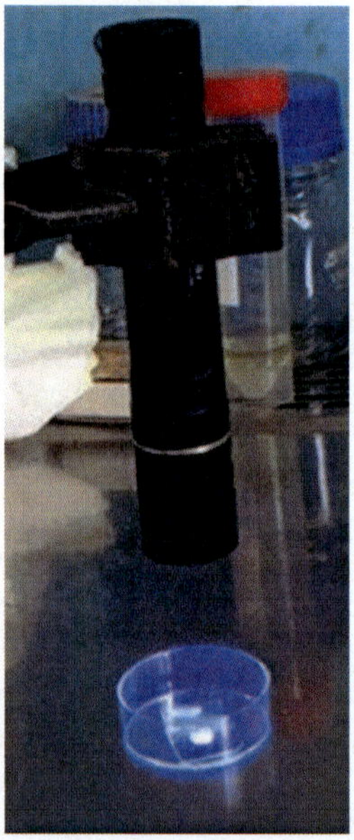

Fig.2. Installation of irradiation experiment

Data analysis was performed using Excel and Graph Pad Prism software.

The installation of the irradiation experiment is illustrated in Fig. 2. The samples were placed at a 4 cm distance from the light source. Drops with *E.coli* suspension were irradiated with the UV-A lamp (1–15 min), ensuring the doses were in a range from 27.5 J/cm^2 to 412 J/cm^2.

3 Results and Discussion

The results of the irradiation experiment are shown in Fig. 3. After 15 min *E.coli* were completely inactivated. The 4 log10 reduction dose was about 342 J/cm^2, corresponding to about 12 min. This dose is much more than required for the inactivation of *E.coli* with UV-C light [1]. However, it is obvious that high-intensity UV-A light destroys the cells, and one must be very careful using intensive UV-A light assuming a non-destructive nature of it.

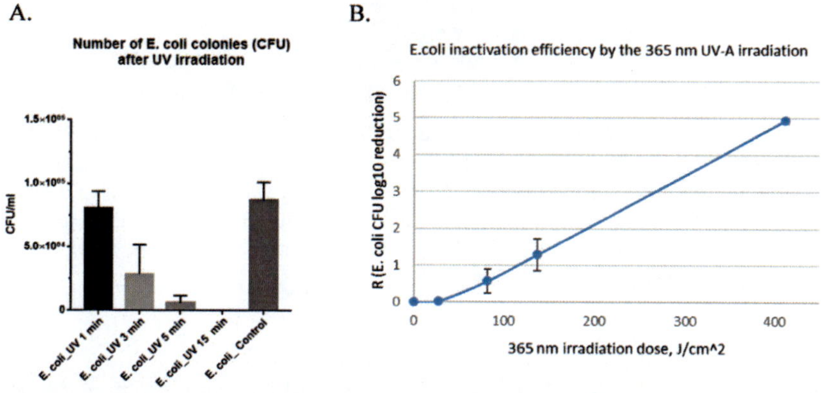

Fig. 3. **A.** Number of *E. coli* CFU/ml after 1, 3, 5, and 15 min of exposure to UV-A; **B.** *E.coli* inactivation efficiency (CFU log10 reduction) as a function of irradiation dose. Error bars represent the standard error of the mean (n = 3).

In Fig. 4, the results of ROS production after the irradiation time of 15 min with ROS-sensitive reagent DCFDA (*E.coli* + UV) in comparison with an untreated sample (*E.coli* - UV) are shown. The inactivation of *E. coli* after 15 min irradiation (412 J/cm^2) caused ROS formation, which gradually increased in time, having indications that it is one of the main inactivation mechanisms of bacteria-killing by irradiation with 365 nm wavelength.

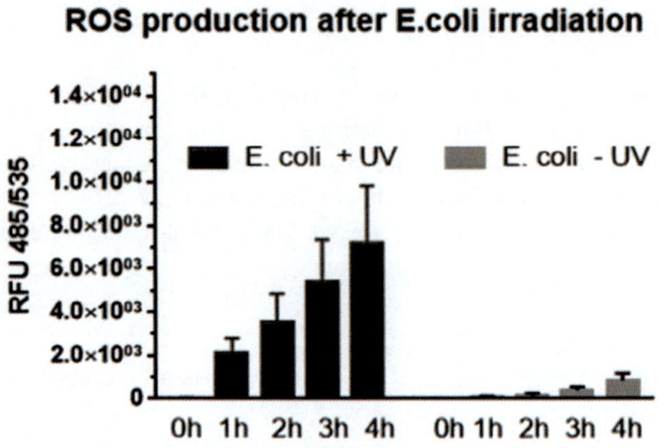

Fig. 4. ROS measurement in *E. coli* cells treated (black) and untreated (grey) with UV 365 nm. Cells were treated or untreated with UV (412 J/cm^2, 15 min), then the cells were transferred into the tube, the fluorescent ROS indicator (DCFDA) was added, and the cells were incubated for up to 4 h to allow the ROS indicator to be activated. The measurement of fluorescence was done at 0, 1h, 2h, 3h, and 4h time points of incubation. RFU - relative fluorescence units. Error bars represent the standard error of the mean (n = 3).

Conclusion

The first results showed effective inactivation (5 log10) of *E.coli* after 15 min of exposure to intense UV-A light at 365 nm (~0.46 W/cm^2), allowing total bacteria elimination. The increasing cellular ROS production was observed during 1–4 h after irradiation. The observed increasing ROS generation after exposure to 365 nm light shows that it is the main factor of cell toxicity at UV-A radiation.

Acknowledgments. The authors acknowledge the State Research Programme of Latvia No. VPP-EM-FOTONIKA-2022/1-0001.

References

1. Skudra, A., et al.: UV inactivation of Semliki Forest virus and E. coli bacteria by alternative light sources. J. Photochem. Photobiol. **10**, 100120 (2022). https://doi.org/10.1016/j.jpap.2022.100120
2. Song, K., Mohseni, M., Taghipour, F.: Mechanisms investigation on bacterial inactivation through combinations of UV wavelengths. Water Res. **163**, 114875 (2019). https://doi.org/10.1016/j.watres.2019.114875

A Prototype Device for Continuous Monitoring of Breathing Mode: Nasal Versus Mouth

Lukas Pipiras[1(✉)] and Vaidotas Marozas[1,2]

[1] Department of Electronics Engineering, Kaunas University of Technology, Kaunas, Lithuania
lukas.pipiras@ktu.edu

[2] Biomedical Engineering Institute, Kaunas University of Technology, Kaunas, Lithuania

Abstract. Breathing is one of the important physiological processes that can occur through the nose or the mouth. Namely, frequent breathing through the mouth can become a bad habit that can cause various difficult-to-cure diseases. As a result, such a breathing habit should be detected as early as possible to change or cure it. A device with discrete and real-time feedback might be helpful in the formation of the right breathing habits.

This paper describes the development of a device for continuous monitoring of the breathing mode that can effectively detect and differentiate between nasal and mouth breathing, providing valuable information on breathing patterns. Assessment of the mode is based on the analysis of Hjorth and sample entropy parameters derived from temperature and acceleration signals. A data registration protocol was developed and a database of 43 examples in nasal, mouth, and speech modes was recorded.

The results of this pilot study demonstrate that the device can detect nasal and mouth breathing cycles for continuous monitoring of breathing mode using NTC thermistors with a magnetic holder and sample entropy as the parameter. However, future development is necessary to achieve device miniaturization and algorithmic accuracy to discern speech from mouth breathing in long-term continuous monitoring.

Keywords: Respiration · sensors · temperature · accelerometer · Hjorth parameters · sample entropy

1 Introduction

Breathing is crucial for life and can occur through the nose or mouth. Habitual mouth breathing can cause challenging conditions such as halitosis, ear infections, fatigue, and sleep disturbances. Children are especially vulnerable to negative effects, including dental issues, reduced blood oxygen, attention disorders, hyperactivity, and jaw development problems, often requiring complex surgical treatment [1]. Mouth breathing also affects skull formation, leading to structural changes such as narrow upper jaws and protruding front teeth, while breathing through the nose fosters proper skull growth [2]. A study on children found a strong correlation between mouth breathing and severe halitosis, with 57.1% of the 35 children studied exhibiting halitosis of grades 3 and 4 [3]. Diagnosing and treating long-term breathing mode issues requires the use of unobtrusive technology.

Y. Dekhtyar and I. Saknite (Eds.): NBC 2023, IFMBE Proceedings 89, pp. 55–62, 2023.
https://doi.org/10.1007/978-3-031-37132-5_8

There are several methods suitable for differentiation between breathing through the mouth and breathing through the nose. In a study by Curran et al. [4], an acoustic sensor was used to classify nasal and mouth breathing. The study showed that a system with such sensors can detect the type of breathing with 90% accuracy [4]. However, the results were not as good when dealing with breath patterns that are mixed. Huang et al. [5] employed thermography to distinguish between mouth and nose breathing by analyzing the relative temperature differences in different facial regions during sleep [5]. Another study with favorable results was carried out by Akre et al. [6], where internal thermistors positioned in the epipharynx, oropharynx, and hypopharynx were used to differentiate between nasal and oral breathing during sleep. The sensors were placed deep in the respiratory tract, so it is not comfortable for the user and thus not suitable for long-term wear. Other less invasive methods such as respiratory plethysmography [7] or respiratory impedance plethysmography [8] were considered for continuous, long-term monitoring of breathing rate. However, these methods cannot distinguish between nasal and mouth breathing.

In order to address the issue of wearability during ambulation, an innovative device has been developed that includes an external thermistor with a magnetic holder and a relevant signal-processing algorithm. This device allows for continuous monitoring of breathing mode and effectively detects and differentiates between nasal and mouth breathing, providing valuable insights into breathing patterns.

2 Material and Methods

2.1 Device, Acquisition Protocol, and Signals

To monitor the human breathing process and determine whether breathing occurs through the nose or mouth, a monitoring system comprising multiple sensors capable of monitoring several main breathing indicators simultaneously is required. Two NXFT15XH103FA2B140 NTC thermistors from Murata Electronics are used to track temperature changes in both nostrils. These thermistors are attached to a magnetic holder, while an MMA7361L XYZ-axis accelerometer from Freescale Semiconductor is used to record the z-axis acceleration on the human chest. The STM32 NUCLEO-L432KC development board from STMicroelectronics is used to control the general prototype device. The device is powered by a 5 Ah power bank, and data are stored on a microSD card. In addition, an external DS3231 Real-Time Clock module from Analog Devices Inc./Maxim Integrated is used for real-time recording.

The block diagram and real implementation of the prototype device for continuous monitoring of breathing mode are shown in Fig. 1.

Data from 5 healthy participants were collected for a biomedical signal database. The protocol, as shown in Fig. 2, was repeated ten times for 30 min each, resulting in 50 examples per mode. However, due to signal quality issues, only 43 examples were suitable for analysis. The database includes temperature signals from the right and left nostrils and a z-axis acceleration signal from neck movements during breathing, all of which were sampled at 30 Hz.

The illustration of raw signals is shown in Fig. 3.

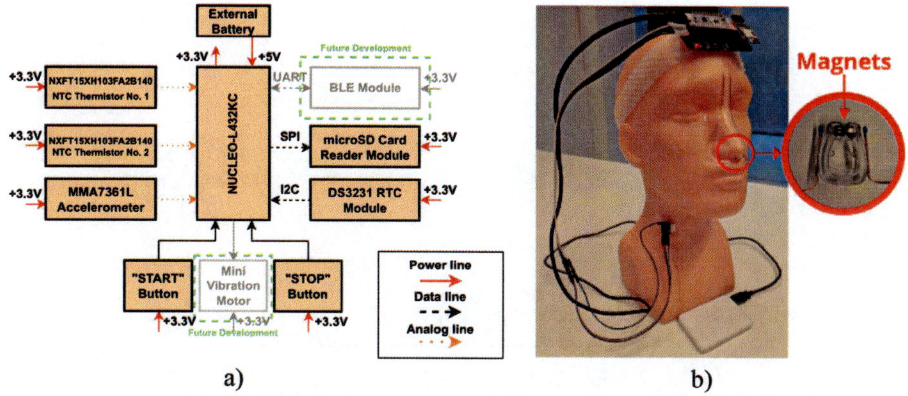

Fig. 1. A prototype: a) block diagram; b) implementation.

Fig. 2. Data acquisition protocol

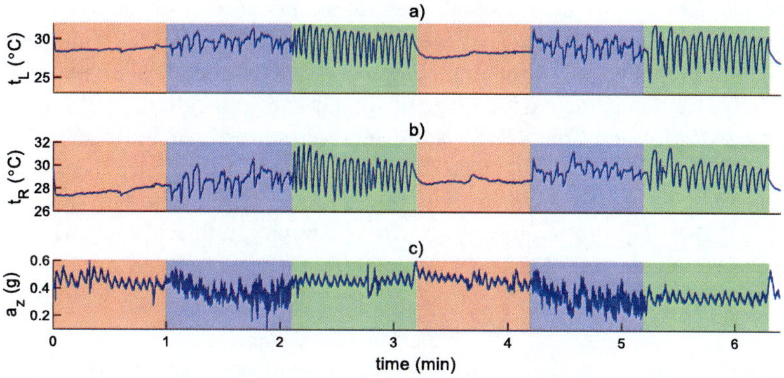

Fig. 3. Example of raw data registered with NTC thermistors and acceleration: a) right nostril; b) left nostril; c) acceleration on the z-axis. Red area – mouth breathing, blue area – speech, green area – nasal breathing

2.2 Signal Processing Algorithm

The recorded signal processing involves two stages: signal pre-processing and estimation of signal features (parameters). The signals were pre-processed using a 2nd-order high-pass (HP) and low-pass (LP) IIR Butterworth filter. The LP filter was used with a lower cutoff frequency of 0.08 Hz to minimize signal noise and remove baseline wander. On the other hand, the HP filter was set with a cutoff frequency of 10 Hz to ensure a good ratio of high-frequency noise rejection and speech detection in the recorded signals. The preprocessed signals, displayed in Fig. 4, were then used for the estimation of Hjorth and sample entropy parameters.

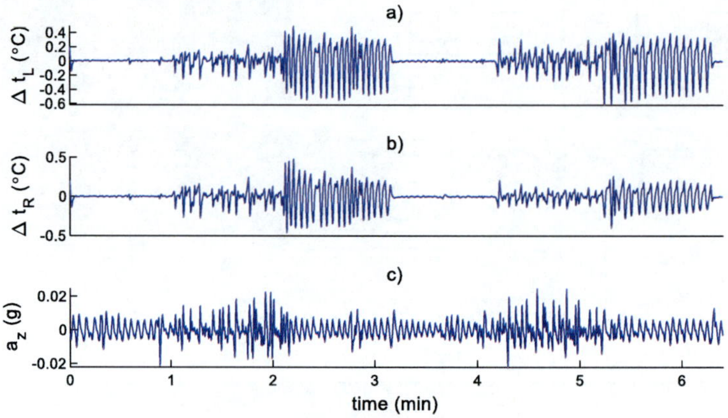

Fig. 4. Filtered NTC thermistors and z-axis acceleration signals: a) left nostril; b) right nostril; c) z-axis acceleration.

Hjorth parameters are statistical measures commonly used in signal processing capable to represent signal regularity. They were introduced by Bo Hjorth [9] and are easy to compute, and suitable for real-time implementation in a microcontroller, making them a popular choice for analyzing a wide range of signals and applications [10]. In this study, the *Activity*, *Mobility*, and *Complexity* parameters were investigated to monitor three different conditions during breathing: nasal breathing, mouth breathing, and speech. These conditions were chosen because of their relevance to respiratory and speech-related disorders. The first Hjorth parameter, known as Activity, refers to the variance of the amplitude of the signal y(t). It can be calculated:

$$Activity(y(t)) = \sigma^2(y(t)) \tag{1}$$

In addition to Activity, the Mobility parameter is another commonly used measure in signal processing. It is a measure of the rate at which a signal changes over time and can provide valuable information about the dynamics of the underlying system:

$$Mobility(y(t)) = \sqrt{\frac{\text{Activity}(y')}{\text{Activity}(y)}} \tag{2}$$

The third Hjorth parameter, Complexity, is another measure that can provide valuable information about the dynamics of the underlying system. It is calculated using formula (3), which describes the relationship between the Mobility of the time derivative of the signal y'(t) and the Mobility of the signal y(t) itself [9]:

$$Complexity(y(t)) = \frac{\text{Mobility}(y'(t))}{\text{Mobility}(y(t))} \tag{3}$$

The sample entropy [11] parameter was also used to analyze breathing under three different conditions. Unlike traditional methods that require large amounts of data to reliably measure complexity, sample entropy was specifically proposed for handling short and noisy time series data [11]. It's less sensitive to changes in amplitude and more focused on the structure within the data. The value ranges from 0 to infinity, with lower values indicating greater regularity and higher values indicating greater complexity or randomness. It was hypothesized here that sample entropy is able to discriminate among nasal, mouth, and speech signals taking into account decreasing regularity of nasal temperature and accelerometer signals.

The results for various parameters were visually presented and analyzed using violin plots, which combine the summary statistics of a box plot with the density estimation of a kernel density plot. To assess normality, the Lilliefors test was employed on the data, and to evaluate statistical differences between the parameters, a multiple comparison based on the nonparametric Friedman's test was utilized.

3 Results and Discussion

The Hjorth activity, mobility, complexity, and sample entropy parameters were analysed to investigate the statistical difference between the parameters mentioned during three different modes of breathing, mouth breathing, nasal breathing, and speech. Violin diagrams illustrating the analyzed parameters are shown in Figs. 5, 6, 7 and 8.

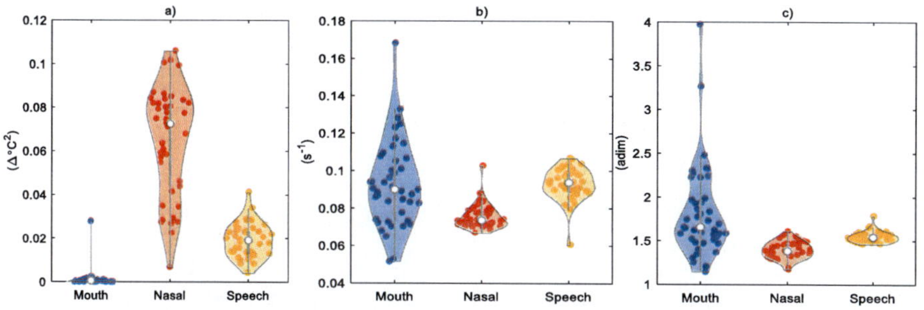

Fig. 5. The violin plots of the left nostril signal Hjorth parameters: a) Activity; b) Mobility; c) Complexity.

The violin diagrams show that the parameters presented in figures Fig. 5 a), Fig. 6 a), Fig. 8 a), and Fig. 8 b) have the least overlap; therefore, it can be assumed that these particular parameters are the most suitable for differentiating the mouth, nasal, and speech breathing modes.

The Lilliefors test has revealed that not all analyzed data groups followed a Gaussian distribution. Only the data groups for the Hjorth Mobility parameter of the left and right nostrils showed a Gaussian distribution, and therefore ANOVA statistical analysis was performed on these two data groups. For the remaining groups, the Friedman non-parametric statistical test was used to determine any statistical differences in the

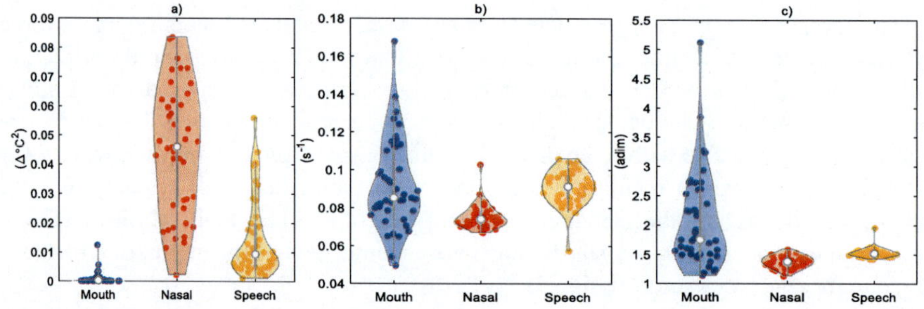

Fig. 6. The violin plots of the right nostril signal Hjorth parameters: a) Activity; b) Mobility; c) Complexity.

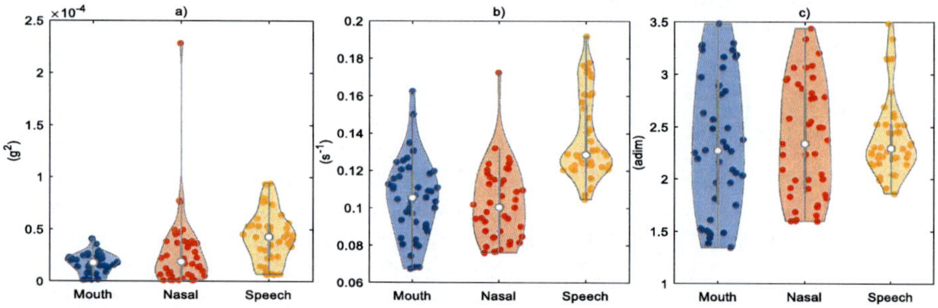

Fig. 7. The violin plots of the z-axis acceleration signal Hjorth parameters: a) Activity; b) Mobility; c) Complexity.

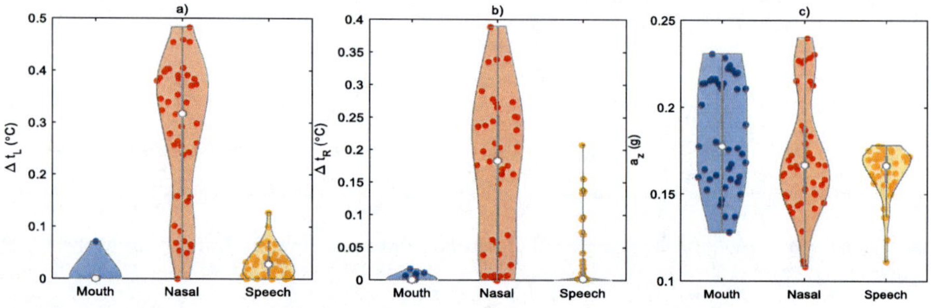

Fig. 8. The boxplots of the calculated signals sample entropy: a) left nostril; b) right nostril; c) z-axis acceleration.

calculated Hjorth and sample entropy parameters among the three different breathing conditions.

Multiple comparisons were made between the MB-NB, MB-SP, and NB-SP groups, with a significance level set at 0.01. Table 1 provides a summary of the calculated *p*-values that estimate the statistical differences between the selected groups.

The results indicate that the Hjorth Activity and sample entropy parameters of the left and right nostril temperature signals exhibit high statistical significance. In contrast, all parameters derived from the z-axis acceleration signal show little statistical significance, suggesting that the accelerometer sensor does not provide useful information. These findings are significant for future device development and sensor selection. For example, the results suggest that a single magnetic nasal temperature sensor can differentiate between three different modes of breathing.

Table 1. Calculated *p*-values between compared groups

Parameter	*p*-values		
	MB-NB	MB-SP	NB-SP
Left nostril (Activity)	<0.01	<0.01	<0.01
Left nostril (Mobility)	<0.01	1	<0.01
Left nostril (Complexity)	<0.01	0.995	<0.01
Right nostril (Activity)	<0.01	<0.01	<0.01
Right nostril (Mobility)	<0.01	1	<0.01
Right nostril (Complexity)	<0.01	0.393	<0.01
z-axis acceleration (Activity)	0.200	<0.01	<0.01
z-axis acceleration (Mobility)	1	<0.01	<0.01
z-axis acceleration (Complexity)	1	1	0.706
Left nostril (sample entropy)	<0.01	<0.01	<0.01
Right nostril (sample entropy)	<0.01	<0.01	<0.01
z-axis acceleration (sample entropy)	0.587	0.156	1

In addition, some of the recorded left and right nostrils temperature signals showed amplitude differences between the nostrils (see Fig. 4 a), b)). These differences could be due to breathing problems and the nasal septum. In future tests, information about the current health conditions of the subjects should be obtained and evaluated.

The study has a few limitations. First, in the pilot study, only a small amount of data was collected, which can limit the generalizability of the results and may not be representative of the larger population. Second, the lack of clinical validation in this study makes it uncertain whether the proposed system can be used effectively to detect a patient's long-term breathing patterns. However, this is the first study proposing a concept system for long-term monitoring of the breathing mode with the potential to be developed as a useful tool to help patients in changing their habits of breathing mode from mouth to nose.

4 Conclusions

The magnetic nasal sensor provides reliable monitoring of temperature fluctuations in the nostrils, and, when combined with the sample entropy and Hjorth Activity parameters, shows statistically significant differences among nasal and mouth breathing modes, as well as speech. In the future, to improve the device's usability, it will be redesigned as a necklace and tested in diverse populations and contexts. A real-time detection algorithm would also be useful in providing immediate feedback to the user.

References

1. Jefferson, Y.: Mouth breathing: adverse effects on facial growth, health, academics, and behavior. Gen. Dent. **58**(1), 18–80 (2010)
2. Harari, D., et al.: The Effect of mouth breathing versus nasal breathing on dentofacial and craniofacial development in orthodontic patients. Laryngoscope **120**(10), 2089–2093 (2010)
3. Motta, L.J., et al.: Association between halitosis and mouth breathing in children. Clinics **66**(6), 939–942 (2011)
4. Curran, K., Yuan, P., Coyle, D.: Using acoustic sensors to discriminate between nasal and mouth breathing. Int. J. Bioinform. Res. Appl. **8**(5–6), 382–396 (2012)
5. Huang, Z., Wang, W., Haan, G.D.: Nose Breathing or mouth breathing a thermography-based new measurement for sleep monitoring. Paper presented at the - 2021 IEEE/CVF Conference on Computer Vision and Pattern Recognition Workshops (CVPRW), pp. 3877–3883 (2021)
6. Akre, H., et al.: Internal thermistors in differentiating between oral and nasal breathing during sleep. Acta Otolaryngol. **119**(8), 934–938 (1999)
7. Miller, K.M., et al.: Long-term tolerability of capnography and respiratory inductance plethysmography for respiratory monitoring in pediatric patients treated with patient-controlled analgesia. Paediatr Anaesth **25**, 1054–1059 (2015)
8. Iqbal, T., Elahi, A., Ganly, S., Wijns, W., Shahzad, A.: Photoplethysmography-based respiratory rate estimation algorithm for health monitoring applications. J. Med. Biol. Eng. 1–11 (2022). https://doi.org/10.1007/s40846-022-00700-z
9. Hjorth, B., Elema-Schönander, A.B.: EEG analysis based on time domain properties. Electroencephalogr. Clin. Neurophysiol. **29**(3), 306–310 (1970)
10. Cocconcelli, M., et al.: Detectivity: a combination of hjorth's parameters for condition monitoring of ball bearings. Mech. Syst. Signal Process. **164**, 108247 (2022)
11. Richman, J. S., Moorman, J.R.: Physiological time-series analysis using approximate entropy and sample entropy. Am. J. Physiol. Heart Circul. Physiol. **278**(6), H2039–H2049 (2000)

Investigation of Electrospun Keratin Mats Containing Biosynthesized Silver Nanoparticles

Akvilė Andziukevičiūtė-Jankūnienė[1(✉)], Ugnė Zasčiurinskaitė[1], Aistė Balčiūnaitienė[2], Jonas Viškelis[2], Erika Adomavičiūtė[1], Carmen Gaidau[3], Maria Rapa[4], Vitalijs Radenkovs[5], Virgilijus Valeika[6], and Virginija Jankauskaitė[1]

[1] Kaunas University of Technology, Studentu St. 56, 51424 Kaunas, Lithuania
akvile.andziukeviciute-jankuniene@ktu.lt

[2] Lithuanian Research Centre for Agriculture and Forestry, Institute of Horticulture, 54333 Babtai, Lithuania

[3] INCDTP - Leather and Footwear Research Institute Division, 93 Ion Minulescu St., Sector 3, 031215 Bucharest, Romania

[4] Politehnica University of Bucharest, Bucharest, Romania

[5] Institute of Horticulture (LatHort), Dobele, Latvia

[6] Kaunas University of Technology, Radvilenu St. 19, 51424 Kaunas, Lithuania

Abstract. Wound healing is a complex mechanism that involves different types of cells to restore damaged skin layers. Therefore, different types of medical bandages are used. Keratin and *Matricaria chamomilla* extract based electrospun medical dressings may improve wound healing process due to promotion of different types of cells that take act in skin regeneration process. Biosynthesized silver nanoparticles can upgrade these bandages due to the strong antibacterial activity. However, the addition of AgNPs to keratin composition may affect electrospinability. The aim of this research is to investigate the possibility to obtain electrospun keratin mats containing biosynthesized silver nanoparticles for medical applications. Therefore, herbal extract and AgNPs were investigated, their influence on keratin compositions viscosity, electroconductivity and electrospinability were evaluated. It was determined that biosyntesized AgNPs improve phenolic and antioxidant activity, however, they lessen keratin composition viscosity, electroconductivity, and electrospinability. Thus, a more viscous keratin solution for mats electrospinning should be used.

Keywords: Green synthesized AgNPs · *Matricaria chamomilla* extract · electrospinning

1 Introduction

Skin healing is a complex process that involves the coordinated action of different types of cells: keratinocytes migrate and proliferate to cover the wound and form a new epithelial layer; fibroblasts, which are responsible for synthesizing and depositing new extracellular matrix (ECM) components, such as collagen, elastin, and fibronectin, endothelial

Y. Dekhtyar and I. Saknite (Eds.): NBC 2023, IFMBE Proceedings 89, pp. 63–71, 2023.
https://doi.org/10.1007/978-3-031-37132-5_9

cells to form new blood vessels, etc. [1]. Improper wound healing may lead to inflammation and scarring, that may increase the time of healing and boost both physical and psychological problems. Therefore, medical dressings with active additives to accelerate the recovery of wounds are on demand. Keratin is a structural protein that is found in many different types of organisms, including humans, fish, and animals. This protein helps to form a strong, protective barrier over the wound helping to keep out harmful bacteria and other pathogens, while also helping to retain moisture and promote healing [2].

Keratin effectiveness for wound healing may be improved with herbal extracts, which can provide additional benefits – anti-inflammatory, antimicrobial, and antioxidant properties that may help to reduce inflammation, prevent infection, and protect the skin from damage, all of which can promote the healing process. [2, 3]. Keratin mats can be obtained by electrospinning technique that has shown great potential for wound healing applications for the ability in simple way to engineer nano-microfibers, which can mimic the extracellular matrix (ECM) of natural tissues and provide a scaffold for cell growth and tissue regeneration. Electrospun keratin fibers can provide a barrier between the wound and the external environment, reduce the risk of infection, and promote wound healing, by promoting cell adhesion and proliferation, regulating inflammatory response, and enhancing angiogenesis [4].

Effectiveness of wound healing can be improved by medicinal herbals addition to keratin hydrolysate solution. Chamomile (*Matricaria chamomilla L.*) is traditionally used for its anti-inflammatory, anti-microbial, and wound healing properties due to bioactive compounds such as flavonoids, terpenoids, and phenolic acids [3]. Combining keratin with *Matricaria chamomilla L. (M. chamomilla)* may enhance the wound healing process by providing structural support to the wound bed and reducing inflammation. Studies have shown that the combination of keratin and *M. chamomilla* may promote the proliferation of fibroblasts and keratinocytes, which are important cells involved in the wound healing process [4]. Additionally, the combination of keratin and *M. chamomilla* may also enhance the formation of new blood vessels (angiogenesis) and the deposition of ECM components such as collagen, which are important for tissue regeneration and remodeling [2–4].

Medical dressings may be improved by incorporation of silver nanoparticles (AgNPs) due to their antimicrobial, anti-inflammatory, cell proliferation-promoting, and circulation-improving properties. New innovative methods for AgNPs synthesis are under investigation. "Green" synthesis of AgNPs refers to the use of environmentally friendly and sustainable methods. This approach involves using natural compounds, such as plant extracts to reduce silver ions into nanoparticles without the use of harmful chemicals or processes. AgNPs biosynthesis is low toxic, cost-effective, scalable, and sustainable. Additionally, the resulting AgNPs are often more stable and biocompatible, making them suitable for use in biomedical applications, including wound healing [5].

The goal of this research is to investigate the possibility to obtain electrospun keratin mats containing biosynthesized silver nanoparticles for medical applications.

2 Materials and Methods

2.1 Compositions Preparation

Keratin hydrolysate (ICPE, Romania) was prepared by solubilization of sheepskin wool preliminarily degreased with 8% NaOH rotulis (Lach-Nersro) at 80 °C for 4 h, followed by filtration [6].

M. chamomilla (Lithuanian university of health sciences, Lithuania) extract was obtained by grinding of dried herb to powder consistency in 10000 min^{-1} speed ("Grindomix GM 200", Germany). Then 30 g of powder were blended with 600 ml 100 °C water and mixed in magnetic stirrer for 10 min and resulted solutions were filtered.

"Green" AgNPs prepared by method, described in [7]. 0.03 g of $AgNO_3$ (Merck, Germany) was dissolved in 2.5 ml distilled water and mixed with 30 ml of *M. chamomilla* aqueous extract under vigorous stirring at room temperature and speed of 400 rpm for 2 h. The mixture was incubated at room temperature for 24 h, and the change in color from yellowish to brown proved the formation of AgNPs reduced by *M. chamomilla* extract (*M. cha*-AgNPs). After that the *M. cha*-AgNPs dispersion was stored in darkness at a 6 °C temperature [7].

The changes of dispersion colour during formation *M. cha*-AgNPs was evaluated using spectrometer MiniScap XE Plus (Hunter Associates Laboratory, Inc., USA). The L*, a* and b* (brightness, red-green and yellow-blue coordinates by CIE L*a*b scale) parameters were investigated. UV light was applied for 2 min., 12 min., and 24 h.

The average of three measurements as test result were taken [8].

M. cha-AgNPs dispersion change colour when exposed to UV light for different durations (Table 1). Change of colour from red/yellow to green/blue identifies the formation of Ag nanoparticles.

Table 1. Changes in the color of the AgNPs mediated with *M.chamomilla* extract

Extract	L*	a*	b*	C*	h*
M.cha extract	28.75 ± 0.06	8.24 ± 0.04	12.55 ± 0.05	15.01 ± 0.04	56.70 ± 0.21
M.cha-AgNPs after 2 min	32.82 ± 0.06	3.71 ± 0.20	10.33 ± 0.24	10.97 ± 0.16	70.25 ± 1.42
M.cha-AgNPs after 12 min	32.41 ± 0.10	3.40 ± 0.11	9.88 ± 0.18	10.45 ± 0.17	71.00 ± 0.76
M.cha-AgNPs after 24 h	32.51 ± 0.05	2.4 ± 0.39	8.62 ± 0.16	8.96 ± 0.16	74.44 ± 2.81

Colour change observed in herbal extracts upon the addition of green-synthesized AgNPs may be explained by the interaction between the nanoparticles and the active compounds in the extract, leading to the phenomenon of surface plasmon resonance, which occurs when light is absorbed by the AgNPs and then scattered in a particular manner due to the collective oscillation of electrons on the surface of the nanoparticles [9].

Keratin hydrolysate was dissolved in distilled water and mixed with water-based polyethylene oxide (PEO, $C = 10\%$) solution in a shaker (KS 4000 in control, Germany) at 150 min^{-1} for 24 h. Keratin composition with concentration of $C = 14.5\%$ was obtained. Before the electrospinning process *M.cha*-AgNPs were added to keratin composition ($C = 5$–15%) and properly mixed.

2.2 Herbal Extract and AgNPs Properties Investigation

Total phenolics content (TPC) was determined using the colorimetric Folin–Ciocalteu method [10]. To perform the assay, 0.2 ml of *M.cha* or *M.cha*-AgNPs was mixed with 1.8 ml double distilled water (DDW) and subsequent vortex mixing (1 min) and ultrasonication (5min) at room temperature (22 ± 1 °C). 0.5 ml of the sample extract was mixed with 2.5 ml of 10-fold diluted Folin–Ciocalteu reagent and 2.0 ml 7.5% Na_2CO_3 with the subsequent vortex mixing for 1 min. Prepared solutions were left to react for 30 min at room temperature. The absorbance was measured using a UV-1800–Visible Spectrophotometer SHIMADZU (Shimadzu Corp., Kyoto, Japan) at 760 nm wavelength.

To evaluate DPPH radical scavenging activity, a 0.15 ml of *M.cha* or *M.cha*-AgNPs was mixed with 2.85 ml of DPPH–EtOH solution (0.039 g DPPH in 1 l EtOH) and left to react for 30 min in the dark at 22 ± 1 °C. The absorbance was measured at 517 nm wavelength (A517) at 0 and 30 min.

Ferric reducing antioxidant power (FRAP) was evaluated by method described by Radenkovs et al. [11] The FRAP reagent was prepared from 100 ml acetate buffer (0.3 mol l^{-1}; pH 3.6), 2,4,6-tripyridyl-s-trizine (TPTZ) solution in 10 mmol l^{-1} of HCl and $FeCl_3$ (20 mmol l^{-1}). The three solutions were mixed respectively at the ratio of 10:1:1 (v/v/v), and heated at 37 °C. Then, 0.15 ml of biosynthesized AgNPs were mixed with 2.85 ml FRAP reagent and vortex mixed for 1 min. The resulting absorbance was measured after 10 min at 593 nm wavelength.

The antibacterial activity was evaluated in vitro using the Agar diffusion test against Gram-positive (*Bacillus cereus, Proteus vulgaris, Staphylococcus aureus* and *Bacillus subtilis*) and Gram-negative (*Klebsiella pneumoniae, Proteus mirabilis, Escherichia coli* and *Pseudomonas aeruginosa*) bacteria strains. 0.5 McFarland unit density suspension (~108 CFU/ml) of bacterial strain was inoculated onto the cooled Mueller Hinton Agar (Oxoid, Basingstoke, UK), using sterile cotton swabs. Wells (Ø = 6 mm) were punched in the agar and filled with 50 μl of investigated solution. Agar plates were incubated at 37 °C for 24 h and zones of the inhibition were measured and tabulated.

2.3 Keratin Compositions Properties Evaluation

The electrical conductivity of keratin solutions with 5 – 15% of *M.cha-AgNPs* was identified (Table 2) by the "HQ40d Portable pH, Conductivity, Dissolved Oxygen, ORP, and ISE Multi-Parameter Meter", measurements were taken at room temperature (23 ± 1) °C.

Table 2. Keratin and AgNPs mediated with *M.chamomilla* composition' viscosity

Composition	Keratin	Keratin + 5% *M.cha*-AgNPs	Keratin + 10% *M.cha*-AgNPs	Keratin + 15% *M.cha*-AgNPs
Viscosity	168.8	168,4	131,7	118,4
Electroconductivity	12.95	12.98	12.99	12.54

The viscosity of the mixtures was determined using the "SMART SERIES ROTATIONAL Viscometer" (Fungilab, USA) device. Measurements were taken at room temperature (23 ± 1 °C), solutions temperature –20 ± 1 °C.

2.4 Nonwoven Material Preparation and Investigation

Electrospun micro/nanofiber mats were formed using electrospinning machine "NanospiderTM" (Elmarco, Check Republic) equipped with rotating spinneret (distance between electrodes 150 mm) at ambient conditions. The operating voltage was 65 kV. Electrospinning duration 10 min.

The structure of micro/nanofibers were analyzed by scanning electron microscope S-3400N (SEM). The diameter distribution was obtained by analyzing 100 nano-microfibers from SEM images by Nis-Elements D 4.50.00 (Nikon) software.

3 Results and Discussion

Chamomile is known to contain various phenolic compounds, which may contribute to its potential health benefits via antiradical activity. Figure 1 shows the antioxidant capacity of *M.cha* and *M.cha*-AgNPs. It was determined that *M.cha* is rich in TPC (3.09 CGA mg · ml^{-1}). Furthermore, this extract shows high antioxidant activity evaluated by FRAP assay. *M.cha*-AgNPs were found to possess better TPC values (up to 27%), free radical scavenging (up to 11.7%) and slight increase of FRAP value (up to 3.9%).

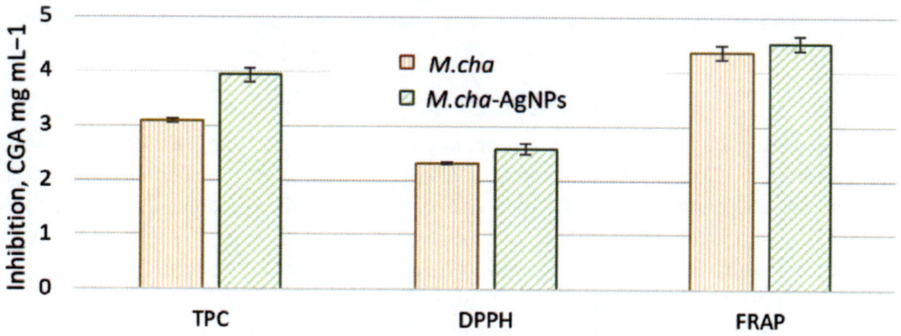

Fig. 1. *M.cha* and *M.cha*-AgNPs antioxidant capacity

Identified TPC, DPPH and FRAP values in *M.cha* extract go in line with the studies [3, 12–14]. Furthermore, an increase of DPPH and FRAP values in green synthesized *M. cha*-AgNPs can be seen. Free radical scavenging improvement may appear due to additional OH- and COOH- groups capped from other bioactive derivatives in plant extracts [15]. Moreover, Salari et al. suggested that the higher antioxidant activity of the silver nanoparticles may be due to their smaller size and increased surface area compared to the plant extract [15].

M.cha extract containing hydroxyl and carbonyl functional groups can also contribute to the antibacterial activity of the nanoparticles by interacting with the bacterial cell membrane, leading to its disruption and ultimately the death of the bacteria [16]. Figure 2 demonstrates *M.cha and M.cha*-AgNPs antibacterial activity.

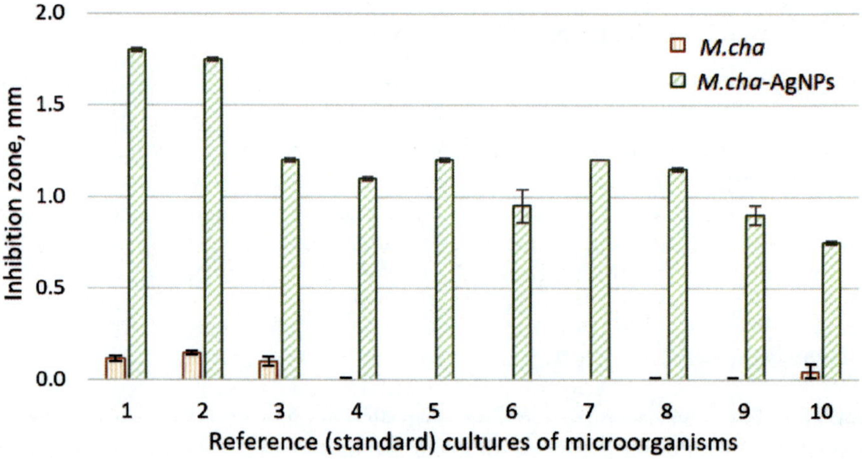

Fig. 2. Antimicrobial activity of *M. cha* ir *M.cha*-AgNPs (1 – S. aureus; 2 – ß- streptococcus; 3 – S. epidermidis; 4 – E. coli; 5 – K. pneumoniae; 6 – P. aeruginosa; 7 – P. vulgaris; 8 – B. cereus; 9 – E. faecalis; 10 – C. albicans)

M.cha extract shows slight antibacterial activity against Gram-positive bacteria, while no effect is observed on Gram-negative bacteria strains. *M.cha*-AgNPs exhibits significantly stronger antimicrobial resistance both to Gram-positive and Gram-negative bacteria.

Keratin compositions modified with AgNPs were investigated.

Electrical conductivity and viscosity are key factors for electrospun nano-microfibers quality. A very high viscosity solution may result in clogging of the needle or spinneret used for electrospinning. This can lead to inconsistent fiber diameter and poor fiber quality [5].

Small amount ($c = 5\%$) of *M.cha*-AgNPs addition to keratin-polymer solution slightly diminished solution viscosity (Table 2). 10–15% of *M.cha*-AgNPs reduced solution viscosity from 28% (at $c = 10\%$) to 43% (at $c = 15\%$).

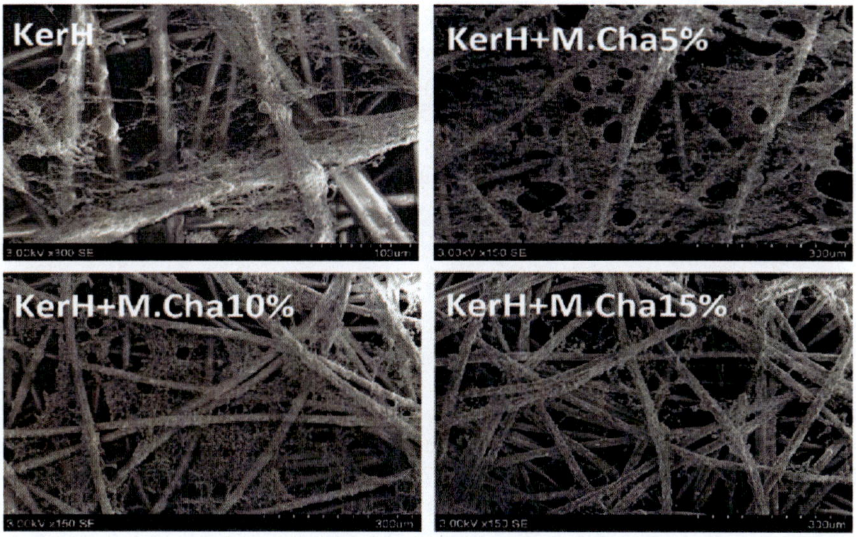

Fig. 3. SEM images of electrospun mats

The addition of 5–10% of green synthesized AgNPs marginally reduced electroconductivity of solution (Table 2). 15% of silver nanoparticles diminished composition's conductivity down to 3.5%.

Electrospun nano-microfibers morphology depends on polymer nature, viscosity, electroconductivity and additive type as well. Figure 3 presents electrospun nano-microfibers SEM images. It is seen that produced fibers are unoriented, inhomogenic, stuck together and have drop-shaped defects (Fig. 3). *M.cha*-AgNps amount increase led to thinner and more even fibers.

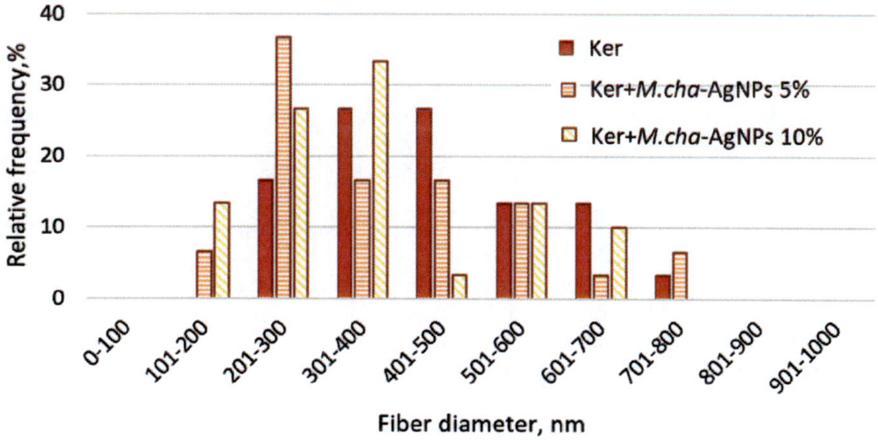

Fig. 4. Polymer composition influence on micro/nanofibers diameter distribution

Keratin compositions micro-nanofibers' diameters vary from 100 nm to 800 nm, when most of them has diameter in the range of 300–500 nm (Fig. 4).

By adding *M.cha*-AgNPs dispersion electrospun nano-microfibers become thinner (44% of them are up to 300 nm at $c = 5\%$ and 73% of them are up to 400 nm when $c = 10\%$). This result can be explained by change of viscosity, slight increase of electroconductivity, polymer and interaction between components [17].

4 Conclusions

Both *Matricaria chamomilla* extract and biosynthesized AgNPs using *Matricaria chamomilla* extract are rich in phenolic and antioxidant compounds. AgNPs show higher antioxidant activity due to their smaller size and increased surface area compared to the plant extract.

While 5–10% of biosynthesised AgNPs have no significant effect on the morphology of the nonwoven material, higher amount ($c = 15\%$) of additives significantly reduce the viscosity of the composition and make it unsuitable for the electrospinning.

Acknowledgements. This work was supported by the Research Council of Lithuania [grant number P-SV-22-73] and by the European Regional Development Fund [grant number 01.2.2-MITA-K-702-10-0004].

References

1. Komi, D.E.A., Khomtchouk, K., Santa Maria, P.L.: A review of the contribution of mast cells in wound healing: Involved molecular and cellular mechanisms. Clin. Rev. Allergy Immunol. **58**, 298–312 (2020)
2. Konop, M., Rybka, M., Drapała, A.: Keratin biomaterials in skin wound healing, an old player in modern medicine: A mini review. Pharmaceutics **13**(12), 2029 (2021)
3. El Joumaa, M.M., Borjac, J.M.: Matricaria chamomilla: A valuable insight into recent advances in medicinal uses and pharmacological activities. Phytochem. Rev. 1–28 (2022)
4. Baseri, S.: Ecological dyeing of cotton fabric with Matricaria recutita L. in the presence of human hair keratins as an alternative copartner to metallic mordants. Sustain. Mater. Technol. **32**, e00405 (2022)
5. Xu, L., et al.: Silver nanoparticles: Synthesis, medical applications and biosafety. Theranostics **10**(20), 8996 (2020)
6. Rapa, M., et al.: New nanofibers based on protein by-products with bioactive potential for tissue engineering. Materials **13**(14), 3149 (2020)
7. Balčiūnaitienė, A., et al.: Eucalyptus globulus and Salvia officinalis extracts mediated green synthesis of silver nanoparticles and their application as an antioxidant and antimicrobial agent. Plants **11**(8), 1085 (2022)
8. Bharadwaj, K.K., et al.: Green synthesis of silver nanoparticles using Diospyros malabarica fruit extract and assessments of their antimicrobial, anticancer and catalytic reduction of 4-nitrophenol (4-NP). Nanomaterials **11**(8), 1999 (2021)
9. Alshehri, A.A., Malik, M.A.: Phytomediated photo-induced green synthesis of silver nanoparticles using Matricaria chamomilla L. and its catalytic activity against rhodamine B. Biomolecules **10**(12), 1604 (2020)

10. Singleton, V., Orthofer, R., Maluela-Raventos, R.: Analysis of total phenols and other oxidation substrates and antioxidants by means of Folin-Ciocalteu reagent. Methods Enzymol. **299**, 152–178 (1999). https://doi.org/10.1007/BF02530903
11. Radenkovs, V., et al.: Wild apple (Malus spp.) by-products as a source of phenolic compounds and vitamin C for food applications. Food Biosci. **38**, 100744 (2020), https://doi.org/10.1016/j.fbio.2020.100744
12. Mahdavi, B., Ghorat, F., Nasrollahzadeh, M. S., Hosseyni-Tabar, M., Rezaei-Seresht, H.: Chemical composition, antioxidant, antibacterial, cytotoxicity, and hemolyses activity of essential oils from flower of Matricaria chamomilla var. chamomilla. Anti-Infect. Agents **18**(3), 224–232 (2020)
13. El Mihyaoui, A., Esteves da Silva, J.C., Charfi, S., Candela Castillo, M.E., Lamarti, A., Arnao, M.B.: Chamomile (Matricaria chamomilla L.): A review of ethnomedicinal use, phytochemistry and pharmacological uses. Life **12**(4), 479 (2022)
14. Qasem, A., et al.: Determination of chemical compounds and investigation of biological properties of matricaria chamomilla essential oils, honey, and their mixture. Molecules **27**(18), 5850 (2022)
15. Salari, S., Bahabadi, S.E., Samzadeh-Kermani, A., Yosefzaei, F.: In-vitro evaluation of antioxidant and antibacterial potential of green synthesized silver nanoparticles using prosopis farcta fruit extract. Iran. J. Pharm. Res. **18**, 430–445 (2019)
16. Philip, P., Jose, T., Prakash, J., Cherian, S.K.: Surface plasmon resonance-enhanced bathochromic-shifted photoluminescent properties of pure and structurally modified electrospun poly (methyl methacrylate) (PMMA) nanofibers incorporated with green-synthesized silver nanoparticles. J. Electron. Mater. **50**(8), 4834–4849 (2021)
17. Al-Hadede, L.T., Hassan, M.I.: Silver nanoparticles synthesis by green method and loading of the enterosein to study its antimicrobial inhibition. In: IOP Conference Series: Materials Science and Engineering, vol. 928, no. 7, p. 072078. IOP Publishing (2020)
18. He, M., et al.: Electrospun silver nanoparticles-embedded feather keratin/poly (vinyl alcohol)/poly (ethylene oxide) antibacterial composite nanofibers. Polymers **12**(2), 305 (2020)

Finite Element Method Modelling of Iron – Oxide Nanoparticle Heat Generation Under Low Radio Frequency Field Conditions

Serhat Ilgaz Yoner[1,2](✉) and Alpay Ozcan[2]

[1] Acibadem Mehmet Ali Aydinlar University, Istanbul, Turkey
ilgazyoner@gmail.com
[2] Bogazici University, Istanbul, Turkey

Abstract. Iron-oxide based magnetic nanoparticles are used as contrast agents in magnetic resonance imaging and as magnetically induced heat generators in magnetic hyperthermia which is a non-invasive cancer treatment method based on delivering nanoparticles to the tumor site. Then, applying alternating magnetic fields generated by specifically designed coils to heat up nanoparticles, thereby killing and/or sensitizing cancer cells to other forms of cancer therapy. Efficiency is measured by heat generation capabilities of nanoparticles under minimum coil power. Therein, measuring temperature rises is more suitable rather than the generated heat. However, nanoparticles' temperature changes are both directly and indirectly effected by many environmental factors, impeding an accurate performance testing. Therefore, evaluations through generated heat, instead of particle temperature should provide more accurate data on system performance. In this work, a single Fe_3O_4 nanoparticle is electromagnetically heated by a 12.6 mT 150 kHz field for 60 min at COMSOL Multiphysics. Time average of total heat generation by a single nanoparticle was found to be 2.43×10^{-41} Wm^3. Fe_3O_4 is preferred due to its great electromagnetic characteristic and widespread use in drug formula Ferumoxytol. Small size of the nanoparticles creates a significant meshing challenge in numerical simulations, resulting in unfeasible computation times when using multiple particles. Accordingly, single particle simulation results were used to theoretically calculate heat generation capacity of 1 ml of Ferumoxytol as 2.62×10^{-18} Wm^3. Combining simulations with theoretical calculations proved to be effective in creating a basis for future studies towards; minimizing total simulation time, computer power and optimizing input waveforms for magnetic hyperthermia systems.

Keywords: fem · simulation · modelling · magnetic · nanoparticle · radio · frequency · heat · hyperthermia

1 Introduction

Magnetic nanoparticles (MNP) are ultra fine particles which are used as part of diagnostic and therapeutic systems [1, 2] such as contrast agents in magnetic resonance imaging (MRI) and as magnetically induced heat generators in magnetic hyperthermia (MH)

Y. Dekhtyar and I. Saknite (Eds.): NBC 2023, IFMBE Proceedings 89, pp. 72–79, 2023.
https://doi.org/10.1007/978-3-031-37132-5_10

which is a non-invasive cancer treatment method based on delivering magnetic nanoparticles to the tumor site prior to applying alternating magnetic fields (AMF) in the range of 80–400 kHz; thereby killing cancer cells through magnetically induced heat [3]. Due to their metal-based structure, MNPs are classified as inorganic materials and they can be used as heating nano-mediators. They are zero-dimensional particles; meaning that their size in any direction is less than 100 nm [4]. The size and shape not only effect their ability to travel in human body but also effect their magnetic properties. Hence, all of these parameters play a huge role in effectiveness of a MNP in MH therapy. Iron oxides and ferrites which are classified as magnetic metal oxide nanoparticles are best candidates for use in MH therapy thanks to their superior biocompatibility and magnetic properties. Out of iron oxides and ferrites, due to presence of Fe2+ and Fe3+ cations in valence states of the inverse spinal structure; iron (II, III) oxide (Fe_3O_4) is widely considered to be the best choice [5, 6]. Efficiency of such particles in MH studies is measured through their magnetically induced heat generation capabilities under minimum coil power [7]. Since it is not practical to track generated heat, temperature rises are tracked instead. However, temperature changes are both directly and indirectly effected by many environmental factors [8], impeding an accurate performance testing. Therefore, evaluations through generated heat, instead of particle temperature should provide more accurate data on system performance. Therein finite element method (FEM) analysis plays an important role in mathematical modeling of physical systems [9]. Nevertheless, relatively small size of the nanoparticles within the modeled environment creates a significant meshing challenge in numerical simulations, resulting in either convergence problems or unfeasible computation times. In a scenario with a single small sized particle, a work around exists. On the other hand, a multiple particle scenario is much more complicated. This study describes relations between complex theories underlying the magnetically induced heat generation mechanisms of MNPs and fundamentals of magnetism, subsequently focusing on steps to structuralize an accurate FEM simulation of a single MNP's heat generation by sticking to fundamentals of magnetism and heat transfer. Finally combining simulation results with theoretical calculations to minimize both required total simulation time and computer power.

2 Theory

In bulk form, iron oxides are paramagnetic materials, therefore posses' weak magnetic properties. This is due to chaotic alignment of bulk iron oxides' multiple magnetic domains (see Fig. 1(a) and (b)). Conversely, iron oxide particles with the size less than 100 nm in all directions are made up of single magnetic domains which yield enhanced magnetic properties under specific conditions and are classified as superparamagnetic particles (see Fig. 1(c) and (d)) [10, 11].

For superparamagnetic particles, magnetically induced heat generation mechanism is affected by two physical phenomena; Néel relaxation (see Fig. 2(a)) and Brownian relaxation (see Fig. 2(b)) [12]. In Nèel relaxations, particles try to align themselves with magnetic field lines through magnetic orientation changes which results in magnetic losses. In Brownian relaxations, particles try to align themselves with magnetic field lines through physical rotations which results in frictional losses [13, 14].

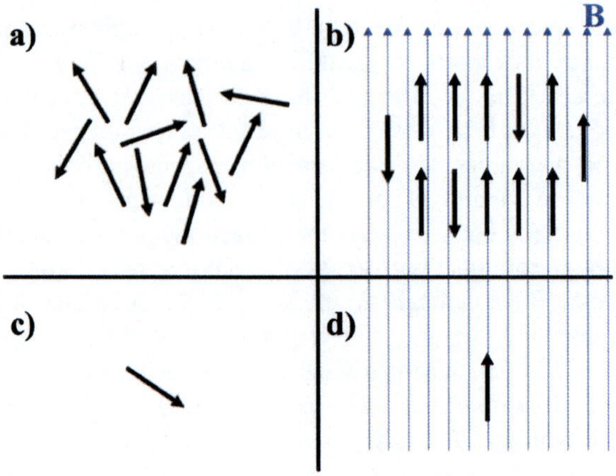

Fig. 1. Alignment of magnetic domains for multi domain paramagnetic materials (a) in the absence of magnetic field B and (b) under magnetic field B; ; for single domain superparamagnetic particles (c) in the absence of magnetic field B and (d) under magnetic field B.

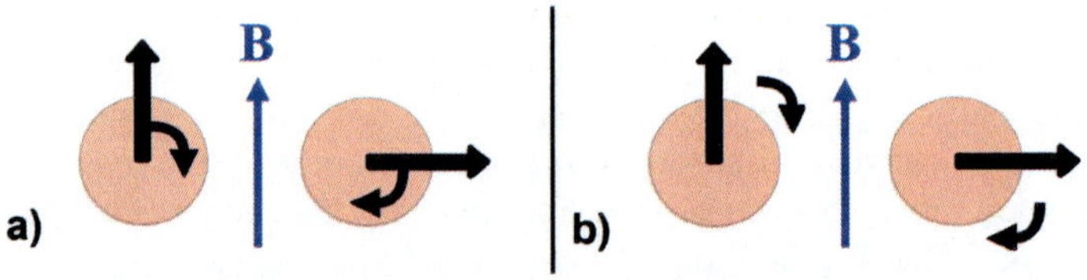

Fig. 2. (a) Mechanism of Nèel relaxation; magnetic orientation change and (b) mechanism of Brownian relaxation; physical rotation both under magnetic field B.

In a study [10], 3 separate theories; equilibrium functions, Stoner–Wohlfarth model-based theories, and linear response theory (LRT) were evaluated and compared with simulations to describe the mechanism of magnetically induced heat generation in superparamagnetic particles. For AMFs with low frequencies, LRT proved to be efficient in calculating magnetically induced heat generation [10]. In this study, low radio frequency (RF) AMFs have been utilized. Magnetically induced heat generation of a MNP is modelled regarding LRT using FEM simulations. In LRT, total affect is called effective relaxation time (τ) and can be calculated as

$$\frac{1}{\tau} = \frac{1}{\tau_N} + \frac{1}{\tau_B}. \tag{1}$$

In (1); τ_N is the Nèel relaxation time and τ_B is the Brownian relaxation time. Nèel relaxation and Brownian relaxation are bounded by (2) and (3), respectively. In (2); τ_0 is the attempt frequency, K is the magnetic anisotropy, V_M is the energy density, k_B is the Boltzmann constant and T is the temperature. KV_M product is the height of energy barrier, which is the minimum amount of magnetic energy required to demagnetize a superparamagnetic particle and k_BT product is the total thermal energy of the system

[10, 11, 15].

$$\tau_N = \tau_0 e^{\left(\frac{KV_M}{k_B T}\right)} \tag{2}$$

In (3); η is the viscosity of the liquid carrier and V_H is the hydrodynamic volume of the particle [10, 11, 15].

$$\tau_B = \frac{3\eta V_H}{k_B T} \tag{3}$$

Fundamental information regarding parameters effecting Nèel relaxation and Brownian relaxation excludes further investigation of (2) and (3). Effective relaxation time directly influences magnetic behavior of a material through complex susceptibility ($\tilde{\chi}$) (4). In (4), χ_0 is the static susceptibility and ω is the angular frequency of the applied AMF. Imaginary part (i) of the equation is bounded by angular frequency and effective relaxation time [10, 11, 15].

$$\tilde{\chi} = \chi_0 \frac{1}{1 + i\omega\tau} \tag{4}$$

LRT states that a magnetic system responds linearly with the magnetic field [10]. (5) describes this linear relationship.

$$M(t) = \tilde{\chi}\, H(t) \tag{5}$$

In (5), $M(t)$ is the magnetization through time and $H(t)$ is the magnetic field strength through time. Complex susceptibility directly effects particle magnetization (5) and is bounded by angular frequency of the applied AMF and the effective relaxation time (4). As previously mentioned, MNPs yield enhanced magnetic properties under specific conditions, due to dynamic hysteresis loop of superparamagnetic particles [10]. If the multiplication of angular frequency of the applied AMF with effective relaxation time is smaller than 1; similar to paramagnetic particles, there can be no considerable delay in the relaxation of the magnetic moment [10, 11]. Hence, required demagnetization energy is the same as magnetization energy which does not result in heat loss (see Fig. 3) [11].

$$\omega\tau < 1, \tag{6}$$

if the multiplication of angular frequency of the applied AMF with effective relaxation time is larger than 1 (7), a considerable delay in the relaxation of the magnetic moment occurs due to increased coercivity which results in requirement of more energy than magnetization energy to demagnetize the particle (ferromagnetic regime). This results in relaxation losses, thereby the system outputs energy in the form of heat (see Fig. 3) [10, 11].

$$\omega\tau > 1 \tag{7}$$

This normally undesirable heat loss forms the essence of MH therapy [3]. Total heat loss is equal to green area (A) enclosed by solid lines in magnetic field strength versus magnetization plot (see Fig. 3) [16];

$$A = \int_{-H_{max}}^{+H_{max}} \mu_0 M(H) dH \tag{8}$$

In (8); μ_0 is the magnetic permeability of free space, $M(H)$ is the magnetization over magnetic field strength, dH is the differential of magnetic field strength. Negative and positive maximum limits of the integral are bounded by $-H_{max}$ and $+H_{max}$, respectively. Maintaining the condition in (7) is essential to trigger the desired magnetic loss (see Fig. 3) for a highly effective and efficient MH therapy. The LRT model is bounded by fundamentals of magnetism. Therefore, fundamentals of magnetism can be utilized to structuralize a basic FEM analysis to model the heat produced during MH therapy.

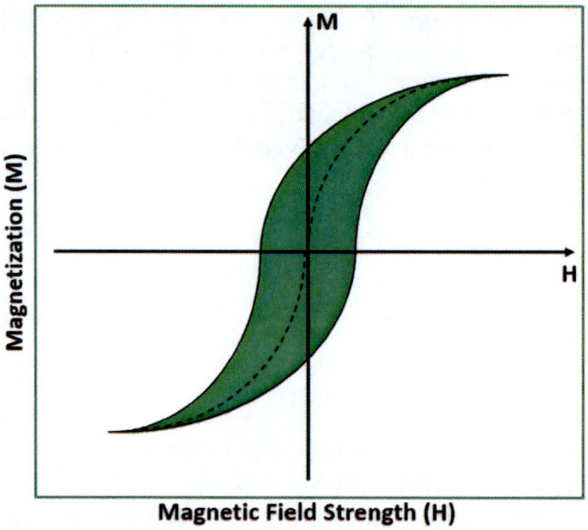

Fig. 3. Simplified plot showing relationship between the magnetic field strength (H) and magnetization (M) of a superparamagnetic particle. Dashed line shows the paramagnetic regime for the condition $\omega\tau < 1$. Solid lines show the ferromagnetic regime (hysteresis loop) for the condition $\omega\tau > 1$. Integration of the green area enclosed by solid lines gives the total heat loss.

3 Methods and Materials

COMSOL Multiphysics 5.5 (COMSOL Inc., USA) has been used for FEM simulations. COMSOL Multiphysics 5.5 is very effective especially in scenarios where modeling of multi-physical phenomena or coupling different physics is required [17]. Whole study has been performed in 3-dimensional (3D) space. Electromagnetic Waves Frequency Domain and Heat Transfer in Solids physics have been coupled in a one-way coupling mode for Electromagnetic Heating multi-physics calculations. First, modelling geometries have been generated using primitive shape options; sphere, block, point and plane geometries. A 14-turn coil geometry with a 22.45 mm radius and 103 mm length has been designed with square wires instead of rounded wires. Primary reason for choosing square wires is to avoid the exponential increase of meshing complexity caused by small rounded geometries such as rounded wires. Since COMSOL Multiphysics 5.5 lacked advanced tools needed to design complex geometries with ease; the coil geometry has been designed in Fusion 360 (Autodesk, USA) and then imported to COMSOL Multiphysics 5.5 environment and positioned at the center of the coordinate system. Then, a square with 30 nm diameter have been built and positioned at the center of the coil as the MNP. 2 different size spheres have been built to reduce the complexity of the meshing process. Material properties are assigned to geometries automatically regarding selected physics which for this study are as follows; relative magnetic permeability, relative permittivity, electrical conductivity, heat capacity at constant pressure, density and thermal conductivity. To model the magnetic loss of the MNP, relative magnetic permeability property can be used in complex form. Since;

$$\tilde{\chi} = \mu_r' - i\mu_r'' - 1, \tag{9}$$

wherein $\mu_{r'}$ is the real part and $i\mu_r''$ is the imaginary part of the relative magnetic permeability. Equation (5) can be rewritten as;

$$M(t) = \left(\mu_r' - i\mu_r'' - 1\right)H(t) \tag{10}$$

Material properties for all geometries have been numerically determined across well established data found in scientific literature. Although properties of Fe_3O_4 have been assigned to MNP; imaginary part of the complex magnetic permeability value could not be found which is essential in modelling of magnetic loss. Therefore, only the real part has been determined across data and the imaginary part has been determined as 65-i20 through reverse engineering process. Multiple simulations have been run to observe the magnetically induced heat output of the Fe_3O_4 and the relative magnetic permeability values of the most efficient simulation run have been adopted. User controlled meshing has been utilized; maximum and minimum mesh element sizes for each geometry have been determined based on experience to avoid any meshing problems. The coil has been fed with 150 kHz (Ferromagnetic regime as in (7), 45 A peak to peak current for 60 min straight. Frequency Domain and Time Dependent studies have been performed to calculate magnetic flux density and heat generation parameters for each mesh element of the 3D space, respectively. Magnetic flux density data have been visualized across a 3D color map (see Fig. 4(a)) and time versus heat generation graph (see Fig. 4(b)) has been plotted for further examination.

To theoretically calculate the total heat generation capacity (Q_{Total}) of a single injection of Ferumoxytol (1 ml) which contains 30 mg of elemental iron, (11) has been utilized.

$$Q_{Total} = \left(\frac{Q_{SingleFe_3O_4}}{3}\right) \times \left(\frac{m_{TotalFe}}{m_{AtomicFe}}\right) \tag{11}$$

In (11), $Q_{SingleFe_3O_4}$ is the heat generated by a single Fe_3O_4 particle. $m_{TotalFe}$ is the mass of total elemental iron and $m_{AtomicFe}$ is the atomic mass of iron. Since $Q_{SingleFe_3O_4}$ consists 3 Fe atoms, it is divided by 3 for correct energy output calculations. 11 evenly distributed data points regarding momentary heat production by a single Fe_3O_4 particle have been averaged to attain $Q_{SingleFe_3O_4}$. Atomic mass of an iron particle ($9.2732796 \times 10^{-23}$ g) is also known from scientific literature.

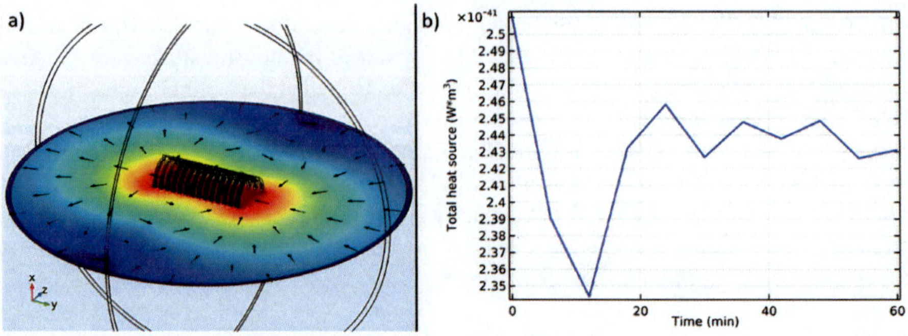

Fig. 4. (a) In COMSOL Multiphysics 5.5, rainbow color map is the magnetic flux density and black arrows are vector field of magnetic effect on YZ plane. (b) Magnetically induced momentary heat production of a single Fe_3O_4 nanoparticle for a duration of 60 min.

4 Results and Discussion

Coil design simulated through FEM produced a maximum of 12.6 mT magnetic flux density on a Fe_3O_4 particle. Time average of total heat generation by a single Fe_3O_4 was found to be 2.43×10^{-41} Wm^3. Single particle simulation results were used to theoretically calculate heat generation capacity of a single injection (1 ml) of Ferumoxytol as 2.62×10^{-18} Wm^3. Combining simulation results of a single particle with theoretical calculations proved to be effective in creating a basis for future studies towards minimizing both required total simulation time and computer power and also optimizing input waveforms for alternating magnetic fields. Practical studies must be performed to measure imaginary magnetic permeability value of superparamagnetic Fe_3O_4 particles. Assigning measured imaginary magnetic permeability values to Fe_3O_4 particle in this model will surely increase accuracy of the structured simulation model.

Acknowledgement. This project is funded by Bogazici University Research Fund with Grant Number 19661.

References

1. Chen, G., et al.: Nanochemistry and nanomedicine for nanoparticle-based diagnostics and therapy. Chem. Rev. **116**(5) (2016)
2. Khalili, S.R., et al.: Charpy impact behavior of clay/basalt fiber-reinforced polypropylene nanocomposites at various temperatures. J. Thermoplast Compos. Mater. (2014)
3. Périgo, E.A., et al.: Fundamentals and advances in magnetic hyperthermia. Appl. Phys. Rev. **2**(4), 041302 (2015)
4. Reddy, L.H., et al.: Magnetic nanoparticles: design and characterization, toxicity and biocompatibility, pharmaceutical and biomedical applications. Chem. Rev. **112**(11) (2012)
5. Ali, R., et al.: Impacts of ni-co substitution on the structural, magnetic and dielectric properties of magnesium nano-ferrites fabricated by micro-emulsion method. J. Alloys Compd. **584** (2014)

6. Tailhades, P., Gillot, B., Rousset, A.: Mixed-valence defect ferrites: a new family of fine powders and thin films of spinel ferrites. J. Phys. IV Colloq. **7** (1997)
7. Tishin, A.M., et al.: Developing antitumor magnetic hyperthermia: Principles, materials and devices. Recent Patents Anti-Cancer Drug Discov. **11** (2016)
8. Çengel, Y.A., Boles, M.A.: Thermodynamics: An Engineering Approach. McGraw-Hill Higher Education (2008)
9. Chen, Z.: Finite Element Methods and Their Applications. Springer (2005)
10. Carrey, J., Mehdaoui, B., Respaud, M.: Simple models for dynamic hysteresis loop calculations of magnetic single-domain nanoparticles: Application to magnetic hyperthermia optimization. J. Appl. Phys. **109**(8), 083921 (2011)
11. Hervault, A., Thanh, N.T.: Magnetic nanoparticle-based therapeutic agents for thermo-chemotherapy treatment of cancer. Nanoscale **6**(20), 11553–11573 (2014)
12. Usov, N.A.: Low frequency hysteresis loops of superparamagnetic nanoparticles with uniaxial anisotropy. J. Appl. Phys. **107**(12) (2010)
13. Oura, S., et al.: Radiofrequency ablation therapy in patients with breast cancers two centimeters or less in size. Breast Cancer **14**(1), 48–54 (2007). https://doi.org/10.2325/jbcs.14.48
14. Seip, R., Ebbini, E.S.: Noninvasive estimation of tissue temperature response to heating fields using diagnostic ultrasound. IEEE Trans. Biomed. Eng. **42**(8) (1995)
15. Dieckhoff, J., et al.: Magnetic-field dependence of Brownian and Néel relaxation times. J. Appl. Phys. **119** (2016)
16. Jiles, D.C., Atherton, D.L.: Theory of ferromagnetic hysteresis. J. Magnetism Magnet. Mater. **61**(1) (1986)
17. Adam, T., Hashim, U.: COMSOL multiphysics simulation in biomedical engineering. Adv. Mater. Res. **832** (2014)

Detection of Physical Stressors in Patients with Frailty Syndrome Using Wearable Devices

Daivaras Sokas[1(✉)] and Andrius Petrėnas[1,2]

[1] Biomedical Engineering Institute, Kaunas University of Technology, Kaunas, Lithuania
daivaras.sokas@ktu.lt

[2] Department of Electronics Engineering, Kaunas University of Technology, Kaunas, Lithuania

Abstract. Frailty is considered one of the most important challenges of the aging population and it is characterized by a decline in physiological reserve and vulnerability to internal and external stressors. Cardiac autonomic imbalance, as reflected by heart rate response measures, may contribute to frailty worsening, which in turn may decrease the capacity to maintain homeostasis when exposed to physical stressors. One approach to improve frailty assessment is the monitoring of the heart rate response to physical stressors unobtrusively using wearables. However, there is a lack of algorithms for detecting physical stressors in wearable-based biosignals obtained from frail adults. Accordingly, the paper proposes and explores a derivative dynamic time-warping-based algorithm to distinguish between walking and stair-climbing, which are the most common physical activities among adults with frailty. The performance of the proposed algorithm was studied on biosignals acquired from 86 patients with frailty who had undergone cardiac rehabilitation after open-heart surgery. The best performance is achieved using the acceleration signal of the vertical axis, resulting in a sensitivity of 84.2% and 81.6%, a positive predictive value of 82.1% and 83.8% for detecting walking and stair-climbing, respectively. Identification of walking and stair-climbing in daily activities opens the possibility to assess the heart rate response to detected physical stressors.

Keywords: Dynamic time warping · Sensors · Accelerometer · Walking · Stair-climbing

1 Introduction

Frailty is one of the most essential healthcare challenges of the aging population and it is characterized by a decline in physiological reserve and vulnerability to internal and external stressors [1]. Cardiac autonomic imbalance, as reflected by heart rate response measures, may contribute to frailty worsening, which in turn may decrease the capacity to maintain homeostasis when exposed to physical stressors [2]. Existing tools for frailty assessment are based on indexes and questionnaires and do not involve heart rate response. Meanwhile, clinical methods, such as veloergometry, require special equipment and are too demanding for severely frail adults. One approach to improve frailty assessment is the monitoring of the heart rate response to physical stressors unobtrusively

Y. Dekhtyar and I. Saknite (Eds.): NBC 2023, IFMBE Proceedings 89, pp. 80–86, 2023.
https://doi.org/10.1007/978-3-031-37132-5_11

using wearables [3]. However, to apply this approach, it is crucial to know both the intensity and type of physical activity. Currently, there is a lack of algorithms for detecting physical stressors in wearable-based biosignals obtained from adults with frailty.

Advances in wearable technology have opened the possibility to distinguish between different physical activity types, even when performed in activities of daily living [4]. However, the proposed algorithms for the detection of physical activities have not yet been tested on elderly frail adults who have distinctive movement patterns. Frail adults are more likely to have uneven and asymmetrical movement which may reduce the reliability of the algorithms developed for use in healthy populations.

To account for the variability in physical activity patterns, introduced by the weakness and slowness of frail adults, this study presents the derivative dynamic time warping-based algorithm for detecting daily physical stressors in adults with frailty. The performance of the algorithm was explored on walking and stair-climbing given that these activities are common in daily living and are often feasible even for adults with advanced frailty.

2 Methods

2.1 Study Population

The patients after open-heart surgery, who performed cardiac rehabilitation at Kulautuva Rehabilitation Hospital of Kaunas Clinics (Kulautuva, Lithuania), participated in the study. The inclusion criteria were age $\geq$ 65 years, a score on the Edmonton Frail Scale (EFS) $\geq$ 4 at admittance to the rehabilitation hospital, and a 6-min walk distance $\geq$ 150 m. The study cohort consists of 86 participants (29 females) assigned to groups of non-frail/vulnerable and frail, based on the EFS, see Table 1.

Table 1. Participant characteristics in the groups of non-frail/vulnerable (EFS $<$ 6 points) and frail (EFS $\geq$ 6 points). Certain parameter values are given as mean (range).

	Non-frail/vulnerable	Frail
Number of participants (females)	35 (12)	51 (17)
Age, years	71.7 (65–83)	73.8 (65–87)
Body mass index, kg/m^2	27.2 (19.9–41.0)	27.4 (18.7–41.8)
EFS	4.47 (3–5)	7.14 (6–10)

The study was approved by the Kaunas Regional Biomedical Research Ethics Committee (approval number BE-2-99). Signed written consent to participate in the study was obtained from all subjects. The study was conducted following the ethical principles of the Declaration of Helsinki (64th WMA General Assembly, Fortaleza, Brazil, October 2013). Personal information was removed from the collected data to ensure participant anonymity.

2.2 Data Acquisition

A textile strap Polar H10 (Polar Electro OY, Kempele, Finland), placed below the chest, was used to acquire triaxial acceleration signals at a sampling rate of 200 Hz. A custom-made software, based on a software development kit provided by Polar, was installed on a Samsung Galaxy A21s smartphone and further used for transferring data from the wearable device to a storage server. The participants carried the smartphone in the holder wrapped around the upper arm.

Acceleration signals were recorded during clinical walking and stair-climbing tests that were conducted at various intervals throughout a monitoring session lasting 2–4 h. During the walk test, participants had to walk in a straight path for a 6-min, whereas, during stair-climbing, they had to climb one flight of stairs. The onset and the end of each activity were set manually by a healthcare specialist. The tests were performed before and after inpatient cardiac rehabilitation. A part of the signal database is freely available at Physionet [5].

2.3 Templates of Physical Activity

Acceleration signals were filtered using 3rd order Butterworth lowpass filter with a cut-off frequency of 15 Hz. The duration of walking was 6 min, whereas climbing one flight of stairs took from 10 to 20 s depending on the participant's functional status. Therefore, the analysis time interval was set to 10 s to cover both activities and was extracted from the middle of each activity in case the activity lasted longer. The extracted intervals were further used to manually detect three individual strides in the magnitude vector of the acceleration signal. Then, the strides were resampled to 200 samples and averaged to create a representative activity for each participant. Finally, the templates for walking and stair-climbing were created by averaging the representative activities of all participants. To explore the effect of the choice of the acceleration axes on the detection performance, the templates were found for the anterior-posterior, mediolateral, and vertical axes, as well as for the magnitude vector.

2.4 Derivative Dynamic Time Warping-Based Algorithm

To distinguish between walking and stair-climbing, a derivative dynamic time warping was applied which nonlinearly assesses a similarity between two time series, i.e., a signal under analysis and a template [6].

The similarity is assessed by finding the optimal warping path from the signal X to the template Y. Minimal distance $w_{i,j}$ is found by matching x_i samples to the samples y_j of template time series as visualized in Fig. 1 (a). The optimal distance W^* of the signal under analysis to the template is found by minimizing the path from (x_0, y_0) to (x_N, y_N), as shown in Fig. 1 (b):

$$W^* = \underset{W \in \mathbf{W}}{\arg min} D_W(X, Y), \tag{1}$$

where D_W is the function that represents the distances of each of the possible warping paths W.

The optimal distance of the signal under analysis is calculated for both walking and stair-climbing templates. The signal under analysis is classified as either walking or stair-climbing, depending on which activity resulted in a lower W^* value.

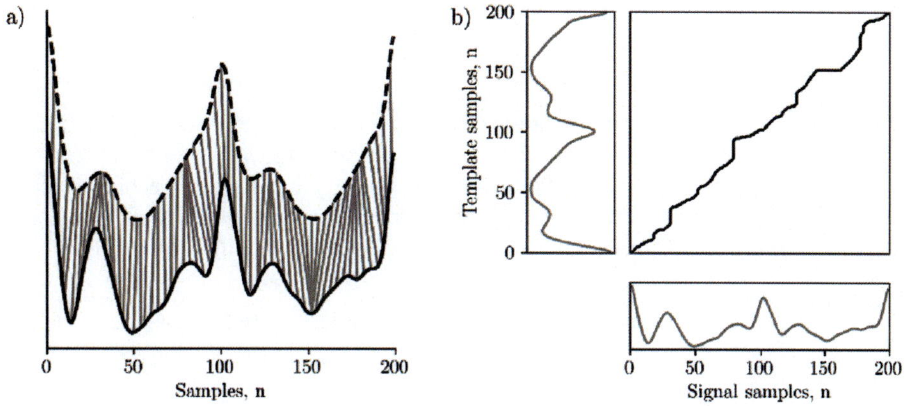

Fig. 1. Template matching using a derivative dynamic time warping: a) alignment of a signal under analysis (solid line) with the template (dashed line), and b) optimal alignment path.

Figure 2 illustrates the calculated distance for each participant during walking and stair-climbing using the acceleration signal of a vertical axis.

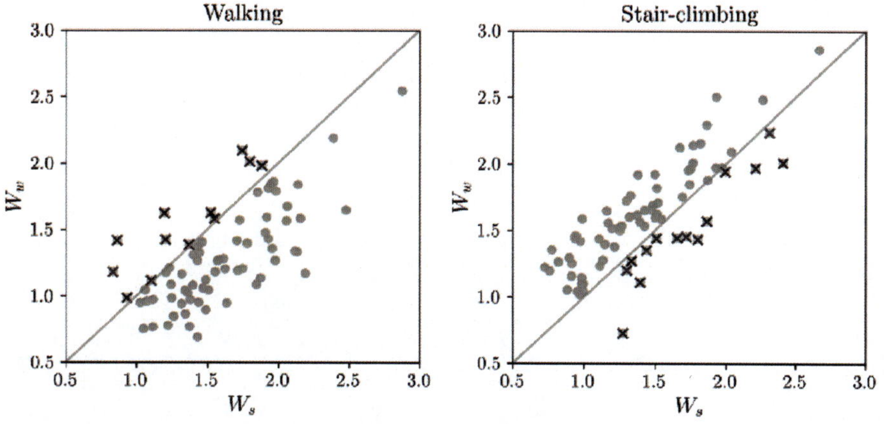

Fig. 2. The distance of the signal under analysis from templates for walking W_w, and stair-climbing W_s. Crossed dots represent incorrectly classified physical activity.

2.5 Performance Measures

Detection performance was investigated in terms of sensitivity (*Se*), positive predictive value (*PPV*), and accuracy (*ACC*). Sensitivity is the ratio of the total number of a particular physical stressor correctly detected and the total number of that stressor, whereas a positive predictive value is the ratio between the number of the correctly detected particular stressor to the total number of that stressor detected including false positive detections. Accuracy refers to the ratio of correctly detected stressors to the total number of both stressors.

3 Results

The performance of a derivative dynamic time warping-based algorithm for acceleration signals of anterior-posterior, mediolateral, and vertical axes, as well as for the magnitude vector of acceleration is given in Table 2. The best overall performance is achieved using acceleration signal of the vertical axis, resulting in *Se* of 84.2% and *PPV* of 82.1% for detecting walking and *Se* of 81.6% and *PPV* of 83.8% for detecting stair-climbing.

Table 2. The performance of a derivative dynamic time warping-based algorithm.

	Walking		Stair-climbing		
	Se, %	*PPV*, %	*Se*, %	*PPV*, %	*ACC*, %
Anterior-posterior	95.1	69.5	57.9	93.6	76.9
Mediolateral	75.0	68.7	65.8	72.5	70.4
Vertical	84.2	82.1	81.6	83.8	82.9
Magnitude vector	97.4	65.5	46.7	94.9	73.0

Table 3 presents the performance for the groups of frail (EFS $<$ 6 points) and vulnerable/non-frail (EFS $\geq$ 6 points) participants using an acceleration signal of a vertical axis. The results show that the algorithm is more sensitive ($Se = 86.9\%$) when detecting walking in frail adults compared to stair-climbing ($Se = 79.1\%$). The opposite tendency is observed for a vulnerable/non-frail group.

Table 3. The performance in the groups of different frailty statuses.

	Walking		Stair-climbing		
	Se, %	*PPV*, %	*Se*, %	*PPV*, %	*ACC*, %
Frail	86.9	81.6	79.1	85.0	*82.5*
Vulnerable/non-frail	80.0	82.8	84.9	82.4	*83.2*

4 Discussion

This study explored the feasibility of detecting physical stressors in adults with frailty which is a challenging issue due to the weakness and slowness of these persons resulting in an inconsistent speed and intensity of physical activities. The proposed derivative dynamic time warping-based algorithm allows to mitigate the influence of inconsistent movements by aligning acceleration signals nonlinearly. The ability to detect physical stressors, such as walking and stair-climbing, in daily living, opens the possibility to provide additional measures about the physiological reserve of frail adults. For example, heart rate recovery expressed as heart rate decrease in 2-min after walking has been considered a useful supplement for the clinical walk test [7]. We previously showed that the estimation of heart rate recovery after stair-climbing using a wearable device is also feasible [8].

Previous studies indicate that people with frailty have increased fatigue and a higher risk of falling. These findings explain additional issues associated with the detection of physical stressors in adults with frailty. Walking, and especially more physically demanding activities such as stair-climbing, tend to be less consistent, often with slowdowns and stops. Additionally, pain in the lower extremities or torso can induce asymmetry of consecutive steps, which further complicates the detection of physical stressors in acceleration signals [9]. These peculiarities hinder the performance of template-matching methods that rely on a correlation between the signal under analysis and the template.

The limitation of this study is that only two types of activities were considered for detection. The main motivation for choosing walking and stair-climbing was that these are common in daily living, including for most adults with frailty. It was observed that 18% of frail adults were unable to walk one bus-stop distance (50 m) and climb one flight of stairs [10]. However, considering the application of the proposed algorithm in daily life, accounting for alternative activity types, such as riding a bicycle, will be necessary. Another limitation is that the physical stressors analyzed were manually extracted, indicating the need for future studies to assess the performance of the proposed algorithm using the entire set of acceleration signals obtained during daily living.

5 Conclusions

The derivative dynamic time warping-based algorithm was proposed to detect walking and stair-climbing in wearable-based biosignals. Identification of physical stressors in daily activities opens the possibility to assess the heart rate response to detected stressors.

Acknowledgment. This work was financially supported by the Research Council of Lithuania (Agreement No. S-MIP-20-54).

Conflict of Interest. The authors declare that they have no conflict of interest.

References

1. Dent, E., Martin, F.C., Bergman, H., Woo, J., Romero-Ortuno, R., Walston, J.D.: Management of frailty: opportunities, challenges, and future directions. The Lancet **394**(10206), 1376–1386 (2019)
2. Parvaneh, S., et al.: Regulation of cardiac autonomic nervous system control across frailty statuses: a systematic review. Gerontology **62**(1), 3–15 (2016)

3. Sokas, D., et al.: Wearable-based assessment of heart rate response to physical stressors in patients after open-heart surgery with frailty. IEEE J. Biomed. Health Inf. In Press (2023)
4. Margarito, J., Helaoui, R., Bianchi, A.M., Sartor, F., Bonomi, A.G.: User-independent recognition of sports activities from a single wrist-worn accelerometer: a template-matching-based approach. IEEE Trans. Biomed. Eng. **63**(4), 788–796 (2015)
5. Sokas, D., et al.: Wearable-based signals during physical exercises from patients with frailty after open-heart surgery (2022)
6. Keogh, E.J., Pazzani, M.J.: Scaling up dynamic time warping to massive datasets. In: Żytkow, J.M., Rauch, J. (eds.) PKDD 1999. LNCS (LNAI), vol. 1704, pp. 1–11. Springer, Heidelberg (1999). https://doi.org/10.1007/978-3-540-48247-5_1
7. van Stel, H.F., Bogaard, J.M., Rijssenbeek-Nouwens, L.H., Colland, V.T.: Multivariable assessment of the 6-min walking test in patients with chronic obstructive pulmonary disease. Am. J. Respir. Crit. Care Med. **163**(7), 1567–1571 (2001)
8. Sokas, D., Petrėnas, A., Daukantas, S., Rapalis, A., Paliakaitė, B., Marozas, V.: Estimation of heart rate recovery after stair climbing using a wrist-worn device. Sensors **19**(9), 2113 (2019)
9. Martínez-Ramírez, A., et al.: Frailty assessment based on trunk kinematic parameters during walking. J. Neuroeng. Rehabilit. **12**(1), 1–10 (2015)
10. Merchant, R.A., et al.: Rapid geriatric assessment using mobile app in primary care: prevalence of geriatric syndromes and review of its feasibility. Front. Med. **7**, 261 (2020)

Expediency of Using a Physical and Mathematical Model in Cell Engineering

Nataliia Moisieieva, Anton Moisieiev, Olga Gorina, and Yuliia Akhatova(✉)

Institute for Problems of Cryobiology and Cryomedicine of the National Academy of Sciences of Ukraine, Pereyaslavska Str. 23, Kharkiv 61016, Ukraine
julija_veselovskaja@meta.ua

Abstract. In order to improve the efficiency of the methods of multicellular objects cryopreservation in the presented work it is proposed to use the method of numerical modeling of the processes that occur during their freezing and affect their viability. It has been proven that the theoretically determined exposure time of spheroids, derived from L929 cells, in 1M DMSO is 10 times less compared to the standard protocol and decreases with increasing temperature. It was established that decreasing the temperature of 1M DMSO cryoprotectant solution from 25 to 10 °C significantly affects the permeability coefficients for water and DMSO molecules into spheroids and their activation energy. It has been proven that the use of theoretically calculated mode leads to an increase in the number of viable spheroids and to a decrease in the number of spheroids with signs of apoptosis compared to standard modes. The permeability and exposure time characteristics, obtained in the work on the basis of theoretical calculations, prove the expediency of taking them into account during cryopreservation process in order to improve the quality of thawed spheroids.

Keywords: Physical and Mathematical Model · Cryopreservation · 3D-cultivation · L929 Cells

1 Introduction

In order to reproduce in vivo conditions, the attention of researchers has recently been focused on multicellular objects (aggregates, organoids, spheroids) [1–5]. The accumulation of the necessary stock and the expansion of the spectrum of the use of multicellular objects as test systems before the clinical approval of drugs requires the creation of biobanks and effective methods of their cryopreservation. Currently there are protocols for obtaining the spheroids from almost all organs of different species origin, but they do not have a thorough cryobiological approach for determining an effective protocol for their cryopreservation. Dehydration and saturation of biological objects with cryoprotectants are one of the most important processes at the cooling stage during cryopreservation of any biological object [6, 7]. The permeability coefficients of the plasma membrane of cells in the composition of spheroids for water and cryoprotectant calculated in this research make it possible to find out the probability of intracellular crystallization, which is a decisive condition for cell survival/death during cooling.

Y. Dekhtyar and I. Saknite (Eds.): NBC 2023, IFMBE Proceedings 89, pp. 87–93, 2023.
https://doi.org/10.1007/978-3-031-37132-5_12

We aimed to determine the optimal cryopreservation protocols for spheroids at different temperatures of the cryoprotectant solution based on physical and mathematical modeling.

2 Materials and Methods

Studies were performed on spheroids obtained under antiadhesive conditions from cells of L929 line on day 7 of cultivation [8]. The L929 cell line (culture of mouse fibroblasts) was kindly provided by D.K. Zabolotny Institute of Microbiology and Virology of the NAS of Ukraine (Kyiv, Ukraine). Cells were cultured in DMEM/F12 nutrient medium (Biowest, France) supplemented with 10% fetal bovine serum (Biowest, France), 200 U/mL benzylpenicillin (Arterium, Ukraine) and 200 μg/mL streptomycin (Arterium, Ukraine), in a humidified incubator with 5% CO_2 at 37 °C.

Changes in the volume of spheroids over time *in toto* were evaluated in 1M DMSO (at temperatures of 10, 15, 25 °C) according to the method [7]. To calculate the exposure time, mass transfer processes, and penetration coefficients of water and cryoprotectant DMSO molecules into spheroids at cryoprotective solution temperatures of 10, 15, and 25 °C, a physical and mathematical model was used [7]. To determine the cell volume, Petri dish with individual cells or spheroids was thermostated LSM 510 META confocal microscope table (Carl Zeiss, Germany). A solution of 1M DMSO in physiological saline at a temperature of 10, 15, 25 °C was added to the samples. Changes in time of the volume of spheroids or attached to the culture surface during their dehydration in 1M cryoprotectant solution and recovery at the stage of penetration of cryoprotectant into cells were studied. Changes in the hourly volume of spheroids *in toto*, attached to the cultural surface in the process of aquaculture in the 1st range of cryoprotectant and reintroduced at the stage of penetration of the cryoprotectant into cells according to the formula V = V(t)/V0, (de V(t) - about 'm cells at time t, V0 - output volume). Morphometric analysis was performed using the "AxioVision Rel. 4.6" software (Carl Zeiss, Germany). Permeability coefficients were determined in single cells and spheroids with a diameter of 80–100 μm after their attachment under adhesive conditions during the day. The volumetric method was used to determine the permeability coefficients of plasma membranes of L929 cells to water molecules (Lp) and DMSO (CPA) and the eff ective values of similar parameters for spheroids from these cells [7, 8]. A sphere approximated the shape of the cells.

The standard mode of spheroids cryopreservation was used as a control [9]. Fluorescent dyes (DNA dyes Hoechst 33342 and propidium iodide (PI)) were used to assess viability and apoptosis in spheroid cells. The level of cytoplasmic cytochrome C was determined as described in studies [8]. Morphometric studies were carried out using a confocal microscope "Axio Observer Z1" (Carl Zeiss, Germany) with the use of computer program "AxioVision Rel. 4.6" (Carl Zeiss, Germany).

Statistical results were processed using the non-parametric Mann-Whitney test in Statgraphics plus for Windows 2.1 ("Manugistics Inc.", USA). All the experiments were repeated 10 times. The results are expressed as mean ± SEM (standard error of the mean). Differences at $p < 0.05$ were considered significant.

3 Results

By changing the volume of spheroids over time after adding 1 M DMSO at different temperatures (10, 15, and 25 °C) in 1M DMSO cryosolution (Fig. 1A), the optimal spheroid exposure time was determined (Fig. 1B). The optimal exposure time was considered the time required to restore the spheroid volume to its original values after the addition of 1M DMSO (Fig. 1A). Figure 1B shows that decreasing the cryosolution temperature from 25 to 10 °C led to an increase in the exposure time of spheroids by 1.5 times. When comparing the theoretically determined in the work exposure time of spheroids at different temperatures with the standard protocol, significant differences were established. Thus, the theoretically calculated exposure time for spheroids at temperatures of 25, 15, and 10 °C was 14, 12, and 9 times shorter respectively compared to the standard mode (600 s) [9].

On the basis of theoretical calculations, the permeability coefficients of water and DMSO molecules into spheroids were determined and their significant decrease with a reduction of temperature from 25 to 10 °C was proved. Thus, the values of the permeability coefficients for water and DMSO molecules at temperatures of 10 and 15 °C decreased by 2 and 1.5 times and 2.8 and 1.4 respectively compared to the temperature of 25 °C (Fig. 2).

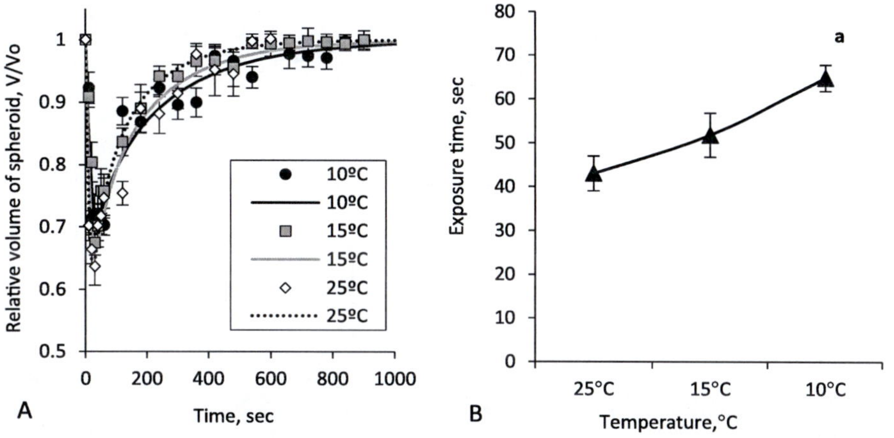

Fig. 1. A - Dynamics of changes in the volume of spheroids on day 7 of cultivation in 1M DMSO solution at temperatures of 10, 15, and 25°C; B - Optimal exposure time of spheroids required to restore the volume to initial values at temperatures of 10, 15, and 25°C. a – the differences are statistically significant in comparison with the temperature of 25°C ($p < 0.05$).

The study of the energy consumption of the permeable substance transfer processes through cell membranes, in particular cryoprotectant and water molecules, was evaluated by the activation energy values. It was established that when the temperature decreased within the range of 10–15 °C, the energy consumption for the penetration of DMSO molecules into the spheroids increased by 2.4 and 8.1 times compared to the temperature ranges of 10–25 °C and 15–25 °C respectively (Fig. 3).

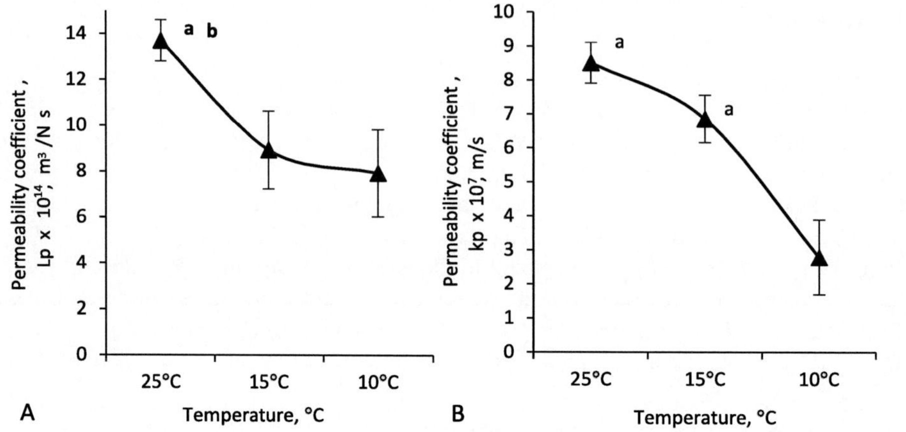

Fig. 2. Permeability coefficients for water (Lp × 10^{14}, m^3/N·s) and DMSO molecules (kp × 10^7, m/s) into spheroids in toto on day 7 of cultivation after incubation in cryoprotective medium at temperatures of 10, 15, and 25 °C; A - for water molecules; B - for DMSO molecules; a – the differences are statistically significant compared to the temperature of 10 °C ($p < 0,05$); b – the differences are statistically significant compared to the temperature of 15 °C ($p < 0,05$).

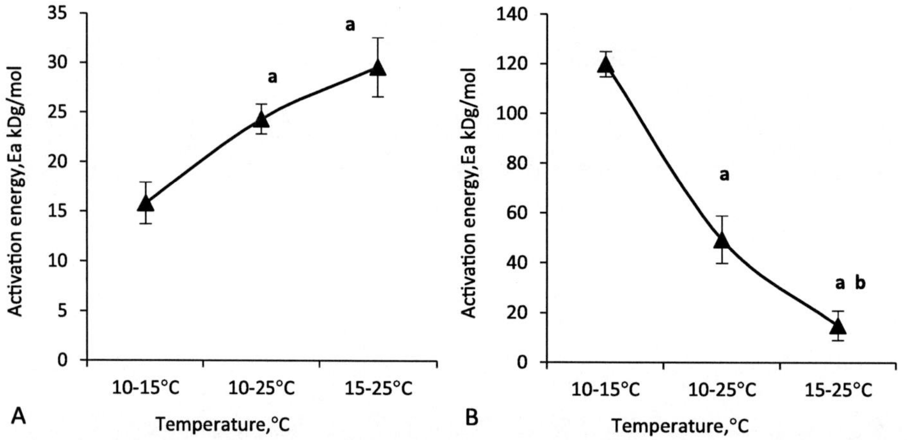

Fig. 3. Activation energy of DMSO molecules penetration (Kp) into spheroids on day 7 of cultivation at different temperature intervals: A – for water molecules, B – for DMSO molecules; a – the differences are statistically significant compared to the temperature range of 10–15 °C ($p < 0,05$); b – the differences are statistically significant compared to the temperature range of 10–25 °C ($p < 0,05$).

Using the obtained values of permeability coefficients in the physical and mathematical model, the degree of spheroids dehydration was predicted. This model describes the kinetics of changes in the relative volume of cells during extracellular crystallization at a certain cooling rate [7]. This made it possible to theoretically determine the optimal cooling rate of spheroids and the final temperature of their freezing before immersion in liquid nitrogen (–196 °C). Thus, for spheroids on day 7 of cultivation, the optimal freezing program includes 65 s of exposure time at the temperature of 10 °C, cooling at a rate of 2 °C/min to –80 °C, followed by immersion into liquid nitrogen.

The effectiveness of the modes determined based on physical and mathematical modeling was confirmed experimentally. The viability of thawed spheroids (the number of Hoechst positive cells with a homogeneous structure of nucleus) was 2.1 times higher compared to the standard mode, which involved 600 s of exposure in cryoprotective solution, cooling at a rate of 2 °C/min to −80 °C followed by immersion in liquid nitrogen (Fig. 4A). Signs of the activation of apoptotic processes significantly decreased after using the theoretically determined mode: the number of cells with nuclear fragmentation at the early and late stages of apoptosis and cytochrome C positive cells in the composition of the thawed spheroids was 17.20 ± 1.85% and 10.38 ± 2.36%, respectively compared to the standard mode of 36.80 ± 2.03% and 40.42 ± 1.67% respectively (Fig. 4B).

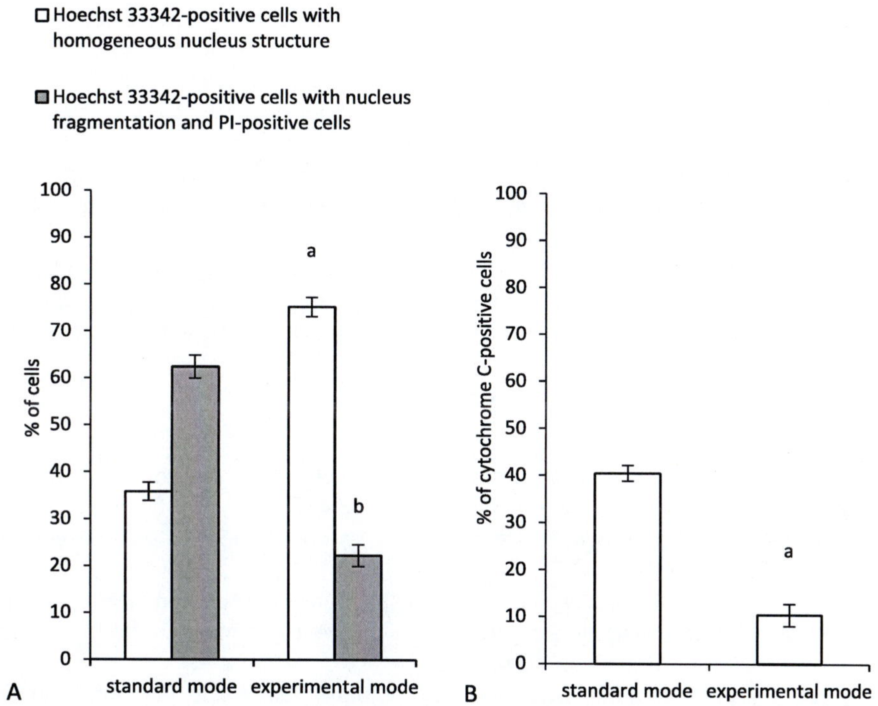

Fig. 4. Cell viability of thawed spheroids according to indicators of disordered membrane permeability (A) and appearance of cytoplasmic cytochrome C (B) after using experimental and standard cryopreservation modes. a, b – differences are significant ($p < 0.05$) in comparison with the corresponding values of the standard mode.

In this way, we have proven that the cryosolution temperature has a significant influence on the permeability indicators, which we believe strongly affect the cryopreservation process.

4 Discussion

Understanding the processes of permeability and mass transfer of compounds into biological objects and the influence of cryosolution temperature on these processes is a necessary component for determining optimal protocols for their cryopreservation. The results presented in the work indicate an ambiguous effect of the cryoprotectant solution temperature on the permeability coefficients and mass transfer processes for water and DMSO molecules into the spheroids. Such discrepancies may be the result of different penetration mechanisms and molecular weights of studied compounds. Diffusion of water and DMSO molecules into spheroids can be influenced by the amount and composition of the extracellular matrix that is synthesized during spheroids formation. The study found that the exposure time of spheroids in cryoprotective medium at different temperatures is 10 times shorter compared to the known conditions (600 s) of spheroid cryopreservation. The reduction of exposure time in the cryoprotective medium, i.e. a reduction in the toxic effect of DMSO significantly increases the safety of biological material, as was demonstrated in a previous work [10] and in the present work. We assume that a 10-fold decrease in the contact time of spheroids with 1M DMSO led to a significant decrease in the number of spheroid's cells with signs of activation of apoptosis and necrosis.

Thus, the results of the study, prove the expediency of using a physical and mathematical model in cell engineering for determining effective protocols for cryopreservation of multicellular objects with different cell composition and cultivation conditions.

5 Conclusions

1. It was established that when the temperature of the cryoprotectant solution decreases from 25 to 10 °C, the permeability coefficients for water and DMSO molecules into the spheroids significantly decrease and the energy consumption of their penetration increases significantly.
2. It has been proven that the exposure time of spheroids in the cryoprotective solution increases with decreasing in temperature from 25 to 10 °C and significantly differs from the values of the standard protocol.
3. After cryopreservation of spheroids according to the theoretically determined mode, the number of viable cells increased significantly while the number of cells with the signs of apoptosis decreased significantly compared to the standard mode.

Acknowledgements. We would like to express our gratitude to our colleagues, Kovalenko I.F. and Gordiyenko O.I., for their assistance in the physical and mathematical calculations of the penetration processes of DMSO and water molecules into the spheroids.

References

1. Gonzalez-Fernandez, T., Tenorio, A.J., Saiz, A.M., Jr., Leach, J.K.: Engineered cell-secreted extracellular matrix modulates cell spheroid mechanosensing and amplifies their response to inductive cues for the formation of mineralized tissues. Adv. Healthc. Mater. **11**(10), e2102337 (2022)
2. Petrenko, Y., Syková, E., Kubinová, Š: The therapeutic potential of threedimensional multipotent mesenchymal stromal cell spheroids. Stem Cell Res. Ther. **8**, 94 (2017)
3. Pinto, B., Henriques, A.C., Silva, P.M.A., Bousbaa, H.: Three-dimensional spheroids as in vitro preclinical models for cancer research. Pharmaceutics **12**(12), 1186 (2020)
4. Pulze, L., et al.: MCF7 Spheroid development: new insight about spatio/temporal arrangements of TNTs, amyloid fibrils, cell connections, and cellular bridges. Int. J. Mol. Sci. **21**(15), 5400 (2020)
5. Kim, T.Y., et al.: Directed fusion of cardiac spheroids into larger heterocellular microtissues enables investigation of cardiac action potential propagation via cardiac fibroblasts. PLoS ONE **13**(5), e0196714 (2018)
6. Kuleshova, L.G., Kovalenko, I.F., Kovalenko, S.Ye., Tsibulko, T.S., Gordienko, O.I.: Transport characteristics of plasma membranes of PK-15 passaged cells. Probl. Cryobiol. Cryomed. **31**(1), 14–22 (2021)
7. Tarusin, D.N., Kireyev, V.A., Kovalenko, S.Ye., Kovalenko, I.F., Rozanov, L.F., Petrenko, A.Yu.: Selection of protocols to cryopreserve mesenchymal stromal cells in suspension and alginate microspheres by studying their osmotic responses in 1M DMSO. Probl. Cryobiol. Cryomed. **26**(2), 133–144 (2016)
8. Bozhok, G.A., Moisieiev, A.I., Gorina, O.L., Bondarenko, T.P.: Morphofunctional features of fibroblasts line L929 in 3D-culture. Fiziolohichnyi zhurnal **65**(3), 34–40 (2019)
9. Ehrhart, F., et al.: A comparative study of freezing single cells and spheroids: Towards a new model system for optimizing freezing protocols for cryobanking of human tumours. Cryobiology **58**(2), 119–127 (2009)
10. Smolyaninova, E.I., Khromenkova, O.B., Zhernoklev, G.V., Pishko, O.V.: Equilibration effect in freezing medium on osmotic resistance and viability of 8-cell mice embryos. Probl. Cryobiol. **1**, 3–11 (2004)

Smart Textile in Post Stroke Patient Rehabilitation Exercises Evaluation for Lower Extremities

Peteris Kozirevs[1](✉), Alexander Oks[2], and Alexei Katashev[1]

[1] Institute of Biomedical Engineering and Nanotechnologies, Riga Technical University, Kipsalas Street 6a, Riga 1048, Latvia
p.kozirevs@gmail.com

[2] Institute of Design Technology, Riga Technical University, Kipsalas Street 6, Riga 1048, Latvia

Abstract. Nowadays stroke remains the second-leading cause of death and the third-leading cause of death and disability combined [1]. In almost all cases post-stroke patients will need rehabilitation under physiotherapist guidelines to handle motor disorders [2]. The role of the physiotherapist includes provision of feedback to the patient to prevent the formation of incorrect movement habits. The alternative approach to provide such feedback, especially for at-home exercises, is based on the use of the smart garment that integrates textile pressure and stretch sensors. To realize this approach a prototype of DAid (Double Aid) leggings system for post-stroke rehabilitation exercises on the lower extremities and a data processing method has been developed. An original method for determining the reference ranges (RR) of exercises for the lower extremities used in post-stroke rehabilitation was proposed, and the corresponding RRs were determined. The classification rules, enabling both the real-time detection of errors of the individual motion of lower extremities during execution of post stroke exercises, as well as estimation of the performance of the entire series, were formulated. The effectiveness of the proposed classification method was demonstrated in a series of test exercises, by comparison of the classification, made using the developed system, with estimation, made by qualified expert - physiotherapist. Depending on the type of exercises, the developed system discovered from 84% to 100% of the errors, identified by the physiotherapist.

Keywords: Stroke · post stroke patients · rehabilitation exercises · smart textile

1 Introduction

Stroke remains the second-leading cause of death and the third-leading cause of death and disability combined (as expressed by disability-adjusted life-years lost - DALYs) in the world [1]. After such of disease person can get complications like paralysis, weakness of one or both sides of the body, difficulties with movements control, thinking, attention, learning, memory, and many other things that makes life more complicated. In almost all cases post-stroke patients will need rehabilitation under physiotherapist guidelines to handle motor disorders [2].

Y. Dekhtyar and I. Saknite (Eds.): NBC 2023, IFMBE Proceedings 89, pp. 94–103, 2023.
https://doi.org/10.1007/978-3-031-37132-5_13

In recent years, the studies of possibilities to apply wearable technologies for applications in personalized and clinical healthcare have been widespread [3]. The information on motor functions provided by body posture/movement monitoring sensors gives possibility to develop effective feedback systems, which boost the motivation of patient's activity during rehabilitation procedures [4] as well as can help doctors to monitor recovery process. Garment with embedded sensors as a part of clothing (Smart garment) is one of widely used wearable solutions for body posture/movement tracking and rehabilitation [5]. The most of proposed smart garment systems are inertial measurement unit (IMU) based [6]. However, in our previous research [7] we demonstrated that developed smart garment system (DAid Shirt) with fully textile strain sensors [8] can be a convenient device for shoulder girdle motion monitoring during advanced motor tasks and in clinical conditions was proven its efficacy for training and rehabilitation of patients with subacromial pain syndrome.

In present study we hypostatize that the same approach can be used for evaluation of correctness execution of post stroke rehabilitation exercises oriented on improvement of patient concentration and coordination skills, by tracking uncontrolled movements of lower extremities. On the present stage of research, the rehabilitation exercises were performed by heathy volunteers which tried to execute exercises correctly as well as simulated typical errors and uncontrolled movements performed by post stroke patients.

2 Materials and Methods

Piezoresistive strain sensors we have used to develop DAid leggings are very soft, plain and have high sensitivity to even small deformations [8]. Due to these properties the applied sensors are highly efficient for evaluation of tiny uncontrolled movements or movement deviations, as shown in previous studies [7, 9].

2.1 Rehabilitation Exercises

On the advice of a physiotherapist, three main exercises for the lower extremities were chosen:

The first is leg abduction and adduction in hip joint with foot dorsiflexion (Fig. 1a.). The rule of correctness is that exercise is performed while the foot is in perpendicular position relative to the floor. If the foot rotates, exercise is performed incorrectly (see Fig. 1a.).

The second: patient lying on the back raises straight leg until it reaches bent leg level and then lowers it, feet are in dorsiflexion position. Rule of correctness is that during exercise working leg does not bend in knee joint (see Fig. 1b.).

The third: patient lying on the back with both legs knees in flexion position performs one leg abduction and adduction with the knee in flexion position. The rule of correctness for this exercise is to save immobility of non-working leg (see Fig. 1c.).

The main goal of all these exercises is to keep static position of a certain region of the body. These exercises are based on concentration and body parts control improvement.

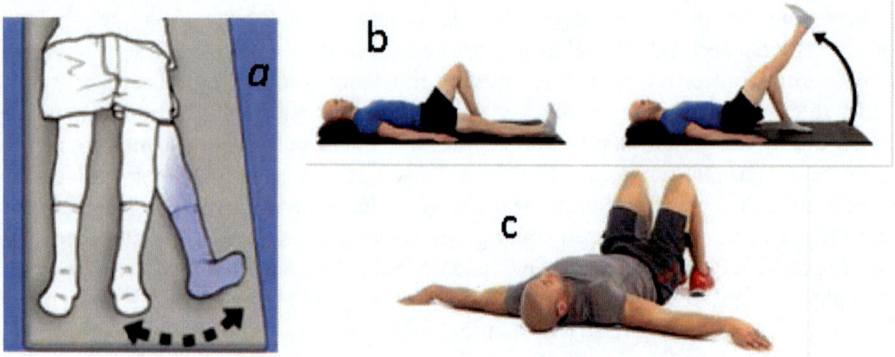

Fig. 1. Exercises visualization: *a* - the first, *b* - the second and *c* - the third exercise.

2.2 DAid Leggings Design

Smart leggings consists of two main parts: textile part – tight elastic leggings with attached knitted strain sensors [8] and data acquisition module [4]. Based on preliminary analysis of body biomechanics during execution of chosen exercises it was hypothesized that it is possible to evaluate quality of exercise performance using three sensors with specific placement. The sensor's placement was defined as follows (see Fig. 2a.):

- for exercise1 "Lying on the back leg abduction/adduction in hip joint with foot dorsiflexion" sensor 1 starts at vastus lateralis bottom part, goes in upper direction through rectus femoris and ends at vastus medialis upper part;
- for exercise 2 "Lying on the back, knees in flexion position, one leg abduction/adduction in hip joint with knee in flexion position" sensor goes through upper lateral part of thigh;
- for exercise 3 "Lying on the back leg flexion and extension with feet dorsiflexion" sensor is positioned in knee joint popliteal region.

Defined zones of sensors placement were transferred and marked on leggings where corresponding sensors were attached (Fig. 2b.). Sensors were connected with acquisition module by conductive pathways made by sewing operation using silver coated nylon yarn Shieldex® 110/2 and snap buttons. Acquisition module provided data transfer via Bluetooth to the laptop for further data processing and feedback.

2.3 Research Methodology

On the present stage of research only healthy volunteers were recruited for the study: in total 7 volunteers (2 female and 5 male). Participated volunteers met the following criteria: they had similar constitution and age (21 ± 2 years old), there were no observable movement or other health disorders, there were no congenital pathologies, no previous lower extremity injuries.

Each participant executed 4 exercise options during which signal from corresponding sensor was acquired.

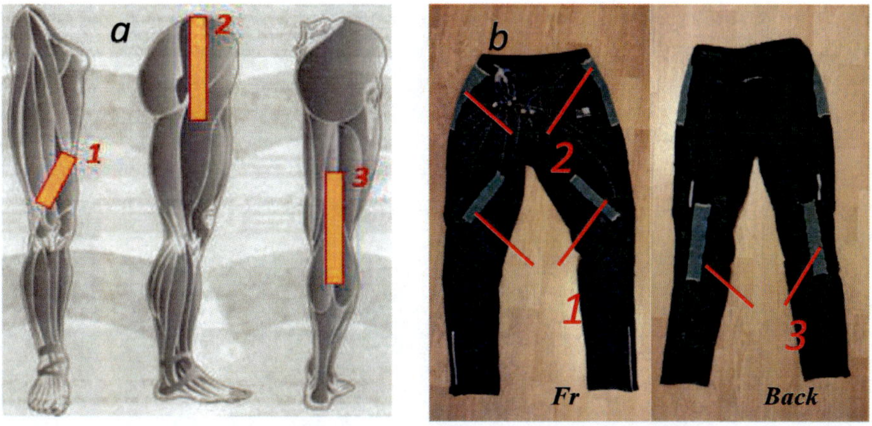

Fig. 2. Sensor's placement: *a*-anatomical position, *b*-leggings with embedded sensors and sewn conductive pathways for sensors connection with acquisition module.

The first option: exercise execution under physiotherapist supervision and control: during exercise performance physiotherapist could physically or verbally correct movements of performer. If during an exercise execution the physiotherapist didn't observe any errors, the signal record was marked as a record with correct performance of the exercise. Otherwise, the record was marked as an incorrect performance of the exercise.

The second option: exercise performance with real time feedback from laptop screen using specially developed software. This software shows as feedback the line generated as a result of processing of data harvested from corresponding sensor on the leggings. Using this line performer could follow up his/her correctness of performance. If the line is straight then exercise was performed correctly, but if peaks were observed - the exercise was executed with errors. Observation of peaks on the line means movements of monitored anatomical area, and therefore incorrect exercise execution. The physiotherapist didn't correct or assist performer: he only observed exercise performance and marked time moments when errors were noted. If during exercise execution errors were not observed, signal record was marked as record with correct performance.

The third option: independent execution of exercise, without any feedback or assistance. Only registration by physiotherapist time moments when errors were observed.

The fourth option: volunteers executed exercises with specially predefined errors shown to them by physiotherapist as mostly common errors performed by real post stroke patients.

Exercises were executed by volunteers on both legs. In this study, recordings for right and left legs were analyzed as separate cases. The potential correlation between data from right and left leg was not studied.

2.4 Data Processing

Analysis of recorded data showed high levels of noise in sensor raw signal records (see Fig. 3a.). To reduce high frequency noise level a median filter with sliding window was used. Numerical tests showed that median filter with window width more than 30 points can essentially reduce the informativeness of a sensor signal otherwise narrow window (less than 30 points) doesn't effectively reduce the level of high frequency noise. Thus, median filter with 30-point width sliding window was used for pre-processing of raw signals. Filtered dataset denoted as *M(i)* is shown on Fig. 3b.

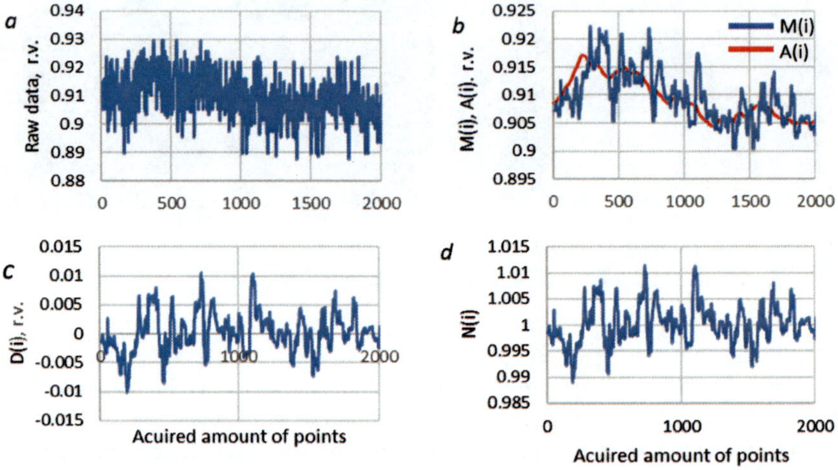

Fig. 3. Stages of signal processing: *a*- raw signal, *b*-signal after noise filtration, *c*-signal after detrending, *d*-final normalized signal.

One can see that filtered data have some trend which can be explained by textile part relaxation and some its sliding relatively exercise performer's skin. So, detrending of filtered signal was made according to the following expression:

$$D_i = M_i - A_i \tag{1}$$

where M_i – i-th signal value after application of median filter, i – point number,

$$A_i = \frac{M_i + M_{i+1} + M_{i+2} + \ldots + M_{i+N}}{N} i \tag{2}$$

where A_i trend data value obtained by application of averaging filter, N - width of window of averaging filter. Numerical tests showed that N is equal to 200 - point sliding window of averaging filter provide the successful signal detrending without loss of useful information (see Fig. 3c.).

To provide possibility to compare signals from different volunteers detrended signal normalisation was used. Calculation of normalized data values was made according to the expression:

$$N(i) = \frac{D_i}{A_v} + 1 \tag{3}$$

where, A_v – average value of first 100 signal points after application of median filter. These first 100 data points had been recorded while volunteer lying motionlessly in a start position before each exercise execution. Such expedient gives possibility to define sensor data baseline which was used for data normalization. Normalized signal, as a result of data processing procedure, is presented on Fig. 3d.

Normalized signal was used in present research as “exercise line” (EL) for monitoring of exercise execution as well as for definition of threshold values of EL fluctuation range to estimate correctness of exercise performance.

2.5 Threshold Definition

Figure 4 Shows an example of ELs, obtained while one of volunteers performed all 4 options of exercise 2. It can be seen that data variation range differs for the options and is much wider for the case of imitation of incorrect performance (Fig. 4d.). The variability of the EL waveform could become a good measure of the correctness of the exercise. To estimate variability of EL recording numerically, we propose to use relative number of outliers – data points that fall out of pre-determined allowed range.

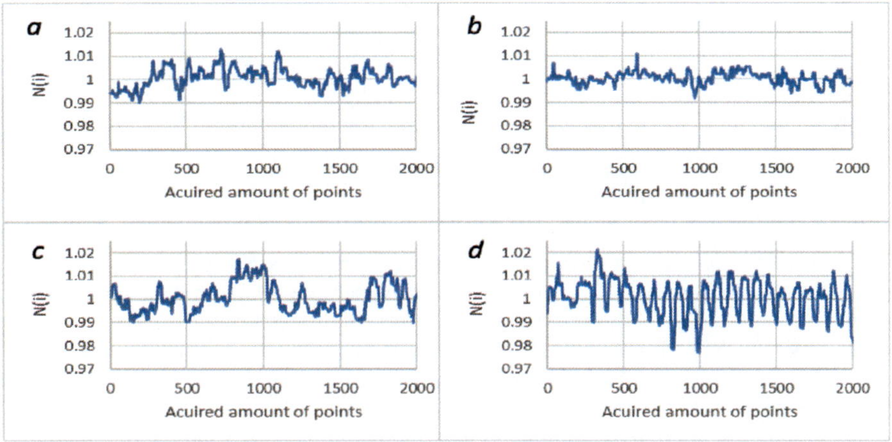

Fig. 4. Examples of “exercise lines” *a* – option 1, *b*-option 2, *c*-option 3, *d*-option 4.

This measurand is convenient for the classification of motions as correct or incorrect. One could demand, for example, that EL waveform of the correct performance should entirely lay within the allowed range. Thus, the EL that falls out of the range, will be classified as incorrect. To set the boundaries of the allowed range, that could provide most accurate classification for each exercise and execution option, the relative number of outliers N_c was calculated as function of the threshold values X, that define the allowed range as [1-X; 1 + X]. The value of x was within range $0.01 \leq |X| \leq 0.04$, the relative number of outliers was calculated as proportion of the total number of recorded values. The calculations were made for every patient, the regularities $N_c(X)$ varied between patients. The outline, that envelopes family of all patients' $N_c(X)$, was constructed for each exercise.

Figure 5a demonstrates as an example the constructed outlines for exercise 1, option 4. The dark red outline corresponds to the data that belongs to incorrect movement simulation. Other outlines are for exercises, performed without simulated errors: blue lined outline area corresponds to exercise execution with DAid feedback, grey one to independent execution, and orange one to execution under supervision of physical therapist. The outlines may be used to select threshold value. For example, demanding that correct EL, obtained under supervision of the therapist, will not have more than 1.3% points outside of the threshold, the threshold value should be selected equal to $|X| = 0.015$. Note, that for performance with simulated errors, the minimal amount of outliers will be approximately 10%. The threshold values, defined in such a way, were $|X| = 0.015$ for exercises 1 and 3, and $|X| = 0.012$ for exercise 2.

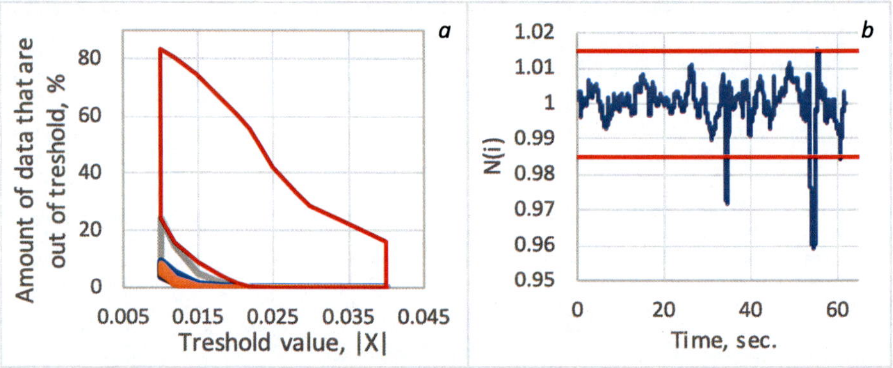

Fig. 5. Threshold definition *a* - Exercise "Lying on the back leg abduction/adduction in hip joint with foot dorsiflexion", all volunteers recorded signals are used, 4 exercise options: area with dark red borders – exercise performance with specially generated errors, area with orange borders – exercise performance under physiotherapist supervision, area with blue borders – exercise performance with real time feedback from leggings, area with grey borders – independent execution of exercise, *b* - validation exercise processed record and defined threshold.

3 Results

To validate system functionality, correctness of applied data processing procedures and threshold definition 3 volunteers were invited. Option 3 of exercise execution was performed: execution without any feedback or assistance with parallel registration by physiotherapist time moments when he/she observed incorrect performance. Collected data were processed and obtained threshold was applied to define DAid legging system's version of incorrectly performed exercises. An example of processed records with threshold lines is shown in Fig. 5b. The graph shows peaks at 34 and 54 s, which exceed the threshold. The same execution errors were noticed by the physiotherapist. The results of error detection using DAid leggings and error detection by a physiotherapist are summarized in Table 1.

Table 1. Validation results.

Exercise	Number of errors noticed by the physiotherapist	Number of errors noticed by system that matched with physiotherapist noticed errors	Number of errors that were noticed by physiotherapist but not noticed by system	Number of errors additionally noticed by system, that were not noticed by physiotherapist	The total number of repetitions performed in the exercise
No. 1	19	16	3	13	120
No. 2	8	8	0	5	90
No. 3	10	9	1	8	108
Total	**37**	**33**	**4**	**26**	**318**

4 Discussion

The present study was devoted to the development of prototype of smart leggings to monitor correctness of execution of post stroke rehabilitation exercises for lower extremities. Monitoring is based on registration of even tiny undesirable movements by highly sensitive knitted piezoresistive sensors [8] embedded into elastic leggings during performance of defined exercises.

System testing using 4 different options of 3 exercises performance gave possibility to develop procedure of data processing and to define threshold values which provide possibility to monitor correctness of exercise performance.

System validation, performed by comparing the execution errors detected by the system and noticed by the physiotherapist, has demonstrated its effectiveness. Table 1 showed the results of validation tests. The developed system noticed most of the same errors as the physiotherapist, with the exception of 3 errors in the 1st exercise and one mistake in exercise 3. The reason of errors missing by system in 1st exercise was poorly developed quadriceps muscle of one of volunteers (noted by a physiotherapist) and accordingly insufficient tightness of leggings and preload of sensor. The latter is necessary for sensor proper functioning and means that smart leggings must fit the exerciser. In case of 3rd exercise data analysis showed that signal value was at threshold level and accordingly was not flagged as an error, but physiotherapist noted a small error. At the same time, the system detected several threshold exceedances that were flagged as exercise errors but were not seen as exercise errors by the physiotherapist. The analysis of the records gave us the possibility to assume that the physiotherapist missed these errors due to their repetition within a short period of time (1–2 s). But this hypothesis needs to be tested in further studies. In present research we assumed all observations of physiotherapist as correct. In accordance with this, graphs of ROC curves were constructed to assess the influence of the selected threshold values on the reliability of error detection. Graphs are presented in Fig. 6.

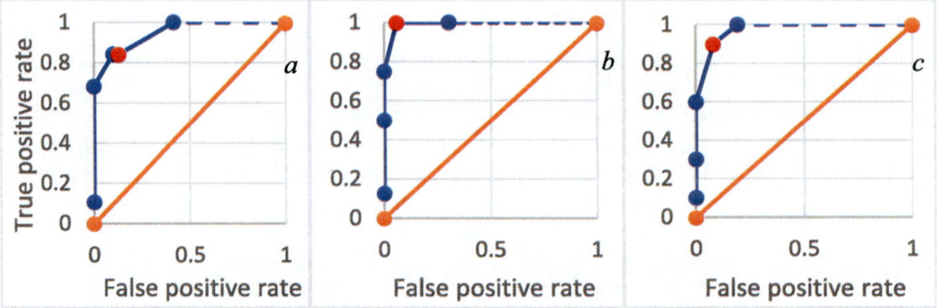

Fig. 6. ROC curves for *a*- exercise 1, *b*- exercise 2, *c*- exercise 3.

The red dots on the curves correspond to chosen threshold values. A good to excellent ratio between the rates of false positive and true positive results is observed for the 2nd and 3rd exercises and satisfactory for the 1st. This confirms the correct functioning of the DAid legging system and its applicability for monitoring rehabilitation exercises of the lower extremities.

The current study has several weaknesses. The first is the recruitment of healthy volunteers instead of real patients, which can significantly affect the repeatability and complexity of the data. The second is to invite only one physiotherapist as an objective alternative in order to notice errors in the performance of exercises, because due to the human factor, he could also make a mistake or miss some errors during the verification tests.

5 Conclusions

DAid leggings system based of textile strain sensors for monitoring of rehabilitation exercises for lower extremities is designed and tested on healthy volunteers. Procedures of data processing and threshold definition are developed and validated. Validation tests showed that the system gives reliable information about quality of exercises performance, correctly define errors of exercise execution (from 84 to 100% of errors were determined during validation tests). Such positive results give the possibility to continue research in clinical conditions involving real patients.

References

1. Feigin, V.L., et al.: World Stroke Organization (WSO): global stroke fact sheet 2022. Int. J. Stroke **17**(1), 18–29 (2022)
2. NIH, Post-Stroke Rehabilitation, National Institute of Neurological Disorders and Stroke, Bethseda (2020)
3. Casselman, J., Onopa, N., Khansa, L.: Wearable healthcare: lessons from the past and a peek into the future. Telematics Inform. **34**(7), 1011–1023 (2017)
4. Okss, A., Kataševs, A., Bernans, E., Aboliņš, V.: A Comparison of the Accuracy of the Smart Sock System to Force Platform and Optical System for Measurement of Temporal Parameters of Locomotion. In: IOP Conference Series: Materials Science and Engineering, vol 254, pp. 1–6. IOP Publishing, United Kingdom, Bristol (2017)

5. Kai-Yu Tong, R.T.: Wearable Technology in Medicine and Health Care. Academic Press, Cambridge (2018)
6. Wang, Q., Markopoulos, P., Yu, B., Chen, W., Timmermans, A.: Interactive wearable systems for upper body rehabilitation: a systematic review. J. Neuroeng. Rehabil. **14**(1), 20 (2017)
7. Semjonova, G., Vetra, J., Oks, A., Katashev, A.: Development of a New Method to Monitor Shoulder Girdle Motion for Ballerina with Shoulder Impingement Syndrome Based on DAid Smart Shirt Application. In: Lhotska, L., Sukupova, L., Lacković, I., Ibbott, G.S. (eds.) WORLD CONGRESS ON MEDICAL PHYSICS AND BIOMEDICAL ENGINEERING 2018. IP, vol. 68/2, pp. 599–602. Springer, Singapore (2019). https://doi.org/10.1007/978-981-10-9038-7_111
8. Oks, A., Katashev, A., Litvak, J.: Knitted resistive fabric: properties and applications. Mater. Sci. Text. Clothing Technol. **9**(21), 28–33 (2014)
9. Semjonova, G., Vetra, J., Cauce, V., Oks, A., Katashev, A., Eizentals, P.: Improving the recovery of patients with subacromial pain syndrome with the DAid smart textile shirt, Sensors (Switzerland) **20**(18), 5277 (2020)

Application of DAid Smart Shirt in Rehabilitation for Subacromial Pain Syndrome Patients Participating in High-Risk for Overuse Injury Sports

Guna Semjonova[1](✉), Janis Vetra[2], Alexander Oks[3], Vinita Cauce[4], and Aleksejs Katashevs[5]

[1] Department of Rehabilitation, Faculty of Rehabilitation, Riga Stradins University, Riga, Latvia
guna.semjonova@rsu.lv
[2] Department of Morphology, Faculty of Medicine, Riga Stradins University, Riga, Latvia
[3] Institute of Design Technologies, Riga Technical University, Riga, Latvia
[4] Statistics Unit, Faculty of Medicine, Riga Stradins University, Riga, Latvia
[5] Institute of Biomedical Engineering and Nanotechnology, Riga Technical University, Riga, Latvia

Abstract. The shoulder rehabilitation challenge faced by athletes participating in sports is characterized by the high risk of overuse injury. Conventional rehabilitation methods for shoulder conditions such as subacromial pain syndrome (SAPS) may not always be effective for athletes who require more targeted and personalized rehabilitation [1]. The DAid smart shirt prototype offers a promising solution to address this issue by providing real-time feedback on movement patterns during physiotherapy, leading to more effective and personalized treatment [2]. This study aimed to investigate the effect of the DAid smart shirt-based feedback on the self-reported DASH (the Disability of Arm, Shoulder, and Hand or DASH) scores during physiotherapy in patients with subacromial pain syndrome who participate in sports, characterized by a high risk for overuse injury. A randomized controlled trial was performed where patients with subacromial pain syndrome had to perform the assigned training exercises while using the DAid smart shirt system. The results showed that the self-reported improvement in DASH scores after the 8-week treatment statistically significantly ($p < 0.001$) differs between the treatment and control groups. The study concludes that the use of the DAid smart shirt-based feedback for the physiotherapy of subacromial pain syndrome in patients, participating in high-risk for overuse injury sports, provides a better self-reported improvement, as compared to the conventional methods.

Keywords: Shoulder Rehabilitation · Subacromial Pain Syndrome · Personalized Rehabilitation for Athletes · Smart Shirt · Self-reported Outcome

Y. Dekhtyar and I. Saknite (Eds.): NBC 2023, IFMBE Proceedings 89, pp. 104–111, 2023.
https://doi.org/10.1007/978-3-031-37132-5_14

1 Introduction

Subacromial pain syndrome (SAPS) is a common condition that affects the shoulder and can cause significant discomfort and disability [3]. Athletes participating in high-risk for overuse injury sports such as overhead sports (basketball, swimming, and volleyball) [4] are more prone to developing shoulder pathologies such as SAPS [5]. Personalized rehabilitation and patient-reported outcomes (PROs) are a crucial part of the management of SAPS for athletes participating in high-risk overuse injury sports [6]. Conventional rehabilitation methods for shoulder conditions such as SAPS may not always be effective for athletes who require more targeted and personalized rehabilitation [2, 4]. And strategies for monitoring shoulder-specific load in overhead athletes call for the development of new technologies, especially at the first stages of shoulder rehabilitation when low-load exercises should be performed [4]. To ensure the correct position of upper extremity training exercises, the use of smart technology in SAPS rehabilitation for more personalized shoulder rehabilitation has been introduced in recent years, for example, Timmermans et al. used sensors on the arm and torso during training for stroke patients [7], for ensuring the correct execution of the task. Beursgens et al. developed a vest for monitoring the patient's posture while playing a game during arm-hand rehabilitation [8]. However, these garments are very loose, and the sensor placement is only approximate. Recently developed DAid smart shirt due to high sensitivity [9] can detect the smallest movements [10] and has shown promising results for recommended integrated approaches of scapular kinematics and rotator cuff strengthening simultaneously during physiotherapy due to its possibility to monitor and provide real-time feedback on movement patterns for shoulder blade during shoulder muscle strengthening exercises [2, 4]. But still little is known about the impact of the DAid smart shirt application in personalized shoulder rehabilitation on PROs for patients with SAPS participating in high-risk overuse injury sports. This study aims to investigate the impact of the DAid smart shirt-based feedback on the patient self-reported DASH (the Disability of Arm, Shoulder, and Hand) scores during physiotherapy in patients with subacromial pain syndrome who participate in sports, characterized by a high risk for overuse injury. The hypothesis is that for patients with subacromial pain syndrome who participate in sports, characterized by a high risk for overuse injury, the physiotherapeutic intervention with the use of the DAid smart shirt prototype as objective feedback, will obtain not worse self-reported outcome measurement results compared with patients with subacromial pain syndrome who participate in sports, characterized by a low risk for overuse injury.

2 Materials and Methods

2.1 Study Design

A randomized controlled trial with a parallel allocation using a 1:1 ratio (experimental group versus control group) was conducted at Orto Clinic (Riga, Latvia), which specializes in traumatology, orthopedics, and rehabilitation of athletes. Participants in the experimental smart shirt group and control group were divided into two subgroups characterized by a qualitative variable:

a. low risk for an overuse injury sports [1, 4]: floorball, table tennis, Nordic walking, track, and field athletics (running).
b. high risk for an overuse injury in sports [1, 4]: basketball, swimming, and volleyball.

The study was conducted from January 2019 to January 2020. A more comprehensive description of the randomization process and the study is presented in [2]. The study was conducted following the Declaration of Helsinki and the study protocol was approved by the Ethics Committee of Riga Stradins University (6–3/39, 27.12.2018).

2.2 Intervention

In the first physiotherapy examination and functional evaluation session, all study participants were required to complete a DASH questionnaire's three modules: Disability/symptom module, Sports module, and Work module. Patient history was collected, and an 8-week session schedule was established for the practical part of the study. Following evidence-based recommendations [11–13], the individual training plan covered 8 weeks, with 2 sessions per week (24–72 h intervals) in the presence of a physiotherapist for 30 min.

The intervention was divided into three stages:

1) The initial stage of training therapy lasted the first 4 weeks. Low load movement control tasks were performed, where the shoulder girdle motion was controlled while performing shoulder joint flexion, and abduction tasks in the frontal, sagittal, and scapular planes [6, 14, 15]. In each plane, the patients performed at their own pace 10 movements in 3 sets with a 1-min break between them [16].
2) Basic stage of training therapy included tailored for each strength and shoulder stabilization tasks specific to the shoulder girdle [6, 15]. The basic principles of the ACSM's Guidelines for Exercise Testing and Prescription [16] were used for individual task progression. The tasks were performed under the supervision of a physiotherapist and a DAid smart shirt.
3) Final stage of training therapy
In the last physiotherapy session, study participants were required to complete a DASH questionnaire's three modules as well as functional clinical tests more comprehensively described in [2].

2.3 DAid Smart Shirt Application in Physiotherapy Sessions

In addition to the training therapy intervention described previously, throughout the intervention period – of 8 weeks – the DAid smart shirt prototype was used for tasks that had to monitor the stability of the shoulder girdle and the quality of movement performance (see Fig. 1.). During the first visit, the patients, who were included in the experimental group, were instructed on the use of the smart shirt prototype during the exercises. The DAid smart shirt was turned on and a laptop was positioned at a comfortable height and position in front of the patients. The real-time reading from both sensors was provided to the patient in the form of a linear graph on the computer screen and the participant was instructed to perform each exercise while actively trying to keep the line as close to the zero point as possible. A more comprehensive description of the DAid smart shirt prototype is presented in [2.10].

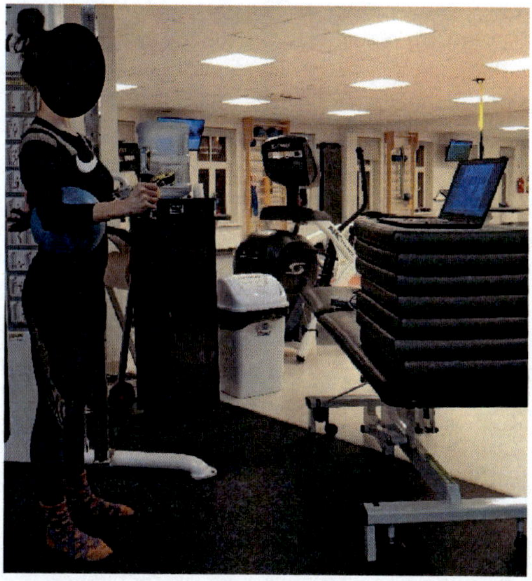

Fig.1. Patient with the DAid smart shirt in clinical settings doing unilateral exercise for rotator cuff simultaneously with shoulder girdle motion control.

2.4 Patient-Reported Outcome Measure

Patient-reported outcomes (PROs) such as DASH (the Disability of Arm, Shoulder, and Hand or DASH) are commonly used to evaluate the effectiveness of personalized shoulder rehabilitation interventions [4, 6]. Upper limb disability was assessed by the patient's self-reported DASH questionnaire, which consists of 30 questions in the disability/symptom module, 4 questions in the work module, and 4 questions in the sports module. Answers are given on a Likert scale of 1 (no symptoms) and 5 (maximum symptomatology), where points are obtained according to the questionnaire methodology. A higher score indicates a greater disability.". Originally, the score interpretation for the DASH outcome measure is where 0 is the best and 100 is the worst outcome. Cutoff scores: less than 15 is interpreted as "no problem," 16 – 40 means "problem, but working," and more than 40 interpretation is "unable to work". In musculoskeletal conditions, the standard error of measurement (SEM) is 5,22, minimal detectable change (MDC) is 10,7, and minimally clinically important difference (MCID) is 10,2 [17].

2.5 Data Statistical Analysis

To determine the effect of the DAid smart shirt-based feedback on the self-reported DASH (the Disability of Arm, Shoulder, and Hand or DASH) scores the following statistical processing and analysis methods were used: descriptive statistical methods to determine mean values, standard deviations, and 95% confidence interval. Determination of normal distribution by the Shapiro-Wilk test. To compare results between the study and control groups and before and after treatment The Mann-Whitney U test and the Wilcoxon Signed Ranks test were used. The Chi-square test and Fisher's exact test were used to determine the difference in the frequency of the qualitative variables of the samples (age, gender, dominant (D) hand, low risk for overuse injury sport, and high risk for overuse injury sport). Results with a p-value of < 0.05 were considered statistically significant. The statistical power of the study was determined to be > 0.8 (80%). Data were analyzed in Microsoft Excel 2016 (Microsoft Corporation, Washington, USA) and SPSS Statistics V22.0 (IBM Corporation, New York, USA).

3 Results

3.1 Characteristics of Patients Involved in the Study

The participant characteristics are provided in Table 1. While 34 participants for each group were selected randomly, there was a similar distribution of age and gender in both groups. The average age of participants in the test group was 38.6 years (with a standard deviation of 12.6), while in the control group, it was 40.8 years (with a standard deviation of 10.1). For the same number of participants in both groups (13 out of 17), the dominant side was affected. The absence of a statistically significant difference between the two groups regarding the baseline characteristics suggests that the groups were comparable for the study.

Table 1. Characteristics of patients involved in the study.

Parameter	Smart shirt group	Control group	*p* values
Age in years (±SD)	38,6 (±12,6)	40,8 (±10,1)	0,053
Gender (fem.) % (n)	58,8 (10)	52,9 (9)	0,730
Low risk for injury sports % (n)	41,2 (7)	47,1 (8)	0,730
High risk for injury sports % (n)	58,8 (10)	52,9 (9)	0,730
Affected dominant (D) hand % (n)	76,5 (13)	76,5 (13)	> 0,999

3.2 DASH Patient-Reported Outcome

All 34 patient DASH outcomes before and after were analyzed in the study, due to conditions "DASH patient–reported outcome may not be calculated if there are greater than 3 missing items in disability/symptom module and work and sports module's score may not be calculated if there are any missing items" were not met. In Table 2 the disability/symptom module, work module, and sports module study group and control group results before and after physiotherapy intervention are shown.

Table 2. DASH module before and after the intervention.

DASH module and study group	Before (±SD) (95% CI)	After (±SD) (95% CI)	Mean difference (±SD) (95% CI)	***p*-value**
Disability/symptom module				
Smart shirt group	54,4 (±2.5) (53,2–55,7)	14,7 (±3,1) (13,1 – 16,2)	39,8 (±4,1) (37,7 – 41,9)	< **0,001**
Control group	52,0 (±4,3) (49,8 – 54,2)	21,5 (±3,0) (20,0–23,1)	28,2 (±5,3) (25,5–31,0)	< **0,001**
***p*-value**	**0,210**	< **0,001**		
Work module				
Smart shirt group	85,2 (±2,7) (83,8–86,6)	17,1 (±3,3) (15,3–18,8)	68,2 (±3,8) (66,2–70,1)	< **0,001**
Control group	86,2 (±3,7) (84,3–88,1)	27,5 (±3,1) (25,9–29,1)	58,2 (±5,3) (55,4–60,9)	< **0,001**
***p*-value**	**0,424**	< **0,001**		
Sports module				
Smart shirt group	83,8 (±2,7) (82,4–85,2)	18,1 (±3,3) (16,4–19,8)	66,3 (±4,7) (63,9–68,8)	< **0,001**
Control group	84,4 (±3,2) (82,7–86,0)	25,7 (±2,9) (24,2–27,2)	57,5 (±3,1) (55,9–59,0)	< **0,001**
***p*-value**	**0,651**	< **0,001**		

In general, for the smart shirt group, the DASH score before the intervention in the disability/symptom module, work module, and sports module did not differ statistically significantly between the DAid shirt group and the control group ($p > 0.05$). Statistically significant differences were observed between the groups in all three DASH modules after the intervention ($p < 0.001$). Obtained statistical power > 0.8 (0.99). The sports module participant's DASH scores in the experimental smart shirt group and control group were divided into two separate subgroups characterized by a qualitative variable: low risk for an overuse injury sports to the shoulder joint and high risk for an overuse injury sports to the shoulder joint (Table 3).

The analysis of the results from the sports module, where participants answered four questions related to physical ability in the past week, revealed statistically significant differences between the smart shirt and control groups after the intervention ($p < 0.001$). But there was no statistically significant difference in the control and the smart shirt subgroups in terms of the risk for overuse injuries in sports after the intervention. The *p*-value for the control group's low risk of overuse injury sports and high-risk of overuse injury sports subgroup was 0,528. The *p*-value for the smart group's low risk of overuse injury sports and high-risk of overuse injury sports subgroup was 0,465.

Table 3. Sports module DASH score in low-risk of overuse injury sports and high-risk of overuse injury sports subgroups before and after intervention

Group	Before (±SD) (95% CI)	After (±SD) (95% CI)	*p*-value
Low risk for injury sports			
Smart shirt group	83,8 (±2,9) (80,9–86,7)	17,2 (±3,7) (13,4 – 20,9)	**0,012**
Control group	85,4 (±3,5) (81,9 – 88,9)	25,2 (±2,9) (22,2–28,2)	**0,011**
High risk for injury sports			
Smart shirt group	83,6 (± 3,9) (80,2–87,0)	14,4 (± 4,4) (10,0–18,7)	**0,012**
Control group	87,4 (± 3,7) (83,7–91,1)	23,4 (± 2,3) (21,1–25,7)	**0,012**

4 Discussion and Conclusions

The present study investigated the effectiveness of the DAid smart shirt-based feedback for the physiotherapy of subacromial pain syndrome (SAPS) in athletes participating in sports characterized by a high risk for overuse injury. The results showed that the use of the DAid smart shirt-based feedback system led to a statistically significant improvement in self-reported DASH scores compared to the conventional methods [12, 13]. The findings of the study are consistent with previous research on the benefits of personalized and targeted rehabilitation methods for athletes with SAPS [1]. The DAid smart shirt system provides real-time feedback on movement patterns during physiotherapy, which allows for more effective and personalized treatment. This is particularly important for athletes who require tailored rehabilitation to address the specific demands of their sport and avoid overuse injuries.

The statistically significant improvement in self-reported DASH scores after the 8-week treatment in the DAid smart shirt group compared to the control group indicates the potential of the system in improving the patient-reported outcomes of rehabilitation for SAPS patients participating in high-risk for overuse injury sports. The results also suggest that the use of the DAid smart shirt-based feedback system may lead to better treatment outcomes and contribute to the prevention of future injuries.

However, it is important to note that while the study demonstrated the effectiveness of the DAid smart shirt-based feedback system, the sample size was relatively small, and the study was limited to athletes with SAPS. Further research with a larger sample size and different patient populations is needed to validate the findings of the present study. And it is important to note, that the DAid smart shirt is still in the prototype phase, so it should be improved as technology by itself towards maturity from a bioengineering perspective and user experience (UX) perspective also.

In conclusion, the results of this study suggest that the use of the DAid smart shirt-based feedback system for the physiotherapy of subacromial pain syndrome in athletes participating in high-risk for overuse injury sports provides a better self-reported improvement in DASH scores, compared to conventional methods. The DAid smart shirt system prototype offers a promising solution to address the challenges faced by athletes with SAPS, and further research is warranted to establish its effectiveness and potential applications in other patient populations.

References

1. Ellenbecker, T.S., Wilk, K.E.: Sport therapy for the shoulder: evaluation, rehabilitation, and return to sport. United States of America: Human Kinetics (2016)
2. Semjonova, G., Vetra, J., Cauce, V., Oks, A., Katashev, A., Eizentals, P.: Improving the recovery of patients with subacromial pain syndrome with the DAid smart textile shirt. Sensors **20**(18), 5277 (2020)
3. Littlewood, C., May, S., Walters, S.: A review of systematic reviews of the effectiveness of conservative interventions for rotator cuff tendinopathy. Should. Elb. **5**(3), 151–167 (2013)
4. Schwank, A., et al.: Athlete Shoulder Consensus Group. Bern consensus statement on shoulder injury prevention, rehabilitation, and return to sport for athletes at all participation levels. J. Orthop. Sports Phys. Ther. **52**(1), 11–28 (2022)
5. Lewis, J.S., Green, A.S., Dekel, S.: The aetiology of subacromial impingement syndrome. Physiotherapy **87**(9), 458–469 (2001)
6. Cools, A., Borms, D., Castelein, B. et al.: Shoulder Rehabilitation. A Practical Guide for the Clinician. Belgium: SKRIBIS (2020)
7. Timmermans, A.A., et al.: Sensor-based arm skill training in chronic stroke patients: results on treatment outcome, patient motivation, and system usability. IEEE Trans. Neural Syst. Rehabil. Eng. **18**(3), 284–292 (2010)
8. Delbressine, F.,Timmermans, A., et al.: Motivating arm-hand use for stroke patients by serious games, In: 34th Annual International Conference of the IEEE Engineering in Medicine and Biology Society (2012)
9. Oks, A., Katashev, A., Litvak, J.: Knitted resistive fabric: properties and applications. Mater. Sci. Text. Clothing Technol. **9**(21), 28 (2014)
10. Eizentals, P., Katashev, A., Oks, A., Semjonova, G.: Smart Shirt for Uncontrolled Movement Retraining.In: IFMBE Proceedings, p. 76 (2020)
11. Reinold, M.M., Escamilla, R., Wilk, K.E.: Current concepts in the scientific and clinical rationale behind exercises for glenohumeral and scapulothoracic musculature. J. Orthop. Sports Phys. Ther. **39**(2), 105–117 (2009)
12. Diercks, R., et al.: Guideline for diagnosis and treatment of subacromial pain syndrome. Acta Orthopaedica **85**(3), 314–322. (2014)
13. Pieters, L., et al.: An update of systematic reviews examining the effectiveness of conservative physical therapy interventions for subacromial shoulder pain. J. Orthop. Sports Phys. Ther. **50**(3), 131–141 (2020)
14. Ellenbecker, T.S., Cools, A.: Rehabilitation of shoulder impingement syndrome and rotator cuff injuries: an evidence-based review. Br. J. Sports Med. **44**(5), 319–327 (2010)
15. Comford, M., Mottram, S.: Kinetic Control : The Management of Uncontroled Movement. Australia: Churchill Livingstone, ELSEVIER (2012)
16. Allen, K., Anderson, M., Balady, G.: ACSM's Guidelines for Excercise Testing and Prescription. Ninth edition. United States of America: Wolters Kluwer Health. Lippincott Williams &Wilkins (2014)
17. Franchignoni, F., et al.: Minimal clinically important difference of the disabilities of the arm, shoulder and hand outcome measure (DASH) and its shortened version (quickDASH). J. Orthop. Sports Phys. Ther. **44**(1), 30–39 (2014)

Effects of Tactical Boots on Foot and Ankle Kinematics

Darja Nesterovica - Petrikova[1(✉)], Normunds Vaivads[2], and Ainars Stepens[1]

[1] Military Medicine Research and Study Center, Rīga Stradiņš University, 14 Baložu Street, Riga 1048, Latvia
darja.nesterovica-petrikova@rsu.lv

[2] Joint Headquarters of the Latvian National Armed Forces Medical Service, 9 Krustabaznicas Street, Riga 1006, Latvia

Abstract. Occupational footwear is designed for personnel protection from injuries. Footwear worn in military populations has been associated with potential injury. During the gait foot and ankle segment is important part of body balance. Lower limb injury incidence, especially ankle injuries, remains common among military populations. Study aim was to compare ankle joint angular velocities when walking barefoot and while wearing tactical boots. $N = 64$ active-duty infantry male soldiers at mean age 30.0 ± 5.5 years participated in this study. Foot and ankle joint angle calculations as well as spatiotemporal gait parameters were evaluated and 2D kinematic assessment was undertaken for barefoot and shod conditions. During the barefoot walk, foot contact angle was reduced comparing with shod conditions. Maximum angular velocities of plantarflexion and dorsiflexion during barefoot walk were elevated comparing with gait using tactical boots. All observed differences were statistically significant ($p < 0.001$). This study demonstrates that tactical boots change foot and ankle kinematics during the gait. Shod walking indicated reduced motion of foot and ankle. Maximum angular velocity values during ankle dorsiflexion and plantarflexion in this study are comparable with reported angular velocities observed in healthy populations while using common and athletic footwear.

Keywords: Angular velocity · Gait analysis · Military boots

1 Introduction

Footwear forms the contact between the body and the standing or walking environment and reduce loading of the lower extremities [1, 2]. Occupational footwear is one of the key components of personal protection in potentially dangerous environments for such professions like fire fighters, construction workers, law enforcement officers and military personnel [3]. Persons working in these occupations use specialized footwear that address different hazards of occupational setting (e.g., heat, irregular terrain, slippery surfaces) and potentially protect the body from injury at the workplace [3]. Tactical boots used by the military personnel are task and environment specific [4]. Occupational boots, as well as tactical boots, appear to be designed for occupational safety on account of comfort [5].

Y. Dekhtyar and I. Saknite (Eds.): NBC 2023, IFMBE Proceedings 89, pp. 112–118, 2023.
https://doi.org/10.1007/978-3-031-37132-5_15

Occupational footwear is being linked to potentially causing musculoskeletal injuries directly, as well as leading to tripping and slipping associated with injuries [3, 6]. For example, ankle and foot region injury that occurred due to falls and slips caused most limited duty days in U.S. Army [7].

Previous studies have reported the footwear ability to maintain postural stability, impact gait and physiological measures (aerobic capacity, heart rates, temperatures, muscle activity), and selected occupational task performance [1, 3]. Footwear also have the potential to serve as a tool that promotes ankle energy storage and release during the gait [8].

During the barefoot gait, two series of dorsiflexion and plantar flexion in the ankle joint occur [9]. Movement at the foot and ankle has been appraised as an important part of balance recovery [3], and fast ankle muscle reaction is essential for fast ankle joint stabilization [10]. Besides, it has been observed previously that changes in peak ankle angular velocity positively correlates with changes in ankle power [11]. Peak ankle joint angles and angular velocities correspond to foot plantarflexion prior to push-off of the gait cycle and ankle dorsiflexion for ground clearance [12]. Occupational footwear significantly restricts the ankle's range of motion and it is proposed that compensation at the knee joint could occur due to a decrease in muscle activity at the ankle [5].

Considering the impact of occupational footwear on lower limb kinematics there is a need to ensure that occupation-specific footwear is designed to stabilize the ankle region and mitigate injuries not to cause them [3, 5]. This study aim was to compare ankle joint angular velocities during the gait with tactical boots and barefoot gait.

2 Materials and Methods

2.1 Study Population

Available active-duty infantry soldiers of Latvian Land Forces were invited to participate in this study during the annual medical check-up at the Latvian National Army Logistic Command Military Medical Support Centre. The study protocol was approved by the Ethics Committee of Rīga Stradiņš University (No.40/26.10.2017). Participation was voluntary and written informed consent was provided by all participants before entering the study. A cross-sectional study was carried out among male soldiers ($n = 64$) at the mean age of 30.0 ± 5.5 years (range 22–40 years) and with an average service time of 5.2 years (range 1–15 years).

2.2 Gait Assessment

Study participants were asked to wear shorts and were instructed to walk comfortably on the 5-m long walkway barefoot and in tactical boots. Standardized tactical boot model with 25 cm height for hot weather conditions was used during the study (see Fig. 1).

Tactical boots could not be used if any visual attrition signs were found by the researcher before gait test. Two gait trials were used for familiarization for both walking conditions and were not included into the analysis [13]. Full ($n = 50$) gait cycles without turns were recorded and included into the analysis [14].

Fig. 1. Tactical boot example.

Before the gait testing, all study participants were instrumented with spherical retroreflective markers using double-sided tape. Markers ($n = 12$) were attached to the middle shank, the lateral and medial femoral epicondyles, the lateral and medial malleoli, and the head of the 1st, 2nd, and 5th metatarsals, and the posterior calcaneus. A single examiner attached all the markers. Marker set is similar as used in the previous studies for barefoot and shod gait [15, 16]. For gait assessment while using tactical boots the marker placement was defined after the palpation of the anatomical landmark through the shoe.

Gait testing and 2D kinematic assessment of the foot and ankle motion was performed in the gait laboratory of Riga Stradiņš University [17, 18]. Two high-speed camera motion capture (100 samples/s) system was used for gait recording. Joint angles and angular velocities were evaluated using software *Quintic v31 Biomechanics* (Quintic Consultancy Ltd, United Kingdom).

2.3 Statistical Analysis

Statistical analysis was carried out using the SPSS 22.0 software package (Statistical Package for the Social Sciences). Data distribution was determined using the Kolmogorov-Smirnov test. The Kruskal–Wallis test by ranks was used to compare shod and barefoot data. Mean gait data from the right side are presented with standard deviation (SD) if not stated otherwise, the left side data was not included in the analysis. The statistical significance level was $p < 0.05$.

3 Results

Gait with tactical boots showed longer stride in terms of covered length and time. Stride length and stride time were statistically significantly different ($p < 0.001$) during barefoot gait and while wearing tactical boots. However, the differences in gait velocity were not significant ($p = 0.45$). See Table 1 for details.

Table 1. Gait characteristics for shod and barefoot conditions

Gait parameter	Barefoot	With boots
Stride length, m (SD)	1.11 (0.33)	1.32 (0.27)
Stride time, m (SD)	1.08 (0.11)	1.21 (0.10)
Gait velocity, m/s (SD)	1.03 (0.29)	1.09 (0.22)

Foot and ankle motion appeared to be different while walking barefoot and with tactical boots. Gait with tactical boots statistically significantly ($p < 0.001$) increased ankle angle at initial contact, but reduced rearfoot eversion and ankle angular velocities. See Table 2 for details.

Table 2. Foot and ankle kinematics

Variable	Barefoot	With boots
Rearfoot eversion° (SD)	5.31 (1.83)	3.08 (1.11)
Ankle angle at initial contact° (SD)	16.72 (5.50)	25.34 (4.66)
Peak angular velocity for plantarflexion°/sec (SD)	249.29 (34.09)	156.78 (24.06)
Peak angular velocity for dorsiflexion°/sec (SD)	153.44 (22.71)	119.61 (33.38)

4 Discussion

To the author's knowledge, this is the first study that describes angular velocities of the ankle joint during a walk with tactical boots. According to the study results, tactical boots significantly restrict and decelerate ankle joint motion. Other studies also showed that walking barefoot decreases ankle range of motion and increases ankle joint motion while shod [19, 20].

Peak angular velocities observed during ankle plantarflexion and dorsiflexion among our study participants appeared similar with reported data among healthy populations [12]. Additionally, walk with tactical boots increased stride time and showed prolonged stride. Similar effects on stride time and length have been previously found during walk with common footwear among adults [21, 22]. However, differences in gait velocities during barefoot and walking with tactical boots were not significant due to laboratory setting and limited walking space. It has been reported previously that ankle angular velocity increase if the barefoot gait velocity increase [12] and future research could focus on different gait speeds.

Walking with tactical boots provides stability of the ankle joint, but shod gait evaluation could mask ankle joint motion restriction. Barefoot gait assessment should be performed in occupational groups with high lower limb injury risks for comprehensive ankle joint function evaluation.

Tactical boots are used not only for walking, but also for running. Running with boots used by the military personnel has been physically more demanding and it has been associated with a greater compensation of hip and knee joints motion due to the ankle joint limitations, which could potentially increase the risk of musculoskeletal injuries [23, 24]. Tactical boot effects on ankle joint motion during the gait in this study appeared to be similar to gait with running shoes, that also showed increased ankle dorsiflexion at initial contact [8]. However, further research of running gait assessment using tactical boots could provide additional data regarding possible lower limb injury mechanisms among military populations.

This study has some limitations. This was a cross-sectional study with male participants only and obtained ankle kinematics data could not be generalized to all military populations. Moreover, gait data measured in laboratory setting could not be easily transferred to outdoor conditions [21].

We did not analyze shoe attrition, however it could influence the kinematics of ankle joint [16]. Study participants used already worn tactical boots that could shift the results, however tactical boots in Latvian Land Forces are changed regularly if visual attrition sign exist; before marker placement no visual tactical boots damage was found. Additionally, ankle motion analysis with markers could be influenced by soft tissue artefacts (STA), but all the markers were placed following standardized marker placement scheme by one examiner and STA at the hindfoot region have been likely small [25, 26]. Despite these limitations, this study shows the positive effects of such occupational footwear as tactical boots, and these findings could be similar to boots used by law enforcement officers.

5 Conclusions

This study showed that walking with tactical boots used in the military populations restricts ankle joint and decelerates motion comparing with barefoot walking. Observed tactical boot effects on gait are similar as walking in common footwear and could not explain high incidence of ankle joint injuries among military personnel.

References

1. Chander, H., Garner, J.C., Wade, C., Knight, A.C.: Postural control in workplace safety: role of occupational footwear and workload. Safety. **3**, 1–8 (2017). https://doi.org/10.3390/safety3030018
2. Zhang, X., Paquette, M.R., Zhang, S.: A comparison of gait biomechanics of flip-flops, sandals, barefoot and shoes. J. Foot Ankle Res. **6**, 45 (2013). https://doi.org/10.1186/1757-1146-6-45
3. Orr, R., Maupin, D., Palmer, R., Canetti, E.F.D., Simas, V., Schram, B.: The impact of footwear on occupational task performance and musculoskeletal injury risk: a scoping review to inform tactical footwear. Int. J. Environ. Res. Public Health. **19**, 10703 (2022). https://doi.org/10.3390/ijerph191710703
4. Hamill, J.B.C.: Biomechanical analysis of military boots: Phase 1. Materials testing of military and commercial footwear. Technical Report. Natick, MA U.S. Army Natick Res. 1992. Rep. No. NATICK-TR-93/006 (1996)

5. Dobson, J.A., Riddiford-Harland, D.L., Bell, A.F., Steele, J.R.: Work boot design affects the way workers walk: a systematic review of the literature. Appl. Ergon. **61**, 53–68 (2017). https://doi.org/10.1016/j.apergo.2017.01.003
6. Andersen, K.A., Grimshaw, P.N., Kelso, R.M., Bentley, D.J.: Musculoskeletal lower limb injury risk in army populations. Sports Med. - Open **2**(1), 1–9 (2016). https://doi.org/10.1186/s40798-016-0046-z
7. Roy, T.C., Faller, T.N., Richardson, M.D., Taylor, K.M.: Characterization of limited duty neuromusculoskeletal injuries and return to duty times in the U.S. Army during 2017–2018. Mil. Med. **187**, e368–e376 (2022). https://doi.org/10.1093/milmed/usaa392
8. Ogaya, S., Okubo, S., Utsumi, T., Konno, F., Kita, S.: Effects of flat-flexible shoes on lower limb joint kinetics and kinematics in gait. J. Biomech. **141**, 111216 (2022). https://doi.org/10.1016/j.jbiomech.2022.111216
9. Haskell, A.: 110 - Foot and Ankle Biomechanics. Elsevier Inc. (2023)
10. Silva, M., et al.: Precueing time but not direction of postural perturbation induces early muscular activation: comparison between young and elderly individuals. Neurosci. Lett. **588**, (2015). https://doi.org/10.1016/j.neulet.2015.01.004
11. Browne, M.G., Franz, J.R.: Ankle power biofeedback attenuates the distal-to-proximal redistribution in older adults. Gait Posture. **71**, 44–49 (2019). https://doi.org/10.1016/j.gaitpost.2019.04.011
12. Mentiplay, B.F., Banky, M., Clark, R.A., Kahn, M.B., Williams, G.: Lower limb angular velocity during walking at various speeds. Gait Posture. **65**, 190–196 (2018). https://doi.org/10.1016/j.gaitpost.2018.06.162
13. Hamacher, D., Hamacher, D., Krowicki, M., Schega, L.: Between-day test–retest reliability of gait variability in older individuals improves with a familiarization trial. Aging Clin. Exp. Res. **29**(2), 327–329 (2016). https://doi.org/10.1007/s40520-016-0536-3
14. Kroneberg, D., et al.: Less is more - Estimation of the number of strides required to assess gait variability in spatially confined settings. Front. Aging Neurosci. **11**, (2019). https://doi.org/10.3389/FNAGI.2018.00435/FULL
15. Peng, Y., et al : Immediate effects of medially posted insoles on lower limb joint contact forces in adult acquired flatfoot: a pilot study. Int. J. Environ. Res. Public Health. **17**, (2020). https://doi.org/10.3390/IJERPH17072226
16. Chen, S.F., Wang, Y., Peng, Y., Zhang, M.: Effects of attrition shoes on kinematics and kinetics of lower limb joints during walking. Front. Bioeng. Biotechnol. **10**, 824297 (2022). https://doi.org/10.3389/fbioe.2022.824297
17. Maykut, J.N., Taylor-Haas, J.A., Paterno, M.V., DiCesare, C.A., Ford, K.R.: Concurrent validity and reliability of 2d kinematic analysis of frontal plane motion during running. Int. J. Sports Phys. Ther. **10**, 136–146 (2015)
18. Zult, T., Allsop, J., Tabernero, J., Pardhan, S.: A low-cost 2-D video system can accurately and reliably assess adaptive gait kinematics in healthy and low vision subjects. Sci. Rep. **9**, 18385 (2019). https://doi.org/10.1038/s41598-019-54913-5
19. Dames, K.D., Smith, J.D.: Effects of load carriage and footwear on spatiotemporal parameters, kinematics, and metabolic cost of walking. Gait Posture. **42**, 122–126 (2015). https://doi.org/10.1016/j.gaitpost.2015.04.017
20. Abdelraouf, O.R., Abdel-aziem, A.A.: Ankle and foot mechanics in individuals with chronic ankle instability during shod walking and barefoot walking: a cross-sectional study. Chinese J. Traumatol. **24**, 174–179 (2021). https://doi.org/10.1016/j.cjtee.2021.02.010
21. Hollander, K., Petersen, E., Zech, A., Hamacher, D.: Effects of barefoot vs. shod walking during indoor and outdoor conditions in younger and older adults. Gait Posture. **95**, 284–291 (2022). https://doi.org/10.1016/j.gaitpost.2021.04.024

22. Franklin, S., Grey, M.J., Heneghan, N., Bowen, L., Li, F.X.: Barefoot vs common footwear: a systematic review of the kinematic, kinetic and muscle activity differences during walking (2015)
23. Shamsoddini, A., Hollisaz, M.T.: Biomechanics of running: a special reference to the comparisons of wearing boots and running shoes. PLoS ONE **17**, e0270496 (2022). https://doi.org/10.1371/journal.pone.0270496
24. Sinclair, J., Taylor, P.J., Atkins, S.: Effects of new military footwear on knee loading during running. Footwear Sci. **7**, 165–171 (2015). https://doi.org/10.1080/19424280.2015.1066879
25. Alcantara, R.S., Trudeau, M.B., Rohr, E.S.: Calcaneus range of motion underestimated by markers on running shoe heel. Gait Posture. **63**, 68–72 (2018). https://doi.org/10.1016/J.GAITPOST.2018.04.035
26. Benoit, D.L., Ramsey, D.K., Lamontagne, M., Xu, L., Wretenberg, P., Renström, P.: Effect of skin movement artifact on knee kinematics during gait and cutting motions measured in vivo. Gait Posture. **24**, 152–164 (2006). https://doi.org/10.1016/j.gaitpost.2005.04.012

The Reverse Engineering of Human Organs Based on the Application of Method of Anatomical Features

Nikola Vitković[1(✉)], Miroslav Trajanović[1], Miloš Stojković[1], Razvan Pacurar[2], Sergiu-Dan Stan[2], and Filip Górski[3]

[1] Faculty of Mechanical Engineering, University of Nis, A. Medvedeva 14, Niš, Serbia
nikola.vitkovic@masfak.ni.ac.rs
[2] Technical University of Cluj-Napoca, Strada Memorandumului 28, Cluj-Napoca, Romania
[3] Poznań University of Technology, Plac Marii Skłodowskiej-Curie 5, Poznań, Poland

Abstract. The human body is often affected by some pathological processes or illnesses. In many cases, it is necessary to perform surgical intervention, sometimes in a very short period. Computer-Assisted Surgery - CAS defines a set of techniques that use computers and other devices for planning, guiding, and performing surgical interventions, and can greatly improve outcome of pre/intra/post-operative procedures. In this paper, the enhancement of the Method of Anatomical Features – MAF, which is defined as an essential part of CAS, is described. The MAF is improved by adding Characteristic Product Feature - CPF method as one of its essential processes, to better describe human organs and to create Feature models as their virtual representation. The created Feature models will greatly help physicians to better prepare and conduct surgical interventions, thus improving patient treatment and recovery processes.

Keywords: MAF · CPF · human organ · geometrical model

1 Introduction

The human body is often affected by some pathological processes or illness. In many cases, it is necessary to perform surgical intervention, sometimes in a very short period. For the success of surgical intervention, it is particularly important to prepare a good pre-operative plan and to ensure that adequate implants are provided. Computer-Assisted Surgery (CAS) [1] defines a set of techniques that use computers and other devices for planning, guiding, and performing surgical interventions. The most important component of CAS is an accurate personalized model of the affected human organ (e.g., tendon, ligament, or bone). In general, such model(s) can be acquired by the application of two different approaches [2].

The first approach implies application of medical imaging technologies, like Computed Tomography (CT) or X-Ray, to provide 3D geometrical models of human organs. Such models can be created in three general ways:

- Application of software, which is part of a medical scanner (e.g. Vitrea)

Y. Dekhtyar and I. Saknite (Eds.): NBC 2023, IFMBE Proceedings 89, pp. 119–128, 2023.
https://doi.org/10.1007/978-3-031-37132-5_16

- Post-processing of medical images in medical-oriented CAD programs (e.g. Materialise Mimics, 3D Slicer)
- Post-processing in one of the CAD software packages (e.g., CATIA, 3D Fusion).

One of the main drawbacks of this approach is the inability to create a geometrical accurate model of whole organ in cases where the scanned bone incomplete due to illness (osteoporosis, arthritis, cancer, etc.) or trauma (multiple fractures, crushed bones, torn ligaments and tendons, etc.), or when medical images are not of adequate quality. Examples of methods used in this approach are given in the works [3, 4].

The second approach for creation of 3D geometrical models of human organs is based on predictive geometrical or statistical model and data obtained from medical images. In predictive models, geometric entities are described by mathematical functions, whose arguments are morphometric parameters that can be read from medical images. With this approach it is possible to create an accurate 3D geometrical model of a patient's organ [4, 5].

Method of anatomical features (MAF) is a methodology which can be used in both approaches [4]. In this paper, the research focuses on improving MAF by adding the Characteristics Product Features (CPF) [6] method. CPF enables the definition of different product features and properties, and thus it should provide MAF capability to define 3D models of other organs, not just bones. The initial approach to MAF enhancement is described in this paper for the definition of the Anterior Crucial Ligament geometrical model, which is based not just on anatomical and morphological properties but also on its functional characteristics.

2 Characteristic Product Features Methodology

The methodology consists of several steps defined as processes [6], and they are: Point Cloud Import (P1), Feature Description (P2), Feature Definition (P3), Feature Extraction and Implementation (P4). The process starts with importing point cloud and finishes with complete definition of selected product features, and their implementation. Point cloud is initially a set of points acquired from the scanning activity, where each point represents a minimal geometric definition of the scanned product. In the next process (P2) each product feature of interest is described semantically, and a list of features is created. This list can be represented using a plain text file or some other type of representation like a visual (sketch). Finally, it is necessary to conduct process P3 to provide a complete definition of feature(s) properties. This process is the most important activity in methodology application. It consists of several feature definitions and is not limited to those presented in [6] (geometrical, mathematical, functional, etc.). To define product feature(s) geometrically, it is required to use a filtered point cloud. Then, each product feature is defined by selecting a group of points that belongs to that product feature, i.e., product feature(s) is geometrically represented by a set of points in the point cloud. The group of points can be further processed in remodeling steps, like creating mesh(es), forming geometrical elements like curves, and forming 3D models (surface or solid) at the end. Each following individual step can produce output and further define product feature(s). For example, the mesh element can be exported to an STL file and then joined to the point cloud file of the product, thus creating a set of files for an individual feature – point cloud (*.txt) and mesh file (*.stl). The whole set of product features of interest can contain many files for each component; if required, they can be used individually or assembled into the whole product and used as a part model, i.e., Feature(s) model. The one important capability of a product feature is the parametrization of its point set, thus creating a parametric point cloud model for a unique feature with the capability to adapt to different parameters, not just geometrical, but also to functional, technological, and so on. The CPF application is demonstrated in [6] for the creation of the ski shoe heel lip 3D printed model which was successfully implemented and used. In [6] the functional, geometrical, topological, and material features were defined, and they influenced the creation of heel lip geometrical and physical model by 3D printing.

3 Method of Anatomical Features

MAF [4] introduces a new approach to describe geometrical entities of human bones, and it enables creation of various geometrical models of the human bones and other organs. Two different types of models can be created by the original MAF:

- 3D geometrical models – These models are standard polygonal, surface and volume models which are used in CAD for many years. They are created by the application of standard CAD technical features in CAD software packages.
- Predictive (parametric) models of the human bones. By the application of the morphometric parameters acquired from medical imaging methods, these models can be adjusted to the geometry and morphology of the human bone of a specific patient.

Both types of models are created on the basis of data (input models) acquired from medical imaging methods (e.g. CT or MRI). MAF is a complex method, composed of basic and additional processes (Fig. 1), and therefore, Structured Analysis and Design Technique (SADT) [7] is used for its description. The main components of SADT diagrams are input elements, recourses, control elements, and output elements, defined in [4] and presented in Fig. 1.

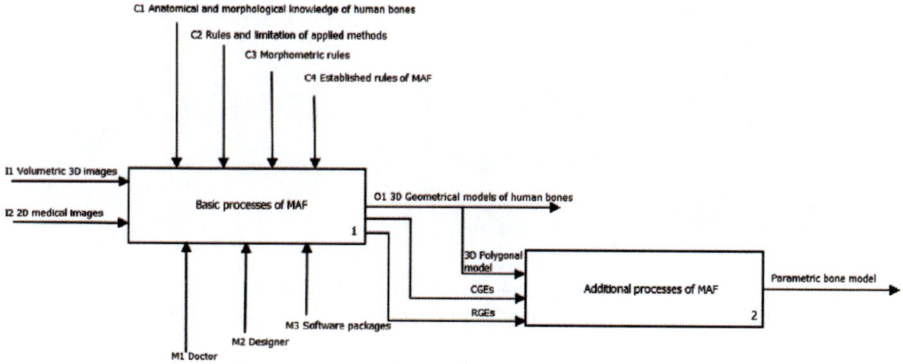

Fig. 1. Basic and additional processes of MAF [4]

Basic processes are defined in [4] and they enable complete geometrical and anatomical definition of the specific human bone. These processes are: Creation of initial polygonal model (A11). The output from this process is a polygonal model of the specific human bone; Anatomical analysis (A12). The outcome of this process is an anatomical model of the specific human bone; Definition of Referential Geometrical Entities (RGEs) (A13). These entities represent basic geometry which is used for the creation of all other geometrical elements like curves or surfaces; Creation of Constitutive Geometrical Entities (CGEs) (A14) – These entities are called constitutive because they are used for the creation of surface and solid models of the human bones, and parts of the bones. Polygonal model, RGEs and CGES are outputs from the basic MAF processes. By the application of these outputs different geometrical models of the human bone can be created.

The output from MAF additional processes is parametric model of the specific human bone [4]. Parametric model is a predictive model which shape, and anatomy are defined by the formed parametric functions, and conditioned by the values of the morphometric parameters acquired from medical imaging methods. Parametric model can be transformed into the personalized model by applying unique values of morphometric parameters acquired from specific patient's medical images.

The MAF is a proven method, and its results are well known [8–12]. The original MAF was developed to create geometrical models of human bones [2, 4, 8–11]. The results presented in these studies demonstrated the geometrical accuracy and anatomical and morphological correctness of MAF application. It is essential to mention that the quality of the models created by MAF can be defined and influenced by the requirement of a clinical case, like the one presented in [9] where a sternum implant was created using initial MAF procedures and additive technologies. The implant was successfully implemented into the patient, and the patient recovered. Additional examples of MAF application are defined in chapters [10–12] in the Springer book "Personalized orthopedics". Besides bones, MAF is also used to create personalized implants for human bones (long and flat) as presented in [8, 11, 12]. The one characteristic application was creating a plastic mandibular reconstructive plate model, for the young male patient with progenia, presented in Fig. 2.

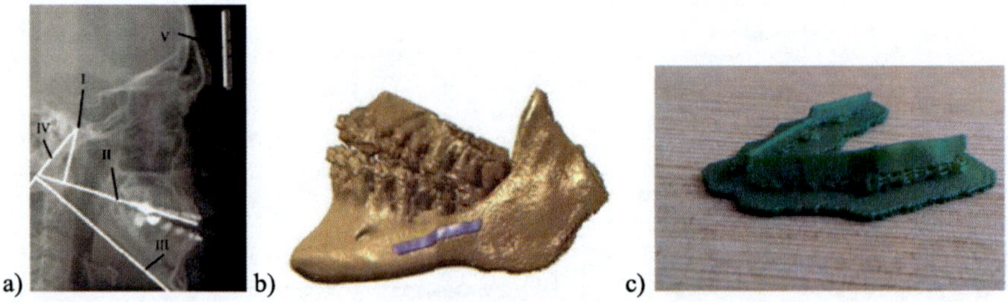

Fig. 2 a) X-ray scan with basic geometry b) Solid model of the personalized plate created over scanned data by using MAF c) printed models of the plate implants (left and right) [8, 11].

The plastic model (Fig. 2c), manufactured by FFF 3D printing, was used to guide real plate bending in the preoperative phase. The bent model was geometrically accurate and fitted the patient perfectly. Also, it is important to note that the presented approach shortened the surgery time, which is always a good thing for a patient. To conclude, MAF can be used to create geometrical models of human bones and implants, and concerning previous applications, results are more than satisfactory. For now, the authors have not found a case in which MAF cannot be applied, because it allows different capabilities to bone remodeling as presented in [4]. Concerning ethical principles, MAF is oriented to digital (geometrical models), and for every real (clinical) application presented in the references and other MAF applications, ethical regulations were and will be followed.

The focus of this research is to enhance MAF, and for that, it is necessary to create specific modifications. The first modification, presented in this study, is an application of the CPF method as one of the basic processes in the methodology definition. The upgraded MAF basic processes with CPF are presented in Fig. 3. CPF method enables various definitions of any product feature with a specific output. As already stated, any product or organ feature can be extracted from the point cloud or polygonal model, and existing or required properties can be defined. In MAF the original procedure is to acquire cloud point from medical imaging methods, then conduct geometrical filtering and transformations and produce some geometry (RGEs, CGEs, and other geometrical entities and models).

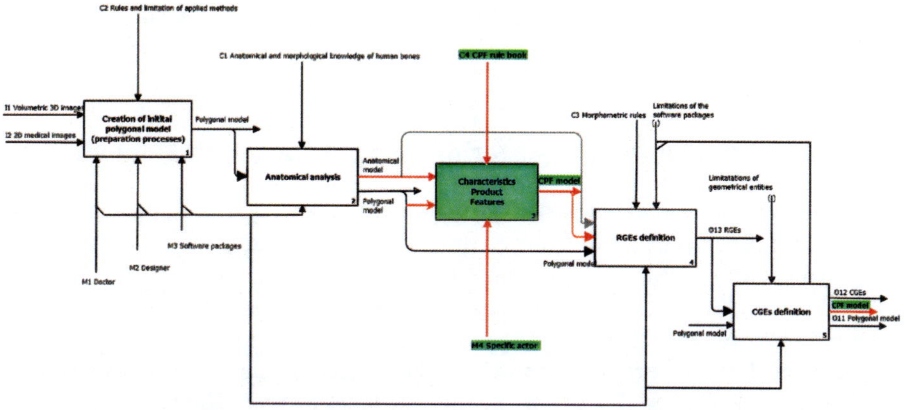

Fig. 3. The addition of CPF to MAF basic processes

CPF (Fig. 4), adds more capability to MAF to define specific features on the human organ and to add more characteristics, not just anatomical ones. Therefore, the resulting model can have additional properties, like required functional or mechanical characteristics. It is important to note that the term model in this context refers to a complex entity composed of different elements (CGE, polygonal model(s), CPF properties), and it will be referred to as Features Model (FM) (specific CPF model). To make a conclusive statement: The resulting FM of the MAF application can reflect various needs and fulfil different requirements from different actors. FM can be used as input model to additional

MAF processes to enable better parameterization according to the novel specifications defined in FM.

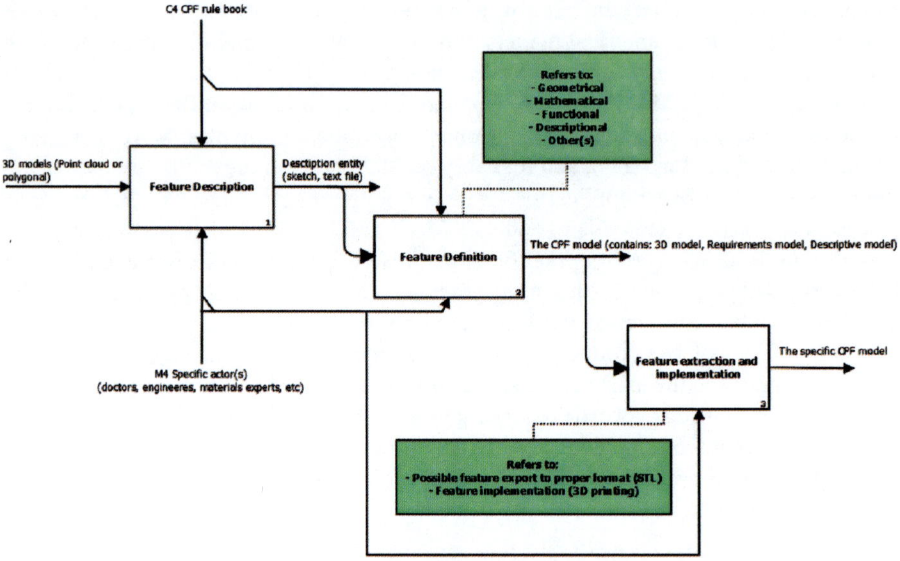

Fig. 4. The CPF basic processes

4 Method Application for Anterior Crucial Ligament Remodeling

In this research, the Anterior Crucial Ligament (ACL) is used as an example for the remodeling of the knee ligament based on different requirements defined in CPF model. Three bones meet to form the knee joint: the femur (thighbone), tibia (shinbone), and patella (kneecap). The kneecap sits in front of the joint to provide some protection. There are four primary ligaments in the knee, and they act like strong ropes to hold the bones together and keep the knee stable. The ligaments connecting the knee bones are: Collateral Ligaments - These are found on the sides of knee; The medial collateral ligament (MCL) is on the inside, and the lateral collateral ligament (LCL) is on the outside. The detailed explanation of ligaments can be found in [13]. One of the most common knee injuries is an ACL sprain, or tear. The main technique for ACL repair is to use tendon graft(s) and to fixate it with screws positioned in the femur and tibia. It is approach used for many years with different results. The other novel approach presented in [14, 15] demonstrates categorically that there is a place for ACL repair with InternalBrace (Fig. 5). Early repair of new joints has an impact on protecting joint health and restoring normal biomechanical function. An ACL repair augmented with the InternalBrace procedure protects the cartilage, reducing the need for resection. Standard repairs with ACL Reconstruction Hamstring Tendon are most prone to resection and arthritis. This approach in ACL reconstruction is due to the concerns regarding the high level of arthritis associated with traditional reconstruction, (reported to be as high as 48% after 10 years).

With InternalBrace, over 80% of the repair patients did not need any form of reconstructive surgery on their knees during the 5 years of follow-up. Those that did still made a tremendous recovery after a second-stage revision procedure using a reduced graft – again supported with the InternalBrace. The difference between the standard and novel approach is that ACL repair with InternalBrace is a lot less traumatic for the joint, as presented in Fig. 6. Repair augmented with the InternalBrace only requires small bone tunnels which take the 2mm InternalBrace. Compared this to the major bone tunnels that need to be drilled during a traditional reconstruction (2.4 mm for femur and 3.5 mm for tibia). There is far less bone trauma in an ACL repair augmented with the InternalBrace, compared to a traditional reconstruction.

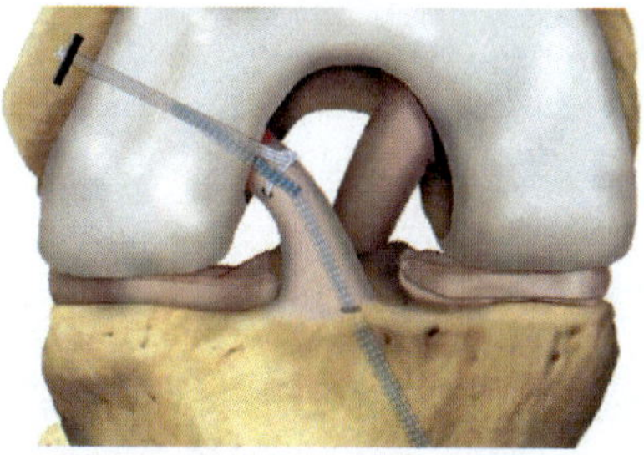

Fig. 5. The ACL InternalBrace reconstruction [15]

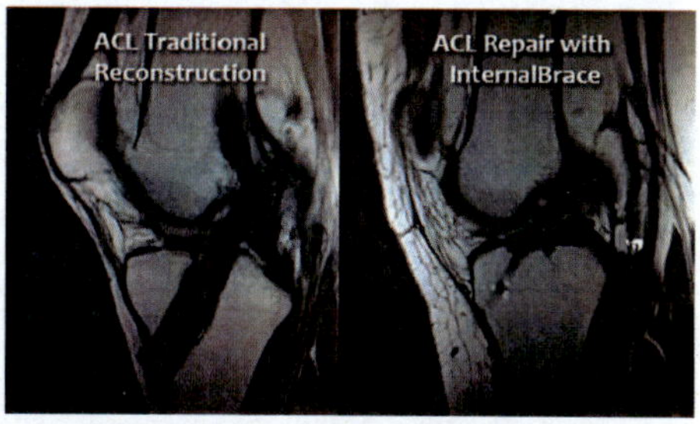

Fig. 6. Comparison of traditional and InternalBrace reconstruction [15]

To create a 3D personalized model of the ACL it is required to define important functional characteristics of the selected treatment case. Because of the novelty of the research and good prognosis, a second case is selected – InternalBrace. The complication of ACL reconstruction may be summarized as the ones due to pre-operative decisions, intraoperative causes, and postoperative causes. The focus of this study will be on pre-operative and post-operative complications which can be reduced with good preoperative analysis of the injured knee, and some of them can be:

- Preoperative assumptions and decisions
- ACL reconstruction is deferred until the swelling has subsided after an acute injury.
- Not performing the ACL surgery when needed and waiting for too long may lead to damage to the articular knee cartilage and meniscus.
- Postoperative complications
- Stiffness is a common post-ACL reconstruction complication that is defined as the incomplete range of motion of the knee following surgery.
- Patella fracture after ACL reconstruction may result from graft harvesting (only in resection for this case).
- Pain, swelling, redness, the rise of temperature, and stiffness may occur because of infection.

- Pain in the front of the knee after the surgery is commonly associated with patellar tendon graft but may also occur in cases of allograft and quadriceps tendon graft.

The stated complications are entry point for the CPG/FM model creation, and they represent the functional and geometrical requirements, which 3D ACL model needs to fulfill. The fulfillment of these requirements will allow better definition, creation, and application of the ACL implant. The personalized 3D model created in the pre-operative procedures with accurate geometry, and with complete positional definition of femur and tibia holes and InternalBrace dimension, will help physician to properly implant the InternalBrace, eliminating the need for tendon graph and thus, reducing potential risks, like patella fracture and infection. Summarized geometrical requirements based on the defined complications for the clinical case are defined as:

- Accurate geometry of femur and tibia. Anatomical axis aligned in Anterior-Posterior (AP) plane.
- Simplified geometrical model of the ACL – Positional accuracy is required.
- Sheet model (3D model with small thickness) of the bracelet.
- Parametrical model of the bracelet (will be completely defined in future research).

The MAF method already provides methods and procedures for the creation of the acurate personalized geometrical models of the femur and tibia (and other long and flate bones) [2, 4]. These models can be used as the basis for the creation of the ACL simplified model. As the example of the model creation in the Fig. 7, different 3D models

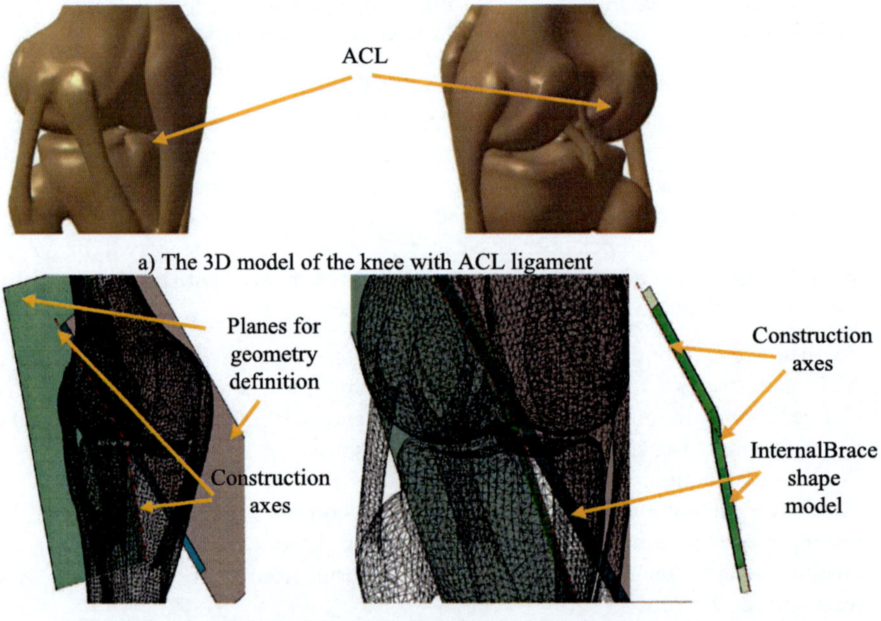

Fig. 7. The InternalBrace and knee 3D model created by using the enhanced MAF

are presented. In the Fig. 7a 3D presentation model of the knee is shown (originated from CT scan, Clinical Center Nis, Serbia, reolution: 512x512,0.5 mm slice thickness), and in the Fig. 7b 3D model of the knee and the braclet is displayed. The models are prepared in CATIA (licensed), but they can be exported to the independet formats like STEP, or IGES and used freely. These models can help surgeons propely prepare surgical interventions using CAS with CAD software and prevent some of the stated complications.

The FM created for the InternalBrace includes geometrical, functional, and material models. The braclet geometrical model is defined by using already defined geometrical models of tibia and femur. Two axes are constructed to define InternalBrace direction through femur and tibia. The width and length of the InternalBrace are parametric values, which mean it can be defined as a feature parameter, and changed accordingly. The functional requirements are stated in the previous section, while material is defined according to the clinical application (FiberTape® suture - polyethylene). By creating this kind of FM, the different modifications are available to the designers and surgeons to prepare and conduct surgical interventions.

5 Conclusion

The human model reconstruction process is a complex one. Therefore, different methods for its creation are presented in this paper. One of these methods is the Method of Anatomical Features (MAF) created by the authors of this research, which focuses on the design of 3D models of human bones and implants. Furthermore, this study demonstrates the enhancement of MAF by adding the Characteristic Product Features (CPF) method as a MAF process. Using this approach, it is possible to create different types of complex models (functional, geometric, material) of human organs and apply them in Computer Assisted surgery (CAS), thus improving patient treatment and recovery processes.

Acknowledgment. This research was financially supported by the ERASMUS+ project "Collaborative e-platform for innovation and educational enhancement in medical engineering" – CALLME. Project Reference: 2022–1-RO01-KA220-HED-000087703.

Initial MAF was financed by Ministry of Education, Science and Technological Development of the Republic of Serbia, and supported by ERASMUS+ project "Boosting the scientific excellence and innovation capacity of 3D printing methods in pandemic period" – BRIGHT. Project Reference: 2020–1-RO01-KA226-HE-095517.

References

1. d'Amato, M., Ensini, A., Leardini, A., Barbadoro, P., Illuminati, A., Belvedere, C.: Conventional versus computer-assisted surgery in total knee arthroplasty: comparison at ten years follow-up. Int. Orthop. **43**(6), 1355–1363 (2018). https://doi.org/10.1007/s00264-018-4114-5
2. Vitković, N., et al.: The parametric model of the human mandible coronoid process created by method of anatomical features, computational and mathematical methods in medicine, ID 574132, 10 (2015)
3. Rathnayaka, K., et al.: Quantification of the accuracy of MRI generated 3D models bones compared to CT generated 3D models. Med. Eng. Phys. **34**(3), 357–363 (2012)

4. Vitković, N., Radović, Lj., Trajanović, M., Manić, M.: 3D point cloud model of human bio form created by the application of geometric morphometrics and method of anatomical features: human tibia example, Filomat, **33**(4), 1217–1225 (2019)
5. Song, D., Jin, Y., Wang, T., Li, C., Tong, R., Chang, J.: A Semantic Parametric Model for 3D Human Body Reshaping. In: El Rhalibi, A., Pan, Z., Jin, H., Ding, D., Navarro-Newball, A.A., Wang, Y. (eds.) Edutainment 2018. LNCS, vol. 11462, pp. 169–176. Springer, Cham (2019). https://doi.org/10.1007/978-3-030-23712-7_24
6. Vitković, N., Korunović, N., Aranđelović, J., Miltenović, A., Perić, M.: Remodeling of complex surface patches by using the method of characteristic features – the ski shoe heel lip example. Innov. Mech. Eng. **1**(2), 96–105 (2022)
7. Minnullina, A., Minnullin, R., Kopytova, A., Savoskina, E.: Production strategy development for an energy company based on SADT. In: E3S Web Conference vol. 217, p. 07015 (2020)
8. Vitković, N., Mladenović, S., Trifunović, M., et al.: Software framework for the creation and application of personalized bone and plate implant geometrical models. J. Healthcare Eng. Article ID 6025935, 11 (2018)
9. Stojkovic, M., et al.: Reverse modeling and solid free-form fabrication of sternum implant. Australas Phys Eng Sci Med. **33**(3), 243–250 (2010)
10. Vitković, N., Trajanovic, M.D., Arsić, S.: Creation of Geometrical Models of Human Bones by Using Method of Anatomical Features. In: Canciglieri Junior, O., Trajanovic, M.D. (eds.) Personalized Orthopedics, pp. 55-78. Springer, Cham. (2022). https://doi.org/10.1007/978-3-030-98279-9_2
11. Mitic, J., Trajanovic, M.D.: Geometrical Model of the Human Mandible: Potential for Application in Personalized Maxillofacial Surgery. In: Canciglieri Junior, O., Trajanovic, M.D. (eds.) Personalized Orthopedics, pp. 79–112. Springer, Cham. (2022). https://doi.org/10.1007/978-3-030-98279-9_3
12. Manic, M., Vitković, N., Mitic, J.: Design and Manufacturing of the Personalized Plate Implants. In: Canciglieri Junior, O., Trajanovic, M.D. (eds.) Personalized Orthopedics, pp. 185–219. Springer, Cham. (2022). https://doi.org/10.1007/978-3-030-98279-9_6
13. Longo, U.G., Viganò, M., Candela, V., de Girolamo, L., Cella, E., et al.: Epidemiology of posterior cruciate ligament reconstructions in Italy: a 15-year study. J Clin Med. **10**(3), 499 (2021)
14. Hopper, G.P., Aithie, J.M.S., Jenkins, J.M., Wilson, W.T., Mackay, G.M.: Satisfactory patient-reported outcomes at 5 years following primary repair with suture tape augmentation for proximal anterior cruciate ligament tears. Knee Surg. Sports Traumatol. Arthrosc. **30**(1), 253–259 (2021). https://doi.org/10.1007/s00167-021-06485-z
15. The Mackay Clinic, https://www.mackayclinic.co.uk/acl-repair-with-the-internalbrace-a-look-at-the-5-year-results/ Accessed 12 May 2023

Synergetic Integration of Electrospinning and Additive 3D/4D Printing Process for Biomedical Applications

Ashok Vaseashta[1,2,3](✉), Didem Demir[4], and Nimet Bolgen[5]

[1] International Clean Water Institute, Manassas, VA, USA
prof.vaseashta@ieee.org
[2] Transylvania University of Brasov, Brasov, Romania
[3] IEEN, Academy of Sciences of Moldova, Chisinau, Moldova
[4] Division of Chemistry and Chemical Process, Tarsus University, Mersin, Türkiye
[5] Chemical Engineering Department, Engineering Faculty, Mersin University, Mersin, Türkiye

Abstract. Electrospinning is a versatile technique and has been used to produce porous fibers ranging from submicron to nanometer in diameter. Using a variety of high-performance polymers and blends, several new configurations are, now, possible for applications in tactile sensing, energy harvesting, filtration, and biomedical technologies. The structures, however, lack desired mechanical conformity, complexity, and single/multi-material three-dimensional rigid constructs essential to mimic specific functionalities. A simple, yet versatile, strategy is by employing a digitally controlled fabrication process of shape-morphing called 3D printing/additive manufacturing process and by conjoining the two promising technologies. Thus, using strategic and hierarchical integration of processes, elaborate shapes, and patterns can be fabricated on mesostructured stimuli-responsive electrospun membranes. The focus of this investigation is primarily on biomedical structures, as part of a large effort of precision and advanced manufacturing for rapid prototyping.

Keywords: electrospinning · 3D printing · prototyping · scaffolds · biomedical

1 Introduction

Three-dimensional (3D) printing and nanotechnology are two advancing frontiers of science and technology that have recently been moving towards strategic integration to leverage their individual advantages and achieve new material characteristics, unique configurations, and hence, applications that will catapult current technology to create highly functional structures. By adjusting processing parameters, variations of lightweight, adaptable, resilient, biodegradable, and highly multifunctional materials are produced in the form of smart structures [1, 2]. However, one of the bottlenecks to smart-structure promulgation is the manufacturing processes which have become increasingly complex, outpacing traditional manufacturing methods. In addition, the traditional methods present challenges in the processing of heterogeneous materials,

Y. Dekhtyar and I. Saknite (Eds.): NBC 2023, IFMBE Proceedings 89, pp. 129–137, 2023.
https://doi.org/10.1007/978-3-031-37132-5_17

hence, the complete structure needs to be divided into multiple components and processed individually and separately. While there are certain restrictions on the minimum size features that can be accurately printed, most of the limitations of additive processing center around how to optimally position a print layer to reduce support dependency and the likelihood of print failure. This allows a large amount of design freedom and enables the creation of very complex geometries with relative ease. Electrospinning is a versatile, efficient, and inexpensive nanofibers processing technique that can be used to produce 1D fibrous materials or composites with a wide range of diameters (from nm to mm) [3, 5]. With the recent advances in nanotechnology, new multifunctional materials are produced by combining nanomaterials with electrospun nanofibers. There are several investigations that show that such nanofibers demonstrate sensitivity to chemical and biological agents and hence nanofibrous structures fabricated using electrospinning are excellent candidates for the fabrication of e-textiles, tactile membranes, and protective clothing due to their adsorptive, lightweight, and functionalization properties [5]. By choosing certain combinations of high-performance polymers and nanomaterials, the newly developed composite materials have new and unique functional properties. Recent developments in electrospinning in hierarchical unification with additive processing are currently used in enhancing defense and security posture by providing force protection [6, 7], sensing/detection of hazardous environments [7], tactile interface for improved functionality [6], comfortable outwear for soldiers and emergency responders [6], biomedical support in the battlefield, energy harvesting patches and devices with communication capabilities [8]. Complex functionalized and hierarchically integrated scaffolds composed of micro-and nanoscale structures are a key objective of conjoining 3D printing with electrospun nanofibers, especially when combined with nanomaterials. Using this framework devices, and design constructs fabricated using 3D printing and electrospinning have great potential in the production of complex structures required for several unique and specialized applications, such as in intervertebral disc tissue engineering, smart scaffolds for abdominal walls, and bi-layer scaffolds for guided tissue regeneration. Complex and hierarchically functionalized structures consisting of micro and nano-scale structures can now be produced by the convergence of the two methods and several commercial instruments are currently designing instruments by the combination of 3D printing and electrospinning. The objective of this investigation is to explore the spectrum of opportunities, challenges, and drawbacks of this combinatorial process. Furthermore, the scope of this investigation is primarily for biomedical applications, although the processes can easily be applied to a wide variety of other fields.

2 Overview of Fabrication Processes

2.1 Electrospinning

Electrospinning is a versatile technique based on applying an electrostatic field to a polymeric solution to yield fibers at the nanoscale. The technique has attracted a lot of attention since it is relatively simple to operate and yields fibers having applications potentially covering various fields. In general, the electrospinning technique involves the fabrication of nanoscale fibers with a very large surface area to diameter ratio employing polymers and their composites with several functional additives, including composites

[3], ceramers [4], drugs [5], nanoparticles [4], nanotubes [4], quantum dots [4], and even plant extracts [9, 10], in the form of melt or solution. The polymer solution is subjected to a high electrical field, thus using the electrohydrodynamic principles whereby a polymeric droplet is electrified to form a jet called a "Taylor cone" [11], the polymer solution undergoes stretching and elongation to form fibers on a grounded collector. The classical electrospinning setup as shown in Fig. 1a basically consists of a syringe pump, spinneret, high voltage power supply, and a grounded collector. The method has been modified over time with the addition of different needle types, collectors, and rings to contain the cone spread. One of the methods of electrospinning, termed coaxial electrospinning, is shown in Fig. 1b, effectively enables the production of ore-shell structure fibers by combining different system materials in the same fiber.

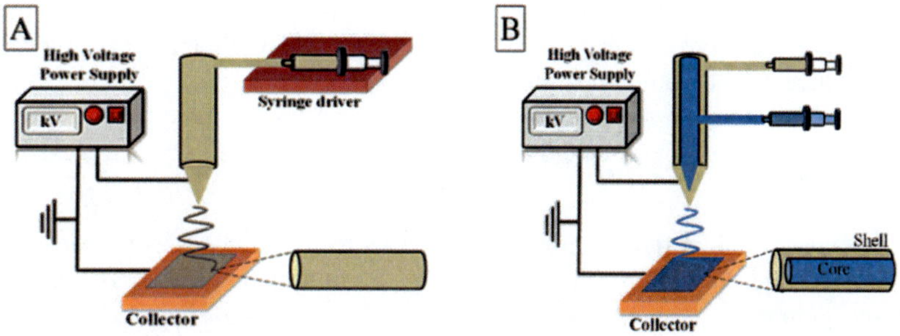

Fig. 1. Electrospinning techniques. a) Conventional method used for basic nanofibers, and b) Coaxial method with core-shell fibers. [1]

2.2 Additive Printing Process

3D printing has developed rapidly in the last few years [12, 13]. The technology is based on the layer-by-layer production of 3D structures directly from computer-aided design (CAD) drawings. Some of the most common methods include fused deposition, stereolithography, digital light processing, and PolyJet photopolymer 3D printing techniques. The method of depositing the polymers by fusion (Fig. 2) is a widely preferred process due to its advantages of being inexpensive and ease of use with which polymer composites containing both thermoplastic polymers and solid particles can be additively printed.

However, the method suffers from disadvantages, such as poor mechanical properties, layer-by-layer appearance, poor surface quality, and the limited number of thermoplastic materials available for this method [14]. Stereolithography (SLA) is an additive manufacturing process that uses liquid UV-curable photopolymer resin and photochemical processes by which light causes chemical monomers and oligomers to cross-link together to form polymers [15]. Yet another technique is called PolyJet printing, which is similar to inkjet technology and is based on the principle of spraying small droplets of liquid resin from nozzles and then solidifying these droplets by exposing them to UV light in each layer. One of the most important advantages of this type of printer is that it

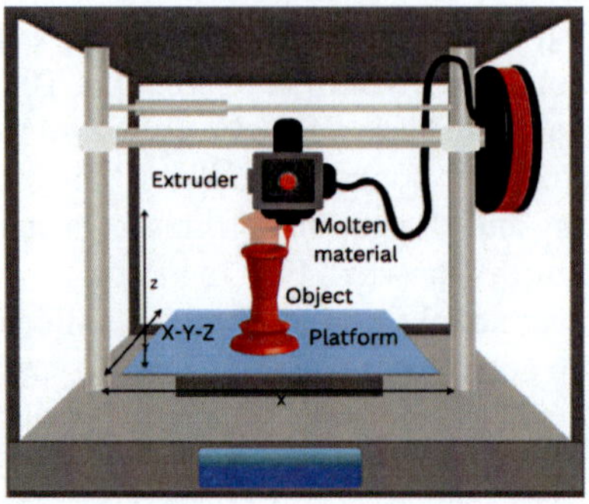

Fig. 2. Commonly used 3D printing techniques

can work in normal room conditions without requiring a special laboratory environment and can produce a wide range of high-resolution materials.

3 Synergetic Integration

The combination of 3D printing and electrospinning technologies is a relatively new and developing methodology and was introduced by the authors to create custom-designed shapes and patterns for several unique and specific applications, as described in the following section. Although preliminary investigations are still in the developing stages, a strategic collaboration of 3D printing methodology by biotech companies has catapulted 3D bioprinting capabilities into the future.

Electrospinning is a promising method for aligning micro/nanofibers due to its simple and versatile fabrication structure. For its applications in scaffolds and tissue engineering, it is known that such structures are categorized as fibrous, porous, or hydrogel. Several methods allow the fabrication of scaffolds in these morphologies by manipulating micro/nanopatterning to provide contact guidance. The electrospun nanofibers have demonstrated considerable potential as scaffolds since these materials exhibit advantageous properties such as mechanical flexibility, high surface area to volume ratio, and the ability to architecturally mimic the native extracellular matrix (ECM) [16]. In addition, electrospinning-based scaffolds with different morphologies can be successfully fabricated for controlled drug delivery systems by controlling the material composition by adjusting the operating parameters. Such combinations are very useful in applying bioactive molecules directly to the bone defect area, which aims to stimulate tissue regeneration while preventing infections.

Despite these advantages, electrospun nanofibers still suffer from drawbacks in their standalone use in regenerative medicine and for scaffolds. Additionally, conventional 2D electrospun membranes use densely packed fibers that can cause inhibition of both cell infiltration and growth along the scaffolds. In bones, the small pore size of electrospun structures can restrict tissue-specific bone cell infiltration, limit tissue ingrowth and simulate the ECM only on a planar structure, which does not allow the repair requirements of large bone defects [17]. Also, as stated earlier, the electrospun structures lack mechanical strength for load-bearing applications of bone. Furthermore, the homogeneous structure, which consists of fibers stacked on top of each other and exhibits a single morphology, is not suitable as a standalone unit to mimic the heterogeneous bone structure of bone. For all these reasons, it is necessary to conjoin these electrospun fibrous structures with rigid scaffold production techniques. In doing so, the objectives then are to improve the mechanical strength, form a larger pore structure that can offer adequate sites for cell adhesion, and to better imitate the compact structure of the bone and other desired configurations. Considering the installation of the combined technologies, a combination system consists of a classical electrospinning apparatus and a computer-controlled x–y motion stage, as shown in Fig. 3.

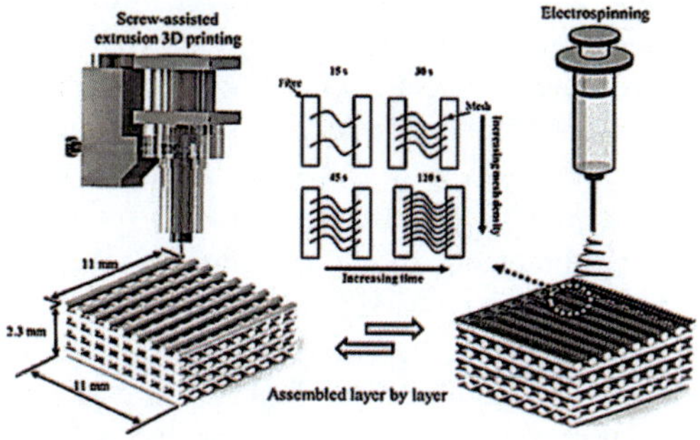

Fig. 3. Schematic of 3D printing and electrospinning to create a dual-scale scaffold. [18]

Hence, strategic, and hierarchical integration of electrospun nanofibers with 3D printing provides desired functionality of several critical structures, which otherwise, would not be readily available.

4 Biomedical Applications of the Combinatorial Processes

So far, there are only very limited investigations that use materials and devices produced by using 3D printing and electrospinning. Realizing that both methods offer limited capabilities and have their own limitations, a conjoined methodology offers a high potential of producing structures, which otherwise would not be possible. Conjoining electrospinning with additive processing is in its very early stages and our initial focus is on the integration strategy of the two. Depending on the application, the first effort has been directed toward developing individual processes accordingly and integrating the two for desired outcomes. The second effort has been towards PVDF-based polymers for use in pressure sensors, tactile systems, structural health monitoring devices, and medical devices that actuate signals induced by pressure and motion. In parallel, efforts are also underway for tissue engineering and bone scaffolds. This leads to prosthetics having a very large number of useful applications. While we continue to explore this subject, we continue to review the literature on the use of this method for additional new and unique applications.

For the second objective, we studied the inclusion of Polyvinylidene fluoride (PVDF) in polymeric solution to produce nanofibers. PVDF is a semicrystalline polymer with alternating monomer units of CH2-CF2, which presents three crystalline phases (commonly α-, β-, and γ-). The α and ε phases are non-polar due to the antiparallel arrangement of dipoles in the cell units, while the β-phase is responsible for the piezoelectric response. By thermal treatment or poling under a high electric field (150kV/mm), the α-phase can be transformed into the γ-phase or β-phase. To increase the β-phase and simultaneously reduce the crystallinity of the α-phase, many methods have been developed during the preparation of PVDF materials. With the addition of PVDF- trifluoro ethylene (PVDF-TrFE), the nanofiber membranes can further align the dipole moment in the nanofiber. The uniaxial stretching technique is used under variable temperatures to obtain PVDF structure in β phase. Also, copolymers of PVDF, such as P(VDF-TrFE), have been synthesized using electrospinning to achieve an intrinsic β-phase crystal. It has been observed that adding Lead Zirconate Titanate (PZT) particles to PVDF improves the dielectric constant, piezoelectric coefficients, and mechanical stability of the films under variable temperature conditions. In addition, such composite films are relatively inexpensive to fabricate functional devices. As a result, studies were undertaken to synthesize nanofiber membranes of PVDF embedded with La+3 ion substituted for Pb+2 ion in PZT, called PLZT, using the electrospinning technique. We have conducted experiments on wearable tactile sensors consisting of electrospun PVDF/PLZT membrane mats as functional layers of biomedical pressure sensors. When put in scaffold format, the devices are likely to yield positive results infant cardiorespiratory monitoring, pediatric dynamometer, respiratory rate monitor, prostate gland stiffness monitor, detection of articular cartilage softening, catheter position sensing, disposable pressure monitoring system, Parkinson's symptoms, and posture monitoring system, just to name a few.

Our next effort has also been toward bone tissue engineering and designing functional and effective techniques for bone regeneration. For this subject, it is critical to understand the existing bone defects, the natural bone regeneration process, and the bone tissue engineering approach. Current modalities include allograft or autograft bones to replace the defect using the Masquelet technique, which uses the body's foreign body response to create a membrane of fibrous tissue around the defect site, and osteogenesis, which uses the natural healing properties of bone to fill in the defective bone. Albeit useful, they have certain challenges and complications. Recently, scaffolds with cells and gene-activating materials have been developed as third-generation bone repair materials. Several processing techniques allow the fabrication of scaffolds in desired configurations, viz. Fibrous, porous, hydrogel, and micro/nano scaffolds, for bone regeneration, such as solvent casting and salt particulate leaching, gas foaming, emulsification, freeze-drying, cryotropic gelation, electrospinning, and 3D printing have been used. Due to the complex architecture of bone consisting of spongy (cancellous bone) and hard (compact bone) tissues, a scaffold in a single morphology does not provide a suitable solution. Thus, a design of multiphase biomaterial structures which is capable of mimicking tissues in complex structures is desired, viz. a scaffold architecture consisting of a dense layer to mimic the compact structure of bone and a porous structure to mimic its spongy nature. Hence, the authors have used a combinatorial approach that combines electrospinning and additive manufacturing to produce bone tissue engineering scaffolds to produce a single material with two different morphologies (nano and micro patterns) having a desired bone anatomy, defect morphology, and size. In this approach, the nanofibers are deposited directly on the 3D-printed patterns by electrospinning. This approach can be divided into two coatings, viz., the outer surface of the final product consisting of layers with electrospinning or coating the interlayers with electrospinning after each layer. Another basic approach is to homogeneously disperse pre-manufactured electrospun fibers in a polymer solution/melt, and then scrape the fiber-polymer combination with 3D printing. Although many other applications are considered, to limit the scope, a review of potential biomedical-related applications is presented below.

Since the aspect described here is relatively new, the authors continue to research ways in which the proposed combinatorial approach can be used for other applications, such as critical infrastructures, defense and security, filtration, and drug encapsulation. Lastly, our efforts are also directed toward 3D-printed tissue engineering constructs developed based on the fundamental characteristics of biodegradability, biocompatibility, and rapid prototyping. In addition, with an aging population, increased traffic worldwide has increased the probability of traffic accidents, sports injuries, and other similar procedures, physical augmentation using prosthetics, Osseointegration, and bionic and robotic prosthetic limbs are gaining much attention in giving amputees full restoration of their lost limbs. Strategic integration of 3D printing coupled with electrospun nanofibers provides a range of applications in various stages of recovery and integration for human performance augmentation efforts. The effort has special application for wounded warriors returning from the battlefield, where injuries are severe and different for different patients. During the COVID-19 crisis, the acquisition of personal protective equipment (PPE) has become ever more critical. While face masks, face shields, hand-held sanitization mounts, etc. have become essential commodities, several designs have appeared in the market and some creative, homemade PPEs have also been observed. In addition, the COVID-19 pandemic also cast a spotlight on ventilators to sustain breathing during surgery or during the condition of hyperinflated lungs, ventilator pumps air—usually with extra oxygen—into patients' airways when they are unable to breathe adequately on their own. Due to immense shortages, 3D-printed designs were widely prototyped and used by several people. Other PPEs are also feasible using 3D printing while electrospun nanofiber-based membranes provide filtration from air-borne contaminants, germs, and viruses.

5 Conclusions and Future Directions

The strategic and hierarchical integration of electrospinning and 3D/4D or additive processing is a relatively new and upcoming field of study with yet unexplored potential. Our research group and several other researchers are continuing to study this straightforward yet comparatively new method to successfully conjoin mats of nanofiber with 3D-printed structures, substrates, or scaffolds to make functional prototypes for a variety of applications as proof-of-concept visualizations. As noted earlier, there is a vast spectrum of applications that span from aviation, automobile, defense, water filtration, sensors for structural health monitoring, and sports, however, the scope of this investigation is within the context of biomedical applications, such as. Tissue-engineered scaffolds, prosthetics, wound dressing, and wearable textiles for health systems monitoring.

References

1. Vaseashta, A., Demir, D., Sakım, B., Aşık, M., Bölgen, N.: Hierarchical integration of 3D printing and electrospinning of nanofibers for rapid prototyping. In: Vaseashta, A., Bölgen, N. (eds.) Electrospun Nanofibers. Springer, Cham (2022). https://doi.org/10.1007/978-3-030-99958-2_22
2. Ashammakhi, N., et al.: Electrospinning and three-dimensional printing for biofab. In: Vaseashta, A., Bölgen, N. (eds.) Electrospun Nanofibers. Springer, Cham (2022). https://doi.org/10.1007/978-3-030-99958-2_20
3. Bölgen, N., Demir, D., Aşık, M., Sakım, B., Vaseashta, A.: Introduction and fundamentals of electrospinning. In: Vaseashta, A., Bölgen, N. (eds.) Electrospun Nanofibers. Springer, Cham (2022). https://doi.org/10.1007/978-3-030-99958-2_1
4. Vaseashta, A., Stamatin, I.: Electrospun polymers for controlled release of drugs, vaccine delivery, and system-on-fibers. J. Optoelectron. Adv. Mater. **9**(6), 1606–1613 (2007)
5. Vaseashta, A.: Loaded electrospun nanofibers: chemical and biological defense. In: Bonča, J., Kruchinin, S. (eds.) Nanostructured Materials for the Detection of CBRN. NSPSSACB, pp. 31–45. Springer, Dordrecht (2018). https://doi.org/10.1007/978-94-024-1304-5_3
6. Vaseashta, A., Karagülle-Bölgen, N.: Loaded Nanofibers: force protection, filtration, decontamination. In: Petkov, P., Tsiulyanu, D., Popov, C., Kulisch, W. (eds.) Advanced Nanotechnologies for Detection and Defence against CBRN Agents. Springer, Dordrecht (2018). https://doi.org/10.1007/978-94-024-1298-7_23
7. Vaseashta, A., Dektyar, Y., Ivanov, V., Klavins, M., Demir, D., Bolgen, N.: Nexus of electrospun nanofibers and additive processing—overview of wearable tactical gears for CBRNE defense. In: Rocha, Á., et al. (eds.) Developments and Advances in Defense and Security. Smart Innovation, Systems and Tech. vol 255. Springer, Singapore (2022). https://doi.org/10.1007/978-981-16-4884-7_11
8. Vaseashta, A., Batra, A.: Electrospun nanofibers of high-performance electret polymers for tactile sensing and wearable electronics. In: Vaseashta, A., et al. (eds.) Proceedings of the Sixth International Symposium on Dielectric Materials and Application (ISyDMA 2006). Springer, Cham (2022). https://doi.org/10.1007/978-3-031-11397-0_3
9. Zhang, W., Ronca, S., Mele, E.: Electrospun nanofibres containing antimicrobial plant extracts. Nanomaterials **7**, 42 (2017)
10. Suryamathi, M., Ruba, C., et al.: Tridax procumbens extract loaded electrospun PCL nanofibers: a novel wound dressing material. Macromol. Res. **27**, 55–60 (2019)
11. Vaseashta, A.: Controlled formation of multiple Taylor cones in electrospinning process. Appl. Phys. Lett. **90**, 093115 (2007)
12. Ngo, T.D., Kashani, A., Imbalzano, G., Nguyen, K.T.Q., Hui, D.: Additive manufacturing (3D printing): a review of materials, methods, applications, and challenges. Compos. B Eng. **143**, 172–196 (2018)
13. Shahrubudin, N., Lee, T.C., Ramlan, R.: An overview of 3D printing technology: technological, materials, and applications. Procedia Manufact. **35**, 1286–1296 (2019)
14. Hossain, N., Chowdhury, M.A., Shuvho, M.B.A., Kashem, M.A., Kchaou, M.: 3D-printed objects for multipurpose applications. J. Mater. Eng. Perform. **30**(7), 4756 (2021)
15. Voet, V.S.D., Schnelting, G.H.M., Xu, J., et. al.: Stereolithographic 3D printing with renewable acrylates. J. Vis. Exp. **2018**(139), 58177 (2018)
16. Chen, S., John, J.V., McCarthy, A., Xie, J.: New forms of electrospun nanofiber materials for biomed. Apps. J. Mater. Chem. B **8**(17), 3733 (2020)

17. Belgheisi, G., Haghbin Nazarpak, M., Solati-Hashjin, M.: Fabrication and evaluation of combined 3D printed/pamidronate-layered double hydroxides enriched electrospun scaffolds for bone tissue engineering applications. Appl. Clay Sci. **225**(February), 106538 (2022). https://doi.org/10.1016/j.clay.2022.106538
18. Vyas, C., Ates, G., Aslan, E., et. al.: Three-dimensional printing and electrospinning dual-scale polycaprolactone scaffolds with low-density and oriented fibers to promote cell alignment. 3D Printing Addit. Manufact. **7**(3), 105–113 (2020)

RGB Laser-Illuminated Spectral Imaging: Applications in Dermatology and Endoscopy

Janis Spigulis(✉), Edgars Kviesis-Kipge, Uldis Rubins, Ilze Oshina, and Madars Mileiko

Biophotonics Laboratory, Institute of Atomic Physics and Spectroscopy, University of Latvia, Jelgavas 3, Riga 1004, Latvia
janis.spigulis@lu.lv

Abstract. Spectral imaging – acquisition of images at specific spectral intervals - is a powerful tool for optical diagnostics, providing objective quantitative data on various clinical parameters, e.g. abnormal content and distribution of chromophores in pathologic tissues. The narrower are spectral bands of imaging, the better is performance of diagnostics; from this point, triple laser line illumination has a potential to ensure excellent spectral selectivity of imaging. This study aims at development and validation of RGB laser-fiber based technologies for high performance spectral imaging of skin and mucosa malformations. Our previous knowledge on skin malformation's spectral line imaging is extended to the whole-body spectral imaging and endoscopic narrowband imaging of mucosa. Design features of two demo-setups ensuring simultaneous illumination of the target tissues by laser lines 450 nm, 520 nm and 638 nm are discussed. Results of laboratory measurements confirm applicability of RGB laser line spectral imaging technology for improved optical diagnostics in dermatology and endoscopy.

Keywords: spectral imaging · skin diagnostics · endoscopy · multi-wavelength lasers

1 Introduction

Spectral image is an image of target within a specific wavelength band. Spectral imaging refers to a group of analytical techniques that collect spectroscopic information and imaging information at the same time. The spectroscopic information tells us about the chemical makeup at the individual points of the image (pixels) allowing a chemical map of the imaged area to be produced [1]. In a general sense "spectral imaging" implies spectrally resolving the process of imaging, either by spectrally refining the illumination source, or detecting the light from the object in a spectrally resolved fashion. The widely used multi-spectral and hyper-spectral imaging technologies are mostly based on sequential capturing of relatively broad-band spectral images, each of them within the range of 10…40 nm [2]. Sequential acquisition of the spectral image set may last from several seconds to a minute or longer [2], so movements of the target area (e.g., *in-vivo* tissue) during the procedure can cause significant image artefacts to be corrected

Y. Dekhtyar and I. Saknite (Eds.): NBC 2023, IFMBE Proceedings 89, pp. 138–144, 2023.
https://doi.org/10.1007/978-3-031-37132-5_18

afterwards. Besides, the shapes of selected spectral bands must be considered during the image processing, which may cause additional problems, e.g. when the distribution maps of absorbing pigments (chromophores) from the sets of spectral images are calculated [3]. Obviously, both acquisition time of the spectral image set and the spectral bandwidth of each image have to be minimized. Ultimate performance can be achieved if the set of narrow spectral line images is obtained by a single snapshot, a technique called "snapshot multi-spectral-line imaging" or SMSLI [4]. In this case spectral line images are obtained under uniform illumination of the target simultaneously by several spectral lines, at a condition that the image sensor system has the corresponding number of spectral sensitivity bands. Standard RGB color cameras, as example, can record three spectral line images by a single snapshot if illumination comprises three spectral lines, each of them positioned within one of the detection bands (R, G or B) [5]. Spectral line imaging under illumination provided by three different lasers has been successfully applied for remote distribution mapping of three main skin chromophores (melanin, oxy-hemoglobin and deoxy-hemoglobin) in a number of cutaneous malformations [6, 7]. A similar approach was recently implemented by using an integrated "3 in 1" RGB laser module that emitted simultaneously 3 spectral lines in the blue, green and red regions of visible spectrum [8]. Uniform illumination of the examined skin region was ensured by means of a laser coupled side-emitting optical fiber loop [9, 10].

This study extends our previous research on local-area skin spectral line imaging [3, 6, 7] to the larger area or whole-body spectral imaging and to the endoscopic imaging. Two prototype setups for those applications are being developed and tested. Design details of the prototypes exploiting simultaneous triple spectral line illumination of the target tissues by means of RGB lasers are presented, along with the first experimental results.

2 Large-Area Skin Imaging Setup with RGB Laser Illumination

Full-body (or whole-, total-body) imaging systems recently have gained attention in dermatology as efficient tools for melanoma detection and monitoring [11]. Contrary to classic investigations focused on a particular lesion, when a single image is taken using a dermoscope or a high-quality camera, such systems capture several photographs of the whole body, which allows not only lesion detection, but also the possibility of comparing images acquired at different times. Several advanced (and correspondingly expensive) whole body imaging systems are commercially available - see [12] as example. Still, the conventional full body imaging provides a lot of clinically unnecessary information thus spending inefficiently the imaging/computing resources, as only specific body location in many cases should be regularly monitored, e.g. the whole back of the patient. Another issue of potential improvements is the spectral content of illumination – most of the existing systems [11] use broadband white light sources so limiting the possibility to extract characteristic spectral features of images for more advanced diagnostics.

We are developing a simpler and less expensive setup for multi-spectral imaging of either full body or a selected large area of patient's skin under triple laser line illumination. The proposed light source enabling uniform illumination simultaneously by the 450 nm, 520 nm and 638 nm wavelengths is a tangled side-emitting optical fiber with 600

a

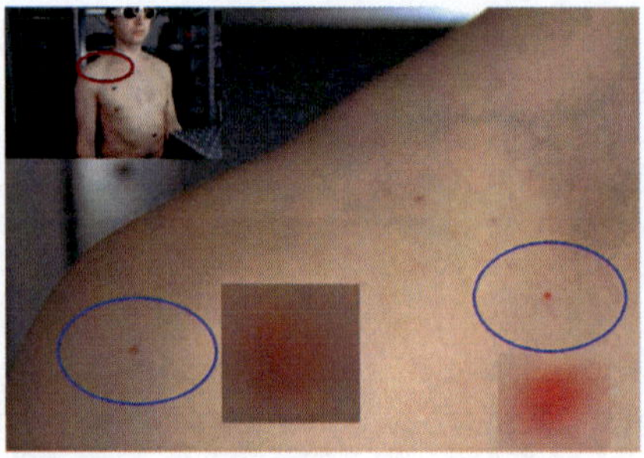

b

Fig. 1. The planar RGB illuminator - tangled side-emitting optical fiber (a) and an example of high-resolution skin color image obtained under such illumination (b).

micron silica core (Fig. 1,a). Fiber is coupled via a standard SMA connector to the 3IN1 RGB High Power 3 W White Laser (*NaKu Technology Co. Ltd, CN*; emission power up to 1W at each wavelength). This light source provided 115 Lx illumination at 65 cm distance, estimated as optimal for the proposed application. Longitudinal uniformity of the side-emitted intensity was enhanced by a micro-reflector mounted at the distal end of fiber [13, 14].

High resolution 61-megapixel RGB camera Sony a7R IVA with the Sony SEL2470GM, F2.8 G Full Frame Standard Zoom Lens was used for skin image acquisition under this specific illumination. After optimizing the camera exposure time and ISO, reasonable quality color images of skin malforations sized down to 1 mm were obtained at the camera-body distance of 70 cm (Fig. 1,b). Three spectral line images at 450 nm, 520 nm and 638 nm were extracted from each of the obtained RGB images [5], with further distribution mapping of three main skin chromophores over the examined malformations by a procedure described in [6] – see two examples on Fig. 2.

In particular, a system of three modified Beer-Lambert equations comprising concentrations and absorption coefficients of these three chromophores at each of the working wavelengths was composed and solved for each pixel of the image. Relative spectral sensitivities of the camera at three exploited wavelengths were determined experimentally [15].

Our single-camera concept of this system proposes capturing several images of the body at appropriate positions of the camera thus obtaining a set of full-body spectral images within tens of seconds. If only one specific large-size area of the body is under interest, the procedure's time is correspondingly reduced.

3 Endoscopy Imaging Setup with RGB Laser Illumination

Filtering relatively narrow (tens of nanometers) spectral bands out of the white light endoscopic illumination facilitates contrast of specific tissue features in endoscopy images; the NBI (narrow band imaging) technique has been widely used over the recent years [16]. Introduction of the SMSLI technology in endoscopy could considerably improve the spectral selectivity (color sensitivity) by narrowing the imaged spectral bands down

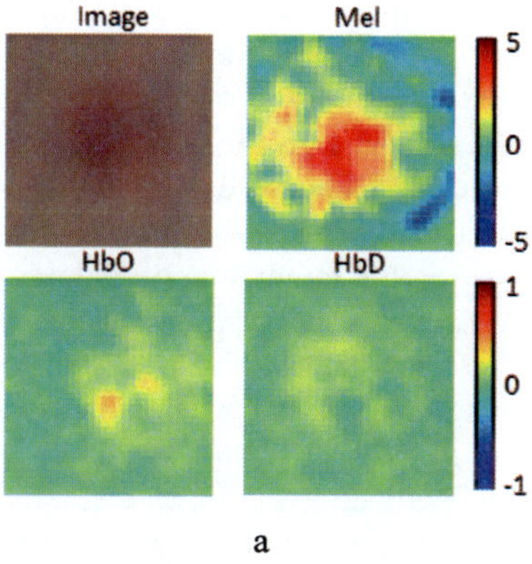

Fig. 2. Distribution maps of melanin (Mel), oxy-hemoglobin (HbO) and deoxy-hemoglobin (HbD) relatively to the surrounding healthy skin for two malformations of a volunteer, presumably nevus (a) and hemangioma (b). The maps were calculated from data of large-area skin color photo taken under the RGB laser-fiber illumination at 70 cm distance. Units of the color scale are milimoles.

to the laser linewidth. It could also open new possibilities of quantitative diagnostics of mucosa by mapping its chromophores. RGB lasers already have been exploited for endoscopic illumination to replace the traditional white light sources for color imaging [17, 18]. To the best of author's knowledge, snapshot acquisition, extraction and analysis of endoscopic spectral line images under RGB laser illumination have not been studied so far.

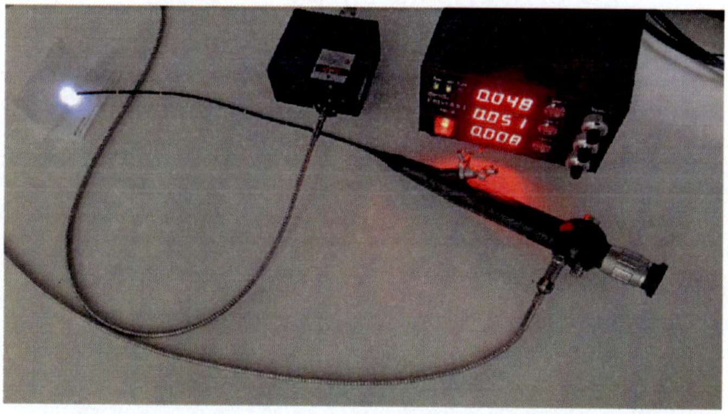

Fig. 3. The endoscopy setup with RGB laser illumination.

To fill this gap, two design options for endoscopic illumination by medium-power RGB laser with multimode fiber output were considered and tested. For this purpose, the RGB 300 mW White Laser (*NaKu Technology Co. Ltd, CN*) emitting manually adjustable power up to 100 mW at each of the 450 nm, 520 nm and 638 nm wavelengths was accessible. The endoscope was STORZ Rhino-Laryngo-Fiberscope, model 11001RDK1 comprising the working channel of 1.5 mm diameter and the RGB camera TH110.

The initial option was to illuminate the target via flat-end 400-micron silica core optical fiber of external diameter 0.8 mm and numerical aperture 0.37 (LU-04.2023, *Lightguide, LV*), attached to the laser by SMA connector and inserted into the working channel of endoscope. Experimental tests confirmed the viability of this approach – bright and relatively uniform illuminated spot was achieved at the target area of endoscope. However, two problems were identified – relatively narrow output beam and risks related to fiber tip sterilization during routine exploitation. They might be solved by deposition of some highly resistant biocompatible light diffusing layer on the polished top of the fiber, which requires an additional study.

Alternatively, the second option – launching the RGB laser light into the input of the built-in illumination channel of endoscope, so avoiding problems with output light shaping and sterilization of the tip – was considered. We used 2 m long metal- armored multimode fiber with 100-micron silica core (NA = 0.22) and SMA connectors at both ends. One of them was connected to the RGB laser output and the other – to the endoscope lighting channel input via a home-made adaptor (Fig. 3). Illumination of 5600 ± 100 Lx at 10 mm distance from the distal end of endoscope was ensured at total output power of 2 mW (equally distributed between the three laser lines).

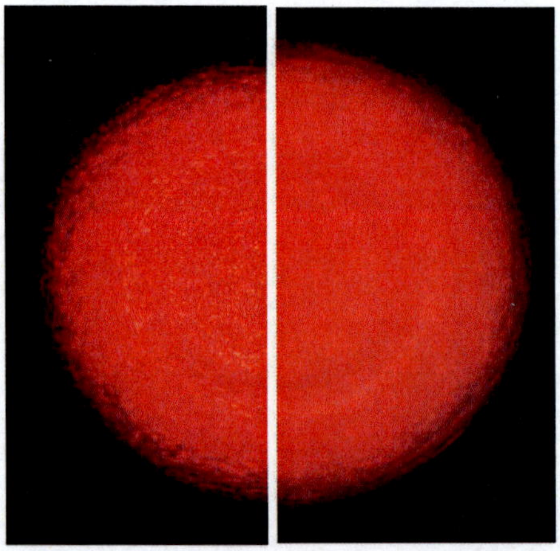

Fig. 4. The red laser beam image without (left) and with (right) vibro-motor attached to the delivery fiber.

This design option appeared advantageous, but one more challenge was faced – grainy laser speckle patterns appeared in the recorded images. As a solution, part of the delivery fiber was trembled by a smartphone vibro-motor with 100 Hz frequency, so "smearing" the speckle grains [19]. Notable improvements were achieved – see Fig. 4 for illustration.

4 Conclusions

To summarize, results of this study confirmed a promising potential of the RGB laser-based illumination systems for high performance spectral imaging in clinical dermatology and endoscopy. We demonstrated that side-emitting fibers are well-suited for uniform large-area skin illumination simultaneously by three laser spectral lines, with subsequent mapping of chromophores in the examined malformations. A simple option for RGB laser illumination of internal cavities via the built-in lighting channel of fibro-endoscope was proposed and tested, allowing further extension of the triple spectral line imaging methodology to endoscopic diagnostics.

The work in this direction will be continued. We aim at the development of completely automated multi-stage full body spectral line imaging system affordable for small and medium-sized dermatology clinics and GP-offices. In the field of spectral line imaging for endoscopy, future studies will be focused on the mucosa laser safety issues and development of specific algorithms for processing of internal cavities' spectral images.

Acknowledgements. This research was funded by the Latvian Council of Science, project "Smart Materials, Photonics, Technologies and Engineering Ecosystem", No. VPP-EM-FOTONIKA-2022/1–0001.

References

1. Spectral imaging. https://www.impopen.com/what-is-spectral-imaging#:~:text=Spectral%20imaging%20refers%20to%20a,imaged%20area%20to%20be%20produced. Accessed 24 Apr 2023
2. Gevaux, L., Perrot, J.L., Hébert, M.: Multispectral and hyperspectral imaging for skin acquisition and analysis. In: Fimiani, M., Rubegni, P., Cinotti, E. (eds.) Technology in Practical Dermatology, pp. 271–279. Springer, Cham (2020). https://doi.org/10.1007/978-3-030-45351-0_26
3. Kuzmina, I., et al.: Skin chromophore mapping by smartphone RGB camera under spectral band and spectral line illumination. J. Biomed. Opt. **27**(2), 026004 (2022)
4. Spigulis, J., Oshina, I., Matulenko, M.: Laser illumination designs for snapshot multi-spectral-line imaging. IEEE Xplore (2019). https://ieeexplore.ieee.org/document/8872998
5. Spigulis, J., Elste, L.: Single-snapshot RGB multispectral imaging at fixed wavelengths: proof of concept. Proc. SPIE **8937**, 89370L (2014)
6. Spigulis, J., Oshina, I., Berzina, A., Bykov, A.: Smartphone snapshot mapping of skin chromophores under triple-wavelength laser illumination. J. Biomed. Opt. **22**(9), 091508 (2017)
7. Oshina, I., Spigulis, J., Kuzmina, I., Dambite, L., Berzina, A.: Three-dimensional representation of triple spectral line imaging data as an option for noncontact skin diagnostics. J. Biomed. Opt. **27**(9), 095005 (2022)
8. Spigulis, J., Rupenheits, Z., Rubins, U., Mileiko, M., Oshina, I.: Spectral line reflectance and fluorescence imaging device for skin diagnostics. Appl. Sci. **10**, 7472 (2020)
9. Spigulis J., Oshina I., Rupenheits Z., Matuļenko M.: Device for uniform surface illumination simultaneously by several laser spectral lines. patent LV 15491 B (2020)
10. Spigulis, J., Pfafrods, D., Stafeckis, M., Jelinska-Platace, W.: The "glowing" optical fiber designs and parameters. Proc. SPIE **2967**, 226–231 (1997)

11. Korotkov, K., et al.: An improved skin lesion matching scheme in total body photography. IEEE J. Biomed. Health Inform. **23**, 586–598 (2019)
12. VECTRA WB360 whole body 3D imaging system. https://www.canfieldsci.com/imaging-systems/vectra-wb360-imaging-system. Accessed 24 Mar 2023
13. Pfafrods, D., Stafeckis, M., Spigulis, J., Boucher, D.: "Side-emitting optical fiber", patent LV 11644 B (1995)
14. Side glowing fibers. http://www.somta.lv/wp30.htm. Accessed 24 Mar 2023
15. Spigulis, J., Oshina, I., Kuzmina, I., Dambite, L.: Method and device for determination of photo-camera relative spectral sensitivities at selected wavelengths. patent application LVP2021000041
16. Barbeiro, S., Libanio, D., Castro, R., Dinis-Ribeiro, M., Pimentel-Nunes, P.: Narrow-band imaging: clinical application in gastrointestinal endoscopy. GE Port J. Gastroenterol **26**, 40–53 (2019)
17. Lu, M.K., Lin, H.-Y., Hsieh, C.-C., Kao, F.-J.: Fiber based illumination for miniature endoscopy. JSAP-OSA Joint Symposia 2016 Abstracts, Optica Publishing Group, paper 15p_C31_9 (2016)
18. Chen, Z.-Y., et al.: Coherent narrow-band light source for miniature endoscopes. IEEE J. Sel. Top. Quant. Electron. **25**(1), 1–7 (2019)
19. Rubins, U., Kviesis-Kipge, E., Spigulis, J.: Device for speckle-free imaging under laser illumination. WO 2018/177565 A1 (2018)

Effect of Age on in Vivo Human Brain Tissue Electrical Conductivity

Juha Latikka[1(✉)] and Hannu Eskola[1,2]

[1] Faculty of Medicine and Health Technology, Tampere University, Korkeakoulunkatu 7, 33720 Tampere, Finland
juha.latikka@tuni.fi

[2] Department of Radiology, Medical Imaging Center, Tampere University Hospital, Tampere, Finland

Abstract. This study aimed to investigate the dependency on age of in vivo human grey and white matter electrical conductivities. Most of the related previous studies were conducted using in vitro animal tissue samples, and it has been assumed that these measurements could indicate age-related dependencies of living human gray and white matter. Electrical conductivity measurements were taken from the surgical channel during 13 brain surgeries from patients aged from 4 to 87 years using a needle electrode with a measurement frequency of 50 kHz. Our results indicated that there was no significant linear correlation between age and gray matter electrical conductivity (R^2 was 0.1095) or white matter electrical conductivity (R^2 was 0.0149) within the patient age range and used measurement frequency. The collated results indicate that age dependency of gray matter conductivity can be observed at a very young age (near newborn age) and that white matter conductivity dependency can be observed at higher frequencies. The conductivity information may have implications for specific absorption rate (SAR) and source localization calculations and simulations with subjects of different ages and with different frequencies.

Keywords: Tissue conductivity · White matter · Gray matter · Age dependency

1 Introduction

Digital head models and simulations having brain tissue electrical conductivity data are used for several purposes, for example, in the planning of treatments using transcranial direct-current stimulation [1] or radiofrequency-induced hyperthermia [2]. In addition to the accuracy of the geometry [3], the accuracy of used brain tissue electrical conductivity values is an important component of simulation models that affects the source localization outcome [4, 5].

It is evident that there are anatomical and geometrical differences between the heads of children and adults, but the possible age dependencies of the electrical conductivities of gray and white matter should also be investigated and, if found, used to improve modeling. The underlying question is whether age dependency needs to be considered,

Y. Dekhtyar and I. Saknite (Eds.): NBC 2023, IFMBE Proceedings 89, pp. 145–152, 2023.
https://doi.org/10.1007/978-3-031-37132-5_19

in addition to the model dimensions, when preparing electrical conductivity models for different age groups.

Most of the related previous studies were conducted in vitro with electrical conductivity measured in tissue samples. Measurements taken from animal tissues have been considered to indicate human brain tissue age-related conductivity. Some researchers have concluded that gray matter conductivity is age dependent [6, 7], others that white matter conductivity is age dependent [8–10], and still others that neither is age dependent [11].

The electrical properties, i.e., the relative permittivity (ε) and the conductivity (σ) of tissues depend on frequency. For a typical soft tissue, different mechanisms dominate at different frequency ranges [12]. The electrical properties change with frequency in three distinct steps, known as alpha (α), beta (β) and gamma (γ) dispersions [13, 14]. This causes variations across the measurement frequency spectrum. Different frequency ranges are of interest when the aim is to model sources of EEG activity or exposure to wireless networks or mobile phones. Since all bioelectrical sources produce frequencies below 1 kHz, a low measurement frequency is necessary for source localization modeling. On the other hand, cellular and other wireless networks operate in the gigahertz region, so a high measurement frequency is necessary for modeling exposure to them.

Electrical conductivity measurements should be taken from living tissues because brain tissue conductivities decrease after death [8, 15]. Physiological changes have a greater effect at the lower frequencies in the α dispersion region [16] than at the higher frequencies in the γ dispersion region, where the measurement results depend more on the tissue water content [17, 18].

In vitro measurements are sometimes taken at room temperature rather than at body temperature. Similar to other tissues, the electrical conductivities of brain tissues are temperature-dependent, and the temperature of the samples affects the results [19]. Thus, the results are affected by whether the measurements are performed in vivo or in vitro, the frequency used and the tissue temperature.

This study aimed to investigate the age dependencies of human white matter and gray matter electrical conductivities by in vivo measurements during brain surgery using a low measurement frequency. In the discussion, we compared our in vivo results to previously reported data on white matter and gray matter conductivities. The collated results and the contradictory statements with their possible reasons are discussed.

2 Material and Methods

Measurements were taken from living brain tissues during surgeries with a monopolar needle electrode. The measurement system used was the FL Fischer Neuro N50 electrosurgical unit. The Neuro N50 has an impedance measurement function and uses a sinusoidal 50 kHz 2 μA current for the impedance measurements. The measurement frequency was low enough to make the assumption that only the real part of the impedance was present in the measurements. The measurement range of the Neuro N50 is 0–2,040 ohms, which is suitable for intracranial tissue measurements. The active electrode was a TCU301 inside a guide cannula with a 2 mm exposed tip. A 130 cm2 disposable reference electrode was usually placed on the patient's hip area. The main advantage of this measurement system is that it can be used in an operating room environment.

Prior to taking measurements, the linearity and possible baseline error of the equipment were tested for calibrated measurement results. For this, liquids of known conductivity and a set of known resistors were used. Then, the conductivity of the saline wash to be used during the surgeries was measured to determine the effect of the surrounding human body on the measurements.

Brain tissue measurements were taken from 13 patients aged 4 to 87 years. The surgeries were scheduled to remove lesions. The tissue conductivity measurements were not part of the standard lesion removal procedure. The number of measurements taken, and the number of different tissues measured in each patient depended on the surgical path. Gray matter conductivity was measured in 13 patients (1 to 6 measurements per patient), and white matter conductivity was measured in 8 patients (1 to 12 measurements per patient). During the operations, the saline wash (with known conductivity from the calibration measurements) was also measured, and this information was later used to calculate the gray and white matter conductivity values. The protocols were approved by the ethical council of Tampere University Hospital.

3 Results

The patient ages and the average values of white and gray matter conductivity in S/m are shown in Table 1. Based on the low 50 kHz measurement current, it was assumed that only the real part of the admittance was present in the measurement.

Table 1. Patient ages and the average of measurement results for white matter (WM) and gray matter (GM) conductivity in S/m. If the surgical path did not go through white matter, no measurements were taken from it; therefore, the white matter value was not available. This is indicated in the table with "-".

Patient #	1	2	3	4	5	6	7	8	9	10	11	12	13
Age (years)	4	32	33	48	49	54	61	61	63	70	73	75	87
GM (S/m)	0.77	0.58	0.21	0.43	0.26	0.27	0.19	0.21	0.39	0.28	0.47	0.51	0.52
WM (S/m)	–	0.29	0.20	0.40	0.32	–	0.23	–	0.21	0.38	0.26	–	–

Linear regression analysis using Excel was performed to test whether age significantly predicted the measured tissue electrical conductivity value. Figures 1 and 2 show the measured values and age with regression best-fit lines for gray matter and white matter, respectively. The figures also show the coefficient of determination (R^2), which is the proportion of the variance in the dependent variable (electrical conductivity of the tissue) that is predictable from the independent variable (age).

The linear regression results indicate that the predictor (age) explains 11% of the variance in gray matter conductivity (R^2 was 0.1095) and 1.5% of the variance in white matter conductivity (R^2 was 0.0149). Based on the measurements and analysis, there was no significant linear correlation between age and gray matter or white matter in vivo conductivity within the age range of 4 to 87 years in humans at a frequency of 50 kHz.

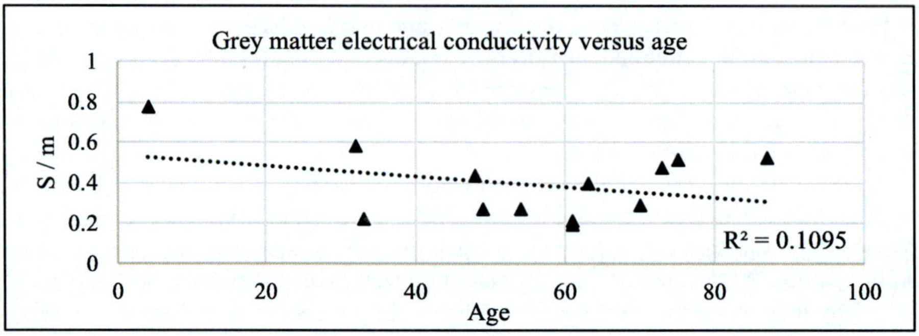

Fig. 1. Gray matter electrical conductivity versus age with linear regression best-fit line and coefficient of determination (R^2).

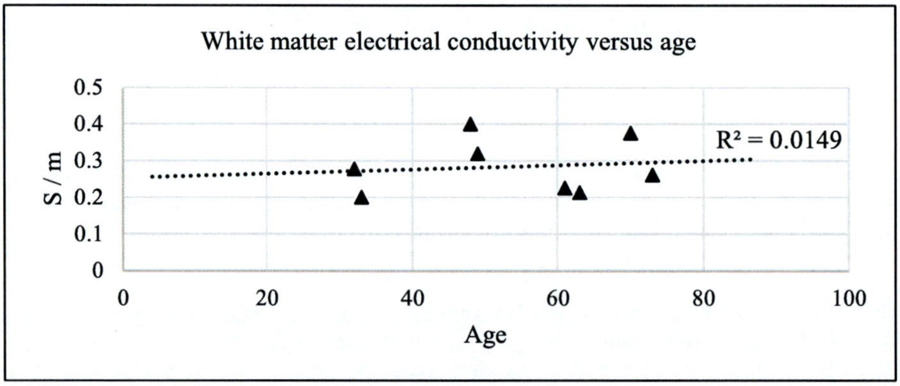

Fig. 2. White matter electrical conductivity versus age with linear regression best-fit line and coefficient of determination (R^2).

4 Discussion

The lack of significant linear age dependency of brain tissues found in this study is consistent with the findings in [11], where the coefficient of determination R^2 was 0.01 for white matter conductivity and R^2 was 0.17 for gray matter conductivity. Compared to the age range of 83 years in this study, the age range in [11] was significantly narrower at 34 years (16 to 50 years old), and all their participants sat well inside our age range. Measurements from very young patients, i.e., from newborn to two years were not taken in either the current study or in [11].

Conductivity measurements have been taken from newborn or nearly newborn gray matter excised from mouse and rat brains [6, 7]. In these in vitro studies, groups of mice aged 3, 5, 19, 26, 33 and 58 days were used [6], and groups of rats aged newborn (less than 24 h old), 10, 20, 30, 50 and 70 days were used [7]. Measurements were taken immediately or two to four hours after excision. The three-day- and five-day-old mouse groups showed different conductivities from each other and the rest of the age groups. The 19-, 26-, 33- and 58-day-old mouse samples showed a similar conductivity, and the authors reported that 19- and 58-day-old mice showed a 95% confidence interval in their dielectric parameters with measurement frequencies below 1 GHz. This is in line with the cerebral cortex of a mouse reaching maturity 16–17 days after birth [6]. In the rat brain, the largest variations were also observed at the younger end of the age range, between the ages of newborn, 10 and 20 days [7]. The measurements were taken with frequencies ranging from 10 MHz to 5 GHz [6] and from 130 MHz to 10 GHz [7].

Other studies have examined brain tissue conductivity age dependence in larger animals and have used age groups from young animals to adults. In [8], female greyhounds were placed into four age groups: 20+ months, 30+ months, 40+ months and 80+ months. The 30+ month and 40+ month groups were considered to be fully grown adult dogs. The measurements were taken immediately after extraction or within 24 h of extraction with a measurement frequency ranging from 300 MHz to 3 GHz. In [9], measurements were taken from freshly excised tissues from two age groups: 16–24 months old bovine (i.e., adult) and 4–6-month-old calves (i.e., young animals). Measurements were taken within eight minutes of the animals' death with a frequency range from 400 MHz to 18 GHz. In [10], measurements were taken from pig tissue samples with approximate ages of 30, 100 and 600 days, an age range considered to cover different developmental stages from piglets to mature animals. Measurements were performed with a frequency range from 50 MHz to 20 GHz. In all three of these studies, the gray matter conductivity did not show a significant change with age across the frequency ranges used. However, in all three studies, the white matter conductivity showed a systematic decrease with age, especially between the youngest and oldest groups in every study. For example, in [9], the mean value of white matter conductivity was 15–22% higher in the young group than in the adult group. In these studies, the measurement frequency range was again mostly within the γ dispersion range, and therefore, the results should correlate with tissue water content.

Table 2 summarizes the results of the previous studies that were reviewed above and the results of the current study. Together, these earlier studies and our measurements reveal that gray matter conductivity is reported to be age dependent in some studies [6, 7], other studies report that there is age dependence in white matter but not in gray matter conductivity [8–10], and some studies report that there is no age dependence in either [11].

The contradictory results may well be due to the different age ranges and brain maturity levels of the subjects. For humans, the brain reaches 80–90% of adult volume in the first two years [20], and the increase in volume is mainly a result of gray matter growth.

According to the collated results, it seems that gray matter conductivity age dependence can be observed in near newborn measurements and white matter conductivity age dependency at higher frequencies where the tissue water-content-related γ dispersion contributes to the result. This means that the age dependency also depends on the measurement frequency. In very young human brains, there could be age dependence of gray matter conductivity based on animal measurements and our one pediatric measurement. However, more human data is needed to verify this.

Most of the measurements in previous studies [6–10] were taken from in vitro tissue samples, and some researchers assume that these measurements give an indication of

Table 2. Observed age dependence in the current study and in selected previous studies. N in the measurement conditions column indicates the number of subjects. It is indicated in the age range column if the subjects were grouped, and in those cases, the average group results are reported instead of individual subject values.

Author and Reference	Gray matter conductivity age dependence	White matter conductivity age dependence	Species	Age range	Measurement conditions and frequency
Thurai et al. 1984 [6]	Yes, conductivity decreased with age	Not measured	Mouse	3 to 58 days, 6 ages with multiple specimens in each age	Tissue samples N = 18 10 MHz to 5 GHz
Peyman et al. 2001 [7]	Yes, conductivity decreased with age	Not measured	Rat	0 to 70 days, 6 ages with multiple specimens in each age	Tissue samples N = 18 130 MHz to 10 GHz
Mohammed et al. 2016 [8]	No significant differences were observed	Yes, conductivity decreased with age	Greyhound (female)	20 to 81 months, divided in 4 age groups	Tissue samples N = 12 300 MHz to 3 GHz
Schmid & Überbacher 2005 [9]	No significant differences were observed	Yes, conductivity decreased with age	Cattle	4 to 24 months, divided in 2 age groups	Tissue samples N = 52 400 MHz to 18 GHz
Peyman et al. 2007 [10]	No significant differences were observed	Yes, conductivity decreased with age	Pig	35 to 600 days, divided in 3 age groups	Tissue samples N not available 50 MHz to 20 GHz
Koessler et al. 2017 [11]	No significant differences were observed	No significant differences were observed	Human	16 to 50 years, individual values	In vivo N = 15 50 kHz
The current study	Inconclusive, more data needed	No significant differences were observed	Human	4 to 87 years, individual values	In vivo N = 13 50 kHz

age-related dependencies of in vivo electrical conductivities of gray and white matter. However, it is not yet evident whether the results obtained in animals accurately reflect the human situation for brain tissue conductivity age dependency. Until such evidence is clearly presented, transferring animal results to human models must be done with caution. Some studies [6–9] have considered different types of age matching, which may be relatively easy with newborn and adult tissues but more difficult with tissues at ages in between (i.e., youth). Different stages of development, such as reaching sexual maturity, have also been used to enable comparison of other animals with humans [9].

More data are needed from aging brain tissues to definitively determine whether there are age dependencies in the brain tissue conductivities at the higher end of the age spectrum. More data are also required from the younger end of the age spectrum to investigate ages that correlate with the developmental stage of humans under two years. Since different brain regions develop at different rates, it is critical that the measurement site is also taken into account when analyzing age dependency.

Unfortunately, previous studies [6–9] often present their results in graphs and do not report numerical data. Several papers [8–10] have also used age groups or averages in reporting the results, which smooths the results and reduces the number of reported data points.

5 Conclusions

Our data derived from living human brain tissue did not show a statistically significant dependency of conductivity on age within the study's age range (i.e., from 4 to 87 years) and with a measurement frequency of 50 kHz. Our results do, however, suggest a slight decrease in gray matter conductivity with age, which is in accordance with earlier human studies. Animal studies using tissue samples and higher frequencies (from MHz to GHz) have suggested that gray matter conductivity is age dependent at a very young age and that white matter conductivity is age dependent when the water content of the tissue decreases with age.

Our methodology has thus far not been applied in other studies. A larger number of human subjects is necessary to definitively determine the conductivities of brain tissues, preferably at different measurement frequencies. Such data would improve the accuracy of for example specific absorption rate and source localization calculations and simulations.

References

1. Datta, A., Baker, J.M., Bikson, M., Fridriksson, J.: Individualized model predicts brain current flow during transcranial direct-current stimulation treatment in responsive stroke patient. Brain Stimul. **4**(3), 169–174 (2011)
2. Johansson, J.D., Loyd, D., Wårdell, K., Wren, J.: Impact of cysts during radiofrequency lesioning in deep brain tissues – a simulation and in vitro study. J. Neural. Eng. **4**, 87–95 (2007)
3. Ramon, C., Schimpf, P.H., Haueisen, J.: Influence of head models on EEG simulations and inverse source localizations. Biomed. Eng. OnLine **5**(1), 55 (2006)
4. Haueisen, J., Ramon, C., Eiselt, M., Braurer, H., Nowak, H.: Influence of tissue resistivities on neuromagnetic fields and potentials studied with a finite element and a boundary element model of the head. IEEE Trans. Biomed. Eng. **4**, 727–735 (1997)
5. Van Uiert, R., Johnson, C., Zhukow, L.: Influence of head tissue conductivity in forward and inverse magnetoencephalographic simulations using realistic head models. IEEE Trans. Biomed. Eng. **51**, 2129–2137 (2004)

6. Thurai, M., Goodridge, V., Sheppard, R., Grant, E.: Variation with age of the dielectric properties of mouse brain cerebrum. Phys. Med. Biol. **29**, 1133–1136 (1984)
7. Peyman, A., Rezazadeh, A., Gabriel, C.: Changes in the dielectric properties of rat tissue as a function of age at microwave frequencies. Phys. Med. Biol. **46**, 1617–1629 (2001)
8. Mohammed, B., Bialkowski, K., Abbosh, A., Mills, P., Bradley, A.: Dielectric properties of dog brain tissue measured in vitro across the 0.3–3 GHz band. Bioelectromagnetics **37**, 549–556 (2016)
9. Schmid, G., Überbacher, R.: Age dependence of dielectric properties of bovine brain and ocular tissues in the frequency range of 400 MHz to 18 GHz. Phys. Med. Biol. **50**, 4711–4720 (2005)
10. Peyman, A., Holden, S., Watts, S., Perrott, R., Gabriel, C.: Dielectric properties of porcine cerebrospinal tissues at microwave frequencies: in vivo, in vitro and systematic variation with age. Phys. Med. Biol. **52**, 2229–2245 (2007)
11. Koessler, L., et al.: In-Vivo Measurements of human brain tissue conductivity using focal electrical current injection through intracerebral multicontact electrodes. Hum. Brain Mapp. **38**, 974–986 (2017)
12. Foster, K.: Dielectric properties of tissues. In: The Biomedical Engineering Handbook, J. D. Bronzino (Ed), pp. 1385—1394, CRC Press, inc. (1995)
13. Schwan H.: Electrical of tissues and cell suspensions: mechanisms and models. Engineering advances: new opportunities for biomedical engineers. In: Proceedings of the 16th Annual Conference of the IEEE engineering in medicine and biology society, vol. 1, pp. A70–A71, IEEE (1994)
14. Gabriel, C., Gabriel, S., Corthout, E.: The dielectric properties of biological tissues: I. literature survey. Phys. Med. Biol. **41**, 2231–2249 (1996)
15. Latikka, J., Hyttinen, J., Kuurne, T., Eskola, H., Malmivuo, J.: The conductivity of brain tissues: comparison of results in vivo and in vitro measurements. In: Proceedings of the 23rd Annual International Conference of the IEEE Engineering in Medicine and Biology Society, pp. 910–912, IEEE (2001)
16. Kraszewski, A., Stuchly, S., Stuchly, M., Smith, A.: In vivo and in vitro dielectric properties of animal tissues at radio frequencies. Bioelectromagnetics **3**, 421–432 (1982)
17. Schmid, G., Neubauer, G., Alesch, F., Illievich, U.: Changes in the dielectric properties of porcine brain tissue in the transition from life to death at frequencies from 800–1,900 MHz. Bioelectromagnetics **24**, 413–422 (2003)
18. Burdette, E., Friederich, P., Seaman, R., Larsen, L.: In situ permittivity of canine brain: regional variations and postmortem changes. IEEE Trans. Microwave Theory Tech. **34**, 38–49 (1986)
19. Foster, K.R., Schepps, J., Stoy, R., Schwan, H.: Dielectric properties of brain tissue between 0.01 and 10 GHz. Phys. Med. Biol. **24**, 1177–1187 (1979)
20. Pfefferbaum, A., Mathalon, D., Sullivan, E., Rawles, J., Zipursky, R., Lim, K.: A quantitative magnetic-resonance-imaging study of changes in brain morphology from infancy to late adulthood. Arch Neurol **51**, 874–887 (1994)

Vision Screening and Training Tool for School-Aged Children

Jelena Slabcova(✉) and Gunta Krumina

Faculty of Physics, Mathematics and Optometry, Department of Optometry and Vision Science, University of Latvia, Jelgavas Str. 1, Riga 1004, Latvia
jelena.slabcova@lu.lv

Abstract. During their school years, modern children experience a significant amount of stress on their visual system due to reading, writing, and extensive use of digital devices at close distance. Therefore, there has been a growing discussion in recent years the importance of assessing near visual functions during vision screening in school-aged children. This is because uncorrected refractive errors and near vision problems can negatively impact children's academic achievements and behaviour. As the first sign of near vision fatigue is an inappropriate functioning of eye accommodation and vergence system, our aim was to develop a vision screening and training tool-prototype that can identify children with near vision problems, provide treatment, monitor progress, and collect and store data on vision training. To achieve this aim, we have developed a prototype of a lens-filter device (LFD), computer program, and methodology to ensure accurate performance of vision tests and training. We conducted a pilot study in a laboratory setting to evaluate the effectiveness of our prototype. A total of 75 subjects (aged 22 ± 5 years) were tested to evaluate repeatability and validity of near vision screening tests performance. Of these subjects, 22 (aged 22 ± 2 years) used our training tool to improve vergence performance. Based on results obtained, our vision screening and training tool provides reliable results and is comparable to optometric tests. Our vision screening and training tool can be useful for vision care specialists who deal with children's vision problem, and it is user-friendly, even for non-specialists.

Keywords: Children · Fusional Vergence Amplitude · Vergence Facility · Vision Screening · Vision Training

1 Introduction

In recent years, there has been increasing amount of the literature on the shortcomings of the vision care system in school-age children [1–3]. Vision screening is crucial for the preventive assessment of visual functions in children who do not complain of visual impairment [4], especially since children under 12–13 may not understand their vision problems and may not complain about them [3, 5]. This issue is of utmost importance, as the visual workload of modern school-aged children is associated with not only reading and writing, but also the use of a wide range of digital devices at close distances, both in

Y. Dekhtyar and I. Saknite (Eds.): NBC 2023, IFMBE Proceedings 89, pp. 153–161, 2023.
https://doi.org/10.1007/978-3-031-37132-5_20

the educational process and for leisure [6–8]. This high near visual effort is associated with an approximately 40% prevalence of myopia in school-aged children worldwide [9, 10]. Furthermore, it is noted that the number of school-aged children with near vision impairment reaches 39% [11]. Since uncorrected refractive errors and near vision problems negatively impact children's academic performance and behaviour [3, 12], there has been a growing discussion in recent years about the importance of assessing near visual functions during vision screening in school-aged children [3, 11, 13].

Technological advances have led to the introduction of computerized screening equipment and programs in eye care practices, which enables a wider range of visual functions to be tested in shorter time [3, 14]. These programs include not only distance visual acuity testing but also screening for near visual function [15], as the first sign of near vision fatigue is an improper functioning of eye's accommodation and vergence system [11, 13]. However, these computerized programs rarely assess full range of accommodation and vergence functions [15, 16]. Vision therapy is the most effective method for treating accommodation and vergence disorders [17]. In recent years, computerized programs have been widely and successfully used to improve accommodative and vergence performance [18–20]. Therefore, our aim was to develop a vision screening and training tool that can identify children with near vision problems, provide treatment, monitor progress, and collect and store data on vision training, while being easy to use by non-specialist.

2 Method

In 2013, our department developed and tested the first computerized prototype and methodology for screening of vision in school-aged children. The manual screening methodology was tested on 11,000 school aged children [21]. The next step was to improve computer-based algorithm (including visual stimuli and additional accessories) was specially designed to identify children with near vision problems. In 2022, our scientific team has upgraded the program and methodology, resulting in the development of a comprehensive vision screening and training tool (see Fig. 1). As a result, we have developed a vision screening and training tool. This tool includes a lens-filter device (LFD) prototype, a computer program, and a methodology that ensures accurate performance of vision tests and training. The LFD synchronizes the generation of visual stimuli with the replacement of lenses or filters (see Fig. 2), while the computer records subject responses, performs data processing, and provides conclusions and recommendations for further monitoring. Our vision screening and training tool is adaptive and can be calibrated to work on a any computer.

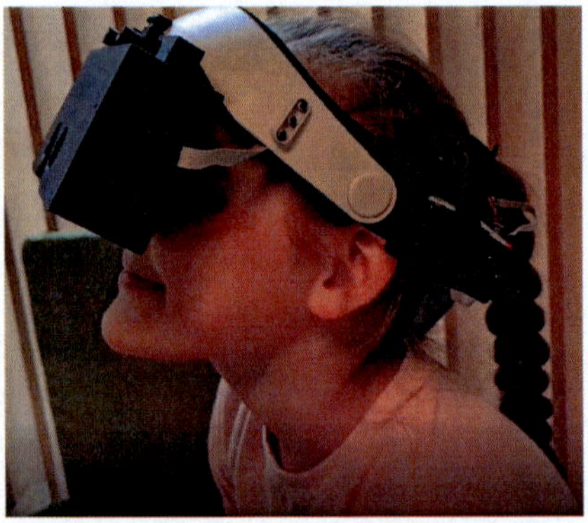

Fig. 1. The head-mounted module of the upgraded vision screening and training prototype.

To evaluate the effectiveness of our prototype vision screening and training tool, we conducted a pilot study in a laboratory setting. Our analysis focused on the repeatability and validity of near vision functions (at 40 cm) including accommodation, heterophoria, stereovision, and vergence performance. We studied a group of 75 subjects (aged 22 ± 5 years), and repeatability was evaluated on three consecutive measurements. Reliability was assessed in comparing our tool's results to those obtained using standard optometric tests. To determine the sensitivity and specificity of the algorithm used by our tool to provide conclusions on the visual system condition, we compared the results to accept norms and recommendations for further monitoring.

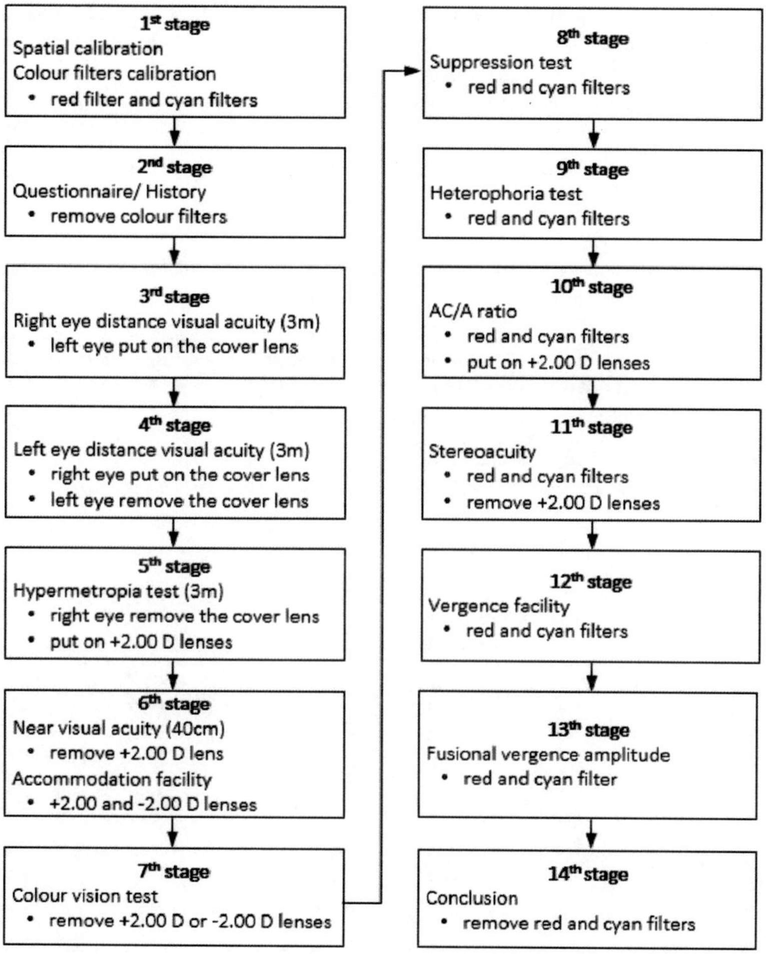

Fig. 2. Stages of operation and sequence of replacement of the lenses and filters of the vision screening tool.

Our developed tool enables users to perform exercises that can improve their accommodation and vergence performance. To test the effectiveness of our vision training tool, we conducted a study 22 subjects with an average age of 22 ± 2 years. Standard optometric tests were used to assess vergence performance before and after training. Data was analysis was carried out using Microsoft Excel Data Analysis ToolPak (version 16.0) and SPSS 28.0 (SPSS Inc., Chicago, IL, USA).

3 Results

3.1 Repeatability and Validity of Vision Screening Tests

Our study found that all computerized tests of near vision functions had good repeatability over three consecutive measurements. In addition, we conducted a Received-Operating Characteristic (ROC) analysis for the developed near tests to determine their sensitivity, specificity, and area under the ROC curve (AUC) ($p < .001$). Table 1 presents the manual test-based norm of each visual function, along with the repeatability and specificity calculated for computerized vision screening using the manual tests norms as reference.

Table 1. Repeatability and validity were assessed for each visual function in the study sample ($n = 75$). Norms for the standard manual optometric tests are cited from Scheiman & Wick [17].

Visual function	Repeatability (ANOVA, *p*-value)	Norms of standard optometric tests (mean ± SD)	Sensitivity (%)	Specificity (%)	AUC
Binocular accommodative facility	.25	10 ± 5 cpm	86	85	.851
Stereoacuity	.32	60 ± 20 arc sec*	37	88	.623
Heterophoria	.76	3 ± 3 pd exophoria	64	77	.703
Vergence facility	.09	15 ± 3 cpm	63	66	.643
Uncrossed fusional vergence amplitude	.87	13 ± 6 pd	100	62	.810
Crossed fusional vergence amplitude	.31	19 ± 9 pd	20	90	.549

* Norm of stereoacuity is cited from Steinman et al. [22].

After analysing the quality of our developed tests using AUC, we have concluded that the uncrossed fusional vergence amplitude and accommodation facility tests have very good accuracy. The heterophoria test has good accuracy, while the stereoacuity and vergence facility tests have medium accuracy. However, the crossed fusional vergence amplitude test showed unsatisfactory accuracy.

3.2 Validity of Vision Screening Algorithm

Based on the varying validity of tests for assessing near vision function, we have developed an algorithm for identifying patients with ocular muscle balance disorders. The algorithm determines whether a patient's results correspond with accepted norms, and recommends a full vision examination if at least of the following conditions is met:

- Decreased accommodative facility
- Heterophoria outside the norm
- Decreased vergence facility and decreased uncrossed or crossed fusional vergence amplitude
- Decreased vergence facility and low stereoacuity
- Low stereoacuity and decreased uncrossed or crossed fusional vergence amplitude

Following the analysis of our findings with a computerised vision screening program, we created our own algorithm to identify cases where a child has failed the vision screening and requires a comprehensive eye examination. Table 1 displays the individual norm for each visual function. In our analysis, we considered the possibility that the child may have failed the screening due to multiple visual functions falling outside the norm. We then evaluated the sensitivity and specificity of the computerised vision screening using manual tests. ROC analysis of our computerized vision screening algorithm shows high sensitivity (91%) and specificity (82%), with a very good accuracy (AUC = .865, $p < .001$). Figure 3 shows the ROC curve of developed vision screening algorithm.

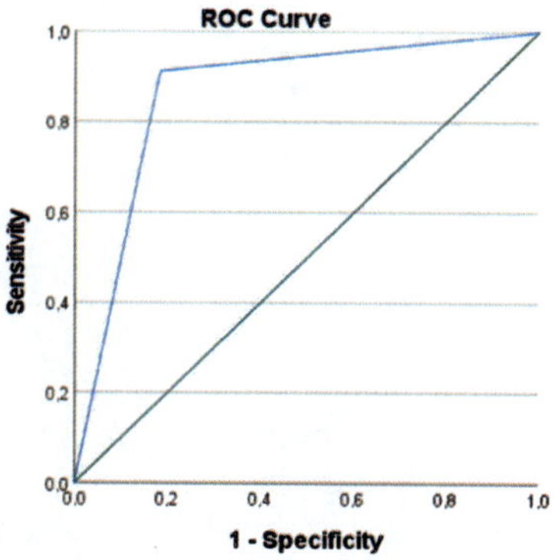

Fig. 3. The ROC curve displays the performance of a vision screening algorithm that compromises several near vision functions. The algorithm is designed to identify patients with suboptimal accommodation and vergence abilities, and it was tested on a sample of 75 subjects. The diagonal segments on the ROC curve corresponds to ties in the data.

3.3 Effectiveness of the Vision Training Tool

Our vision training tool has been shown to effectively improves vergence performance. We have found that subjects with impaired vergence abilities experienced a statistically significant increase in their vergence facility ($M = 18.5$ cycles per minute (cpm), $SD = 2.9$) compared to their pre-treatment state ($M = 15.2$ cpm, $SD = 2.0$), $t\ (21) = 3.74$, $p = .002$. A statistically significant difference was observed in the uncrossed fusional vergence amplitude after treatment (mean = 19.79 pd, standard deviation = 8.33) compared to before treatment (mean = 14.86 pd, standard deviation = 5.46), t(21) = 2.43, p = .030. In addition, the statistically significant increase of crossed fusional vergence amplitude was observed ($M = 34.86$ pd, $SD = 7.15$) compared with the state before treatment ($M = 31.57$ pd, $SD = 8.27$), $t\ (21) = 2.89, p = .013$. Figure 4 displays the mean values of the vergence performance tests before and after vision therapy.

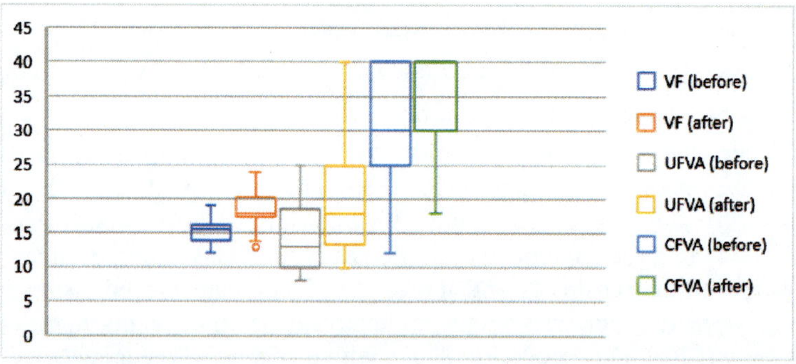

Fig. 4. This study measured the mean values of vergence performance tests before and four weeks after vision therapy, with sample size of n = 022. The results show a statistically significant improvement in all tests of vergence Abbreviations on the chart: VF (before) – vergence facility (cpm) before vision therapy; VF (after) – vergence facility (cpm) after vision therapy; UFVA or CFVA (before) – respectively uncrossed or crossed fusional vergence amplitude (pd) before vision therapy; UFVA or CFVA (after) – respectively uncrossed or crossed fusional vergence amplitude (pd) after vision therapy.

4 Discussion

The widespread use of digital devices among preschool and school-aged children, along with the drastic increase in screen time during the COVID-19 pandemic, has impacted vision habits and the visual system [23]. Research has found that increased screen time and limited outdoor activities [6, 24] are associated with myopia progression [9] and dysfunction in accommodation and vergence [11, 25]. Moreover, vision impairments can negatively affect children's academic performance and behaviour [3, 12]. Vision screening is a crucial tool for assessing visual functions, and recent advancements in digital technology have made computerized programs and automated devices for vision screening widely available [3, 13, 14].

When designing the computer vision screening program, we took into account the resolution of the computer screen and also the reproduction of visual stimuli, as well as the use of colour filters. Unlike manual optometric tests, where real objects are used and only prisms or lenses are used as additional accessories to assess visual function, the previously developed norms cannot be 100% transferred to computer vision screening, knowing that they affect visual parameters, e.g., stereoacuity is worse when assessed by a test using colour filters [26, 27]. Our upgraded vision screening algorithm and training tool provides reliable results and is comparable to optometric tests. The screening tool showed high sensitivity (91%) and specificity (92%) in identifying subjects with accommodative and eye muscle cooperation problems. Compared to the previous version of our computerized vision screening device, we have improved the diagnostic quality of our tool (sensitivity 82%, specificity 77%) [21]. Furthermore, our developed tool is more accurate in detecting visual skills problems than the popular VERA (Visual Efficiency Rating) computerized vision screening protocol (sensitivity 45%, specificity 83%) [4, 16]. Our tool evaluates a variety of visual functions in about 15 min using a lens filter device (LFD) consisting of two disks with lenses and colour filters, connected to a computer and controlled by a computer program. The high repeatability and validity of our vision screening tool are ensured by the LFD and computer program, which synchronize the generation of visual stimuli with the replacement of lenses or filters. Additionally, our instructions and visual information have been designed to make the testing process easy to understand, and our prototype is user-friendly, even for non-vision specialists.

Our tool's innovation lies in its ability to combine vision screening with the option of conducting vision training. Computerized vision training programs have been widely and effectively used for many years [18–20], and our vision training tool can significantly improve vergence performance (vergence facility and fusional vergence amplitudes) in as little as four weeks of training.

This study focused on testing the upgraded prototype, validating the method, and testing the vision training module. As the developed prototype has only been tested on 75 participants and under laboratory conditions, the next steps would include testing the prototype in an appropriate setting, such as a school or a practice of a vision specialist, and accordingly testing it with non-specialists, i.e., assessing whether the screening is easy to perform. It can be used as a standalone training device, but our main goal is to use it to identify near vision functions' disorders, conduct training, and monitor progress through subsequent screenings. The system records and stores answers, processes data, and provides conclusions and recommendations. Our current efforts are focused on establishing standards for computer-based testing methodology and developing a reliable and robust head-mounted tool for use in school settings. This requires rigorous research and testing to ensure that the technology is accurate, efficient, and practical for widespread implementation. In the future, we plan to develop a secure, cloud-based database for vision care specialists to access subject visual function data.

5 Conclusions

In conclusion, our study demonstrates that after four weeks of vision therapy, individuals with impaired vergence abilities showed a statistically significant increase in both their vergence facility and their crossed and uncrossed fusional vergence amplitudes. Our enhanced vision screening and training tool provides reliable results and is comparable to optometric tests. Our screening protocol accurately identifies individuals with vergence issues, and our training program has been shown to effectively vergence performance.

Acknowledgements. The study was supported by the UL and the IDAL Project No. KC-PI-2020/10 and the UL Projects No Y5-AZ77 and No Y9-B003. We would like to express our gratitude for helping with prototype development and data collection to Evita Kassaliete, Karola Panke, Kristine Kalnica-Dorosenko, Zane Jansone-Langina, Renars Truksa, Sergejs Fomins, and Aiga Svede, and anonymous reviewer for their helpful comments.

References

1. Chen, A.-H., Bakar, N.F.A., Arthur, P.: Comparison of the pediatric vision screening program in 18 Countries across five Continents. J. Curr. Ophthalmol. **31**(4), 357–365 (2019)
2. Griffiths, H., Carlton, J., Mazzone, P.: Summary report: Childhood vision screening in Europe. University of Sheffield, United Kingdom, EUScreen (2019)
3. Ali, Q., Heldal, I., Helgesen, C.G., Krumina, G., Costescu, C., Kovari, A., Katona, J., Thill, S.: Current challenges supporting school-aged children with vision problems: a rapid review. Appl. Sci. **11**(20), 9673 (2021)
4. Gallaway, M.S.: The need for better school vision screening: the use of VERA vision screening in a community setting. Optom. Vis. Dev. **41**(4), 232–239 (2010)
5. Lin-Fu, J.S.: Vision screening of children. U.S. Maternal and Child Health Service (1971)
6. Schmidt, S.C.E., Anedda, B., Burchartz, A., Eichsteller, A., Kolb, S., Nigg, C., et al.: Physical activity and screen time of children and adolescents before and during the COVID-19 lockdown in Germany: a natural experiment. Sci. Rep. **10**(1), 21780 (2020)
7. Francis, K., Scholten, H., Granic, I., Lougheed, J., Hollenstein, T.: Insights about screen-use conflict from discussions between mothers and pre-adolescents: a thematic analysis. Int. J. Environ. Res. Public Health **18**(9), 4686 (2021)
8. Przybylski, A.K., Weinstein, N.: Digital screen time limits and young children's psychological well-being: evidence from a population-based study. Child Dev. **90**(1), 56–65 (2019)
9. Matsumura, S., Ching-Yu, C., Saw, S.-M.: Global epidemiology of myopia. In: Ang, M., Wong, T.J. (eds.) Updates on Myopia. A Clinical Perspective, pp. 27–52. Springer Verlag, Singapore (2020). https://doi.org/10.1007/978-981-13-8491-2_2
10. Grzybowski, A., Kanclerz, P., Tsubota, K., Lanca, C., Saw, S.-M.: A review on the epidemiology of Myopia in school children worldwide. BMC Ophthalmol. **20**(1), 27 (2020)
11. Falkenberg, H.K., Langaas, T., Svarverud, E.: Vision status of children aged 7–15 years referred from school vision screening in Norway during 2003–2013: a retrospective study. BMC Ophthalmol. **19**(1), 180 (2019)
12. Mathers, M., Keyes, M., Wright, M.: A review of the evidence on the effectiveness of children's vision screening: effectiveness of children's vision screening. Child Care Health Dev. **36**(6), 756–780 (2010)
13. Metsing, I.T., Hansraj, R., Jacobs, W., Nel, E.W.: Review of school vision screening guidelines. Afr. Vis. Eye Health **77**(1), a444 (2018)

14. Ali, Q., Heldal, I., Helgesen, C.G., Krumina, G., Tvedt, M.N.: Technological supporting vision screening: A protocol for a scoping review. BMJ Open **11**(9), e050819 (2021)
15. Metsing, I.T., Jacobs, W., Hansraj, R.: A review of vision screening methods for children. Afr. Vis. Eye Health **77**(1), a446 (2018)
16. Gallaway, M., Mitchell, G.L.: Validity of the VERA visual skills screening. Optom. J. Am. Optom. Assoc. **81**(11), 571–579 (2010)
17. Scheiman, M., Wick, B.: Clinical Management of Binocular Vision: Heterophoric, Accommodative, and Eye Movement Disorders, 4th edn. Lippincott Williams & Wilkins, Philadelphia (2013)
18. Dusek, W.A., Pierscionek, B.K., McClelland, J.F.: An evaluation of clinical treatment of convergence insufficiency for children with reading difficulties. BMC Ophthalmol. **11**(1), 21 (2011)
19. Nehad, T., Salem, T., Elmohamady, M.N.: Combined office-based vergence therapy and home therapy system for convergence insufficiency in Egyptian children. The Open Ophthalmol. J. **12**, 12–18 (2018)
20. Munsamy, A.J., Paruk, H., Gopichunder, B., Luggya, A., Majola, T., Khulu, S.: The effect of gaming on accommodative and vergence facilities after exposure to virtual reality head-mounted display. J. Optom. **13**(3), 163–170 (2020)
21. Krūmiņa, G., et al.: Skolas Vecuma Bērnu Redzes Skrīnings un tuvuma Redzes Funkciju Novērtēšanas Metode (In English: School-Aged Children Vision Screening and Method for Assessing near Visual Functions). Latvijas Universitāte, Rīga (2013)
22. Steinman, S.B., Steinman, B.A., Garzia, R.P.: Foundations of Binocular Vision: A Clinical Perspective. McGraw-Hill Education, Maidenhead (2000)
23. Rhodes, A.: Screen Time and Kids: What's Happening in Our Homes? Detailed Report. Australian Child Health POLL, Melbourne (2017)
24. Nagata, J.M., et al.: Screen time use among US adolescents during the COVID-19 pandemic: findings from the adolescent brain cognitive development (ABCD) study. JAMA Pediatr. **176**(1), 94–96 (2022)
25. Mohan, A., Sen, P., Shah, C., Datt, K., Jain, E.: Binocular accommodation and vergence dysfunction in children attending online classes during the COVID-19 pandemic: digital eye strain in kids (DESK) study-2. J. Pediatr. Ophthalmol. Strabismus **58**(4), 224–231 (2021)
26. Momeni-Moghadam, H., Kundart, J., Ehsani, M., Gholami, K.: Stereopsis with TNO and Titmus tests in symptomatic and asymptomatic university students. J. Behav. Optom. **23**(2), 35–39 (2011)
27. Garnham, L., Sloper, J.J.: Effect of age on stereoacuity as measured by different types of stereotest. Br. J. Ophthalmol. **90**(1), 91–95 (2006)

Hemodialysis Optical Monitoring Toward Greener Technology: A Potential for Water Saving Dialysis Treatment

Liisi Leis[1], Annika Adoberg[1,2], Joosep Paats[2], Jana Holmar[2], Jürgen Arund[2], Deniss Karai[2], Merike Luman[1,2], Kristjan Pilt[2], Paul Taklaja[2], Risto Tanner[2], and Ivo Fridolin[2(✉)]

[1] North Estonia Medical Centre, J. Sütiste tee 19, 13419 Tallinn, Estonia
[2] Tallinn University of Technology, Ehitajate tee 5, 12616 Tallinn, Estonia
{jana.holmar,ivo.fridolin}@taltech.ee

Abstract. Hemodialysis (HD) treatment of end-stage chronic kidney disease patients is one of the most "resource-intensive" fields of medicine. To mark the significance of the matter, the European Kidney Health Alliance has called for action in connection with "the mutual impact of climate change on kidney health and kidney care on ecology". This study aimed to explore the potential for reducing water consumption during hemodialysis treatments via patient-tailored treatment using optical assessment of the essential markers - uremic toxins' removal from the patients' blood.

Based on the comparison of standard and high flow hemodiafiltration treatments, intradialytic uremic solutes' reduction ratio improved from 9% (urea) up to 16% (indoxyl sulfate) after increasing purified water consumption by 85% on average. Personalizing HD treatment by utilizing the optical dialysis adequacy monitoring feedback could potentially reduce water consumption up to 131 L and save up to 1.22 kWh of electricity per dialysis based on the mean results of the sample of 22 patients. Multiplying it by 5 million dialysis patients predicted for 2030 worldwide receiving three treatments per week, the decrease of the water consumption may be 102.2 million cubic meters and electricity 951.6 million kWh per year.

A specially designed compact and robust on-line dialysis monitoring device for on-line optical monitoring of uremic retention solutes in the spent dialysate offers an opportunity to personalize dialysis through better adaptation of the treatment according to the needs of the patients and considerably reduce the ecological footprint of the treatment.

Keywords: personalized hemodialysis · optics of spent dialysate · uremic toxins · urea · uric acid · β2-microglobulin · indoxyl sulfate · ecological footprint

Y. Dekhtyar and I. Saknite (Eds.): NBC 2023, IFMBE Proceedings 89, pp. 162–171, 2023.
https://doi.org/10.1007/978-3-031-37132-5_21

1 Introduction

Hemodialysis treatment of end-stage kidney disease (ESKD) patients is one of the most "resource-intensive" fields of medicine. Up to 500 L of water, 7 kWh of energy are needed, and more than a kilogram of medical waste is produced during each hemodialysis treatment [1]. Considering that one hemodialysis (HD) patient needs the treatment procedure three times per week (156 per year) and the prediction for HD patients' prevalence in 2030 is 5 million [2], dialysis appears to significantly contribute to global warming while saving patients' lives. To mark the importance of the matter, the European Kidney Health Alliance has called for action in connection with "the mutual impact of climate change on kidney health and kidney care on ecology" [3]. Series of recommendations are proposed to reduce the ecological footprint of hemodialysis treatment [4]. Among these, reducing water consumption is under discussion at the level of everyday treatment practice, considering that the quality requirements for dialysis water are much stricter in comparison with those for drinking water [5], accompanied with the higher costs and energy consumption. This study aimed to explore the potential for reducing water consumption during hemodialysis treatments via patient-tailored treatment using optical assessment of the essential markers - uremic toxins' removal from the patients' blood.

2 Material and Methods

2.1 Patients

In total, 22 ESKD patients were enrolled in the study from the Centre of Nephrology at the North Estonia Medical Centre (NEMC), Tallinn, Estonia (Table 1).

2.2 Treatment Settings

Each patient was observed during three different treatment settings using Fresenius 5008 dialysis machines with water supply from reversed osmosis (RO) machine AquaB-plus 1000, both from Fresenius Medical Care, Bad Homburg v. d. Höfe, Germany. To include large variety of treatment settings, the following dialysis modalities, dialyzers and machine settings were applied (Table 2): (1) hemodiafiltration with standard settings (stHDF), as used routinely for clinical care in NEMC; (2) low flux HD (lfHD) with minimal dialysis settings; (3) high flow hemodiafiltration (hfHDF) dialysis settings presuming maximum elimination of uremic toxins of all three main groups.

2.3 Sampling and Analyses

Sera samples were collected from the patients before their standard treatment. The levels of urea and β2-microglobulin (B2M) in serum and dialysate samples were determined with routine laboratory methods (SYNLAB Eesti OÜ, Tallinn, Estonia), uric acid (UA) and indoxyl sulfate (IS) in both types of samples were determined using high performance liquid chromatography as described earlier [6]. Patients were classified based on the

Table 1. Clinical data of the end stage kidney disease (ESKD) patients. Numerical values are given as a mean ± SD or as a median and interquartile range (Q1–Q3).

Entity of the Data	Specification
Cause of ESKD	Diabetes (4); Hypertension (8); Glomerulonephritis (3); Tubulointerstitial nephritis (3); Renal carcinoma (2); Other (2)
Age (years)	55 ± 17
Gender	M (17), F (=5)
Race, Caucasian (%)	100
BMI, kg/m2[a]	26.8 ± 5.8
BW, kg[a]	81.5 ± 21.3
Ultrafiltration volume, mL	2565 ± 1190
Urinary volume, mL	0 (14 patients) 700 (335–825) (8 patients)
Serum total protein, g/L	62.8 ± 5.5
Hematocrit, %[a]	34.4 (3.5)
Serum calcium, mmol/L[a]	2.25 (0.16)
Serum phosphorus, mmol/L[a]	1.92 (1.63–2.29)
Dialysis access	native fistula (15); graft (7)
Dialysis vintage, months[a]	23 (11–83)
spKt/Vurea[a]	1.47 (1.23–1.67)

[a] Assessed during standard treatment prescribed to the patients

Table 2. Dialysis treatment settings of standard (stHDF) and high flow hemodiafiltration (hfHDF) and low flux hemodialysis (lfHD). The numerical values are given as mean ± SD.

Entity of the Data	stHDF	lfHD	hfHDF
Substitution volume (Vs, L)	21.1 ± 3.1	0 ± 0	25.3 ± 2.8
Dialysis time, min	240	240	240
Blood flow, mL/min (Qb)	300.8 ± 12.7	200 ± 0	364.2 ± 27.1
Dialysate flow, ml/min (Qd)	470.8 ± 105.4	300 ± 0	800.0 ± 0.0
Dialyzer area[a], m^2	2.0 ± 0.2	1.5 ± 0.0	2.2 ± 0.0
eKt/V urea	1.37 ± 0.31	0.94 ± 0.20	1.70 ± 0.35
Number of dialyses (N)	22	22	22

[a] Specification of dialyzers: Standard: FX800 (N = 8), FX1000 (N = 14), Low flux: Lo15 (N = 22), High: FX1000 (N = 22). All dialyzers had polysulfone-based membranes with the following effective membrane area: 1.8 m^2 (Helixone®, FX800), 2.2 m^2 (Helixone®, FX1000), 1.5 m^2 (Amembris®, Xevonta Lo 15)

presence of uremic toxins in their serum and levels of elevated risk for all-cause mortality as the following: IS the reference level 23 μg/mL by Yamamoto et al. [7], and B2M the reference level 32.2 mg/L by Okuno et al. [8]. Spent dialysate samples were collected from the dialysate outlet of the dialysis machine at 7 min after the start of the session and at the end of the session (240 min) during each dialysis session. UV spectra were recorded with the UV-3600 spectrophotometer (Shimadzu, Japan) in the wavelength range of 200–400 nm with the increment of 1 nm using a quartz cuvette with optical path length of 5 mm. Fluorescence spectra were recorded with the spectrofluorometer RF-6000 (Shimadzu, Japan) using the excitation wavelength range of 200–400 nm with the increment of 5 nm and the emission wavelength range of 210–600 nm with the increment of 1 nm. The bandwidths of 5 nm were used in both monochromators and the used quartz cuvette had path length of 4 mm.

2.4 Data Processing

The results of the fluorescence measurement of dialysate samples were corrected considering the sample's self-absorption of the exciting light (the primary inner-filter effect [9, 10]. Optical models were trained based on the optical properties of spent dialysate samples to estimate concentrations of the UA, IS, and B2M in spent dialysate samples. Forward stepwise regression was used to choose up to three optical parameters – UV absorbance in the wavelength range of 200–400 nm and fluorescence in the region of excitation at 220–400 nm and emission at 220–600 nm– to generate regression models with the best coefficients of determination. Concentrations of UA, IS and B2M were determined in dialysate samples taken 7 and 240 min after the start of dialysis based on the trained models and optical data. Removal of toxins during dialysis sessions was described by a reduction ratio % (RR) in dialysate:

$$RR = 100 * \frac{C_0 - C_t}{C_0} (\%)$$

where C_0 is in laboratory or optically estimated concentration of a toxin in sample taken after 7 min from the start the dialysis and C_t in laboratory or optically estimate concentration in the end of the dialysis; to estimate RR of urea optically, UV absorbance values of spent dialysate samples were used instead of concentration to calculate RR.

Relative removal efficiency of uremic toxins and water consumption of different treatment settings were analyzed by comparing stHDF treatment with other treatment settings. The differences between related samples of the same patients were compared via a paired t-test. A p-value of <0.05 was considered significant. The stHDF modality was considered as benchmark in this study and all results of other dialysis modalities were divided by the corresponding stHDF value for comparison. Resource efficiency of different treatment settings was discussed in relation to the different initial levels of toxins in the blood of different patients, considering removal of uremic toxins and the amount of dialysis water consumed.

3 Results

3.1 Assessment of Optical Data

There was generally strong correlation between the laboratory and optically estimated RR values (Pearson correlation coefficient > 0.798). The correlation was higher for UA ($r = 0.998$, $N = 56$), IS ($r = 0.985$, $N = 56$) and urea ($r = 0.954$, $N = 53$), and lower for middle molecule marker B2M ($r = 0.798$, $N = 36$).

Comparison of mean reduction ratios based on laboratory analyses of dialysates and optical data is presented on the Fig. 1.

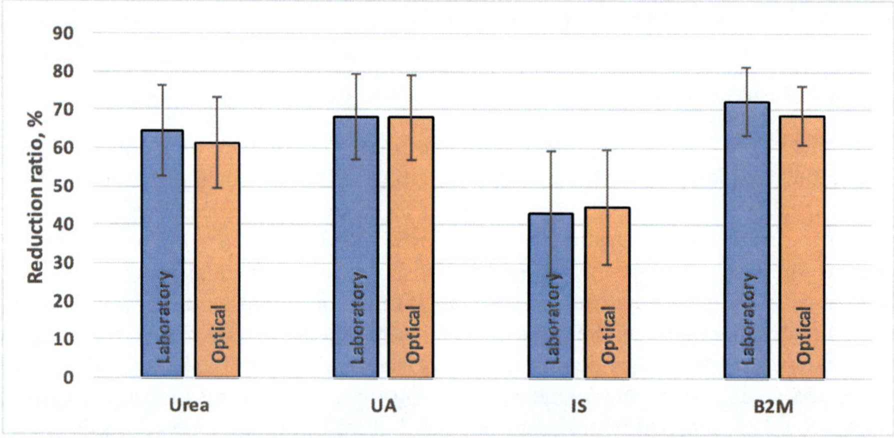

Fig. 1. Comparison of mean (±SD) reduction ratios based on laboratory analyses of dialysates and optical data. UA - uric acid, IS - indoxyl sulfate, B2M - β2-microglobulin.

Corresponding Bland-Altman plots of this comparison presenting statistical accuracy of comparisons are shown on the Fig. 2. It can be seen from the Bland-Altman plots that the random error between laboratory and optically estimated RR values remains similar over the whole range and that the systematic error is small.

3.2 Comparisons of Treatment Settings: Toxins' Removal Efficiency *Versus* Water Consumption

Optically estimated removal efficiencies of toxins and corresponding water consumption for different treatments modalities are presented on the Fig. 3.

Figure 3 shows that hemodialysis treatment with higher blood and dialysate flow rates removes all of the uremic toxins slightly more efficiently, while the water consumption increases significantly. Economy of water consumption in the case of lfHD leads to remarkable reduction of removal of toxins. Corresponding detailed data of comparison with stHDF as reference are presented on in the Table 3.

Removal efficiencies of hfHDF and lfHD were statistically different from corresponding stHDF values for all uremic toxins presented in the Table 3 ($p < 0.05$). Spent

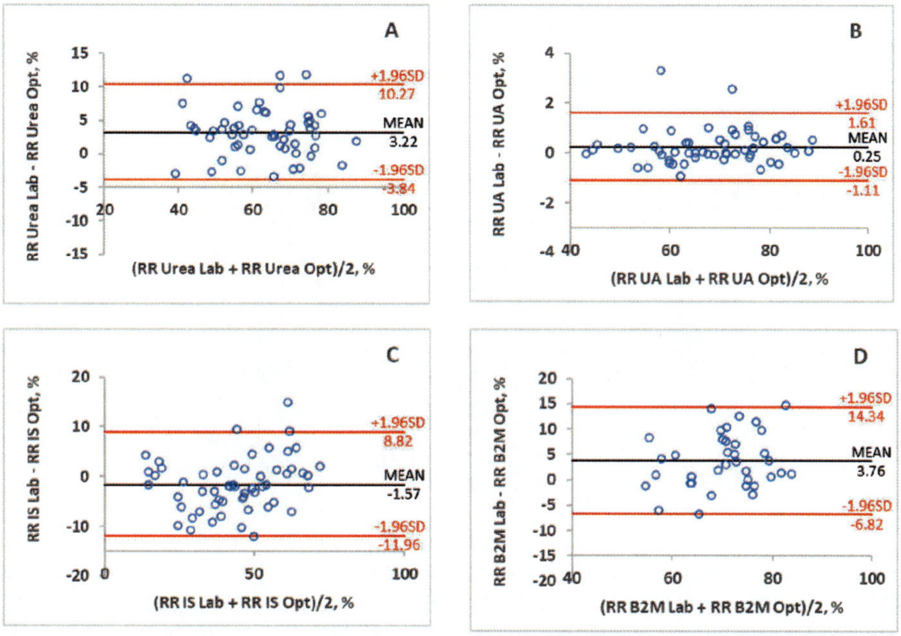

Fig. 2. Bland-Altman plots comparing laboratory (Lab) and optically (Opt) estimated reduction ratios (RRs) of: (A) urea, (B) uric acid (UA), (C) indoxyl sulfate (IS) and (D) β2-microglobulin (B2M).

dialysate based optical measurers indicate a slightly better efficiency of hfHDF modality concerning small molecules urea and UA removal (109.1 and 109.9% compared to 100% of stHDF, respectively) and confirms the benefit of the modality for elimination of IS and B2M (116.4 and 112.6% compared to 100% of stHDF, respectively). However, this improvement of removal efficiencies was achieved with 185 % increase in the consumption of purified (RO) water. This relationship was reversed in the case of lfHD modality of treatment: decreasing RO water consumption down to 69.2% compared to 100% of stHDF reduced elimination of urea and UA down to 76.9 and 80.2%, respectively. Elimination of IS with lfHD reduced nearly equally to the reduction of water consumption in comparison with stHDF. The elimination of B2M was not ratable in the case of lfHD due to low permeability of the low flux membrane for B2M and negligible convective transport.

While classifying the patients according to their toxins' levels, 9 patients had content of IS over the reference level 23 μg/mL, 1 patient had content of B2M over the reference level 32.2 mg/L, and 12 patients had both levels lower than the mentioned levels of elevated risk for all-cause mortality.

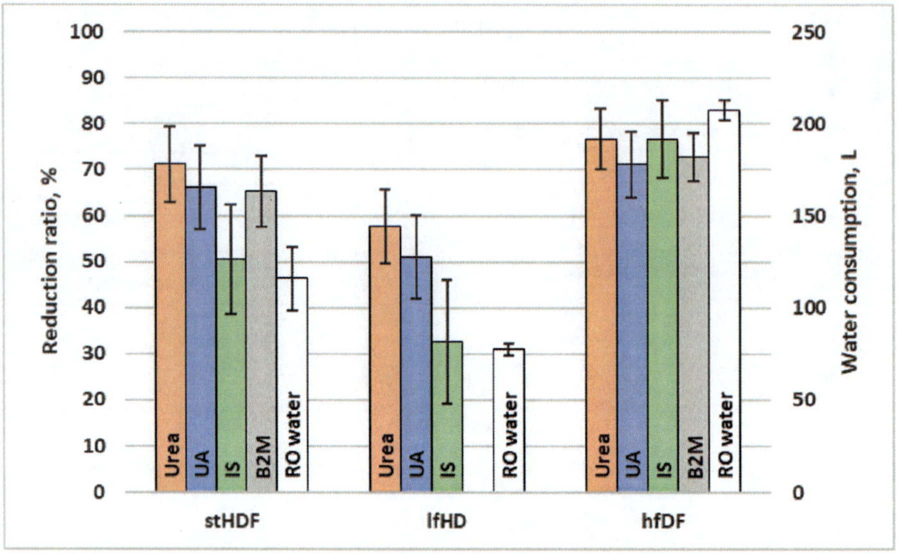

Fig. 3. Reduction ratios of uremic toxins in different dialysis settings and corresponding water consumption (based on optically estimated reduction ratios). Treatment settings: stHDF – standard HDF, lfHD – low flux HD, hfHDF – high flow HDF; see Table 1 for details). UA – uric acid, IS – indoxyl sulfate, B2M – β2-microglobulin, RO – reversed osmosis.

Table 3. Removal ratios of uremic toxins and water consumption during low flux HD and high flow HDF compared with standard HDF treatment.

Group of toxins	Marker compound	Low flux HD (%%)		High flow HDF (%%)	
		Mean	±SD	Mean	±SD
Small water soluble	Urea	76.9	10.6	109.1	7.3
	Uric acid	80.2	8.2	109.9	5.3
Protein bound	Indoxyl sulfate	67.2	28.7	116.4	25.4
Medium molecules	β2-microglogulin	-[a]	-[a]	112.6	8.0
Water consumed		69.2	9.1	185.0	28.0

[a] Not detectable in dialysate

4 Discussion

In this study, including 22 patients, the comparison of laboratory and optical methods revealed that complicated laboratory analyses of selected uremic toxins' markers can be replaced with simple optical measurements in spent dialysate to estimate removal efficiency of toxins with satisfactory results. These results are compatible with earlier findings of Lauri et al. [11] that optical dialysis monitoring, based on UV absorbance and fluorescence in the spent dialysate, can simultaneously provide the removal patterns of marker molecules from all main groups of uremic toxins (urea, B2M and IS) during low-flux HD, high-flux HD and hemodiafiltration treatments without any need for blood sampling. A specially designed compact and robust on-line dialysis monitoring device for on-line optical monitoring of uremic retention solutes in the spent dialysate during dialysis treatments has been proposed to solve the problem of improvement of the quality of dialysis through better adaptation of the treatment to needs of a patient [12–14]. Furthermore, the observations above indicated that the same optical technology has potential to make a substantial contribution also into reduction of the 'ecological footprint' of the treatment at the same time. The results of this work showed that increasing blood and dialysate flow it is possible to achieve substantially better removal of markers of protein bound solutes (IS) and middle molecular toxins (B2M), but not in the degree comparable with the increase of water consumption. As specified in the results section, 9 patients in this study had content of IS over the reference level of elevated risk for all-cause mortality (23 μg/mL by Yamamoto et al. [7]) and 1 patient had content of B2M over the reference level of elevated risk for all-cause mortality (32.2 mg/L by Okuno et al. [8]). Based on the uremic toxins' levels in the blood, these patients evidently need the maximum efficiency of the treatment. However, other 12 patients had levels of both toxins below the levels of elevated risk. For these 12 patients the simple lfHD may be sufficient. By treating those patients with lfHD instead of hfHDF, it could be possible to save up to 116 L of RO purified water per dialysis (as difference between average water consumption of hfHDF and lfHD treatments, Table 3). Taking into account requirements of reversed osmosis purification technology and proportion of the patients with lower risk of mortality (12 of 22 patients), the mean economy of raw (drinking) water and electricity in this group of patients may reach up to 131 L and 1.22 kWh per dialysis, respectively. Potentially, multiplying this water saving by 5 million dialysis patients predicted for 2030 worldwide [2] receiving 3 treatments per week (156 sessions per year), the total economy of environmental loading may be up to 102.2 million of cubic meters less of rough water requirement and 951.6 million of kWh less electricity per year only at the expense of water purification for dialysis. Personalizing the therapeutic prescription is considered to improve a patient's outcome with a focus on reducing the global cardiovascular burden in dialysis patients [15]. Our results that are based on monitoring of optical characteristics in the spent dialysate strongly support this idea and open a chance considerably reduce the 'ecological footprint' of the treatment at the same time.

5 Conclusion

It can be strongly concluded from the results of this work and cited observations, that on-line optical monitoring in the spent dialysate is a promising direction of development of dialysis hardware aimed to reduce 'ecological footprint' of dialysis treatments of ESKD patients.

Acknowledgments. The authors wish to thank all dialysis patients who participated in the experiments and clinical and technical assistance personnel. Research was supported by the Estonian Research Council grant PSG819, by Estonian Centre of Excellence in IT (EXCITE) funded by

European Regional Development Fund, and by the European Commission project 767572 "On-line Dialysis Sensor Phase2" (OLDIAS2).

References

1. Wieliczko, M., Zawierucha, J., Covic, A., Prystacki, T., Marcinkowski, W., Małyszko, J.: Eco-dialysis: fashion or necessity. Int. Urol. Nephrol. **52**, 519–523 (2020). https://doi.org/10.1007/s11255-020-02393-2. Accessed 23 Mar 2023
2. Liyanage, T., et al.: Worldwide access to treatment for end-stage kidney disease: a systematic review. Lancet (London, England) **385**(9981), 1975–1982 (2015). https://doi.org/10.1016/S0140-6736(14)61601-9. Accessed 23 Mar 2023
3. Vanholder, R., et al.: The European Green Deal and nephrology: a call for action by the European Kidney Health Alliance. Nephrol. Dial. Transplant. **38**, 1080–1088 (2022). https://doi.org/10.1093/ndt/gfac160. Accessed 23 Mar 2023
4. Vanholder, R.: Green nephrology. Kidney Dial. **2**, 454–458 (2022). https://doi.org/10.3390/kidneydial203004. Accessed 23 Mar 2023
5. Kasparek, T., Rodriguez, O.E.: What medical directors need to know about dialysis facility water management. CJASN **10**(6), 1061–1071 (2015). https://doi.org/10.2215/CJN.11851214. Accessed 23 Mar 2023
6. Arund, J., Luman, M., Uhlin, F., Tanner, R., Fridolin, I.: Is fluorescence valid to monitor removal of protein bound uremic solutes in dialysis? PLoS ONE **11**(5), 1–12 (2016). https://doi.org/10.1371/journal.pone.0156541
7. Yamamoto, S., et al.: Serum total indoxyl sulfate and clinical outcomes in hemodialysis patients: results from the Japan dialysis outcomes and practice patterns study. Clin. Kidney J. **14**(4), 1236–1243 (2020). https://doi.org/10.1093/ckj/sfaa121. PMID: 33841868; PMCID: PMC8023193. Accessed 23 Mar 2023
8. Okuno, S., et al.: Serum beta2-microglobulin level is a significant predictor of mortality in maintenance haemodialysis patients. Nephrol. Dial. Transplant. Official Publ. Eur. Dial. Transplant Assoc. – Eur. Renal Assoc. **24**(2), 571–577 (2009). https://doi.org/10.1093/ndt/gfn521. Accessed 23 Mar 2023
9. Fonin, A.V., Sulatskaya, A.I., Kuznetsova, I.M., Turoverov, K.K.: Fluorescence of dyes in solutions with high absorbance. Inner filter effect correction. PLoS ONE **9**, e103878 (2014)
10. Wang, T., Zeng, L.-H., Li, D.-L.: A review on the methods for correcting the fluorescence inner-filter effect of fluorescence spectrum. Appl. Spectrosc. Rev. **52**, 883–908 (2017)
11. Lauri, K., et al.: Removal of urea, beta-2-Microglobulin, and Indoxyl Sulfate assessed by absorbance and fluorescence in the spent dialysate during hemodialysis. ASAIO J. **66**, 695–705 (2020). https://doi.org/10.1097/MAT.0000000000001058. Accessed 23 Mar 2023
12. Pilt, K., Arund, J., Adoberg, A., Leis, L., Luman, M., Fridolin, I.: Online urea concentration estimation from spent dialysate using optical sensor. In: Henriques, J., Neves, N., de Carvalho, P. (eds.) MEDICON 2019. IP, vol. 76, pp. 1459–1464. Springer, Cham (2020). https://doi.org/10.1007/978-3-030-31635-8_180
13. Adoberg, A., et al.: TO015 Intradialytic on-line multicomponent total removed solute monitoring in spent dialysate by a novel miniturizwd optical sensor. Nephrol. Dial. Transplant. **35**(Issue Supplement 3), gfaa141.TO015 (2020). https://doi.org/10.1093/ndt/gfaa141.TO015. Accessed 23 Mar 2023

14. Leis, L., et al.: Intradialytic online multicomponent total removed solute monitoring in spent dialysate by a novel miniaturized optical sensor. In: JASN Kidney Week 2020 Abstract Supplement, PO1093 (2020). https://www.asn-online.org/education/kidneyweek/2020/program-abstract.aspx?controlId=3444979. Accessed 23 Mar 2023
15. Canaud, B., Collins, A., Maddux, F.: The renal replacement therapy landscape in 2030: reducing the global cardiovascular burden in dialysis patients. Nephrol. Dial. Transplant. **35**(Suppl. 2), ii51–ii57 (2020). https://doi.org/10.1093/ndt/gfaa005. PMID: 32162663; PMCID: PMC7066547

The Impact of Different Lighting Conditions on the Neural Processes Underlying Relative Depth Perception in 3D Visualization Using Volumetric Multiplanar Display

Mehrdad Naderi(✉), Albina Abdullayeva, Tatjana Pladere, and Gunta Krumina

Department of Optometry and Vision Science, Faculty of Physics, Mathematics and Optometry, University of Latvia, Riga, Latvia

mehrdad.naderi@lu.lv

Abstract. Advancements in technology have led to the emergence of new three-dimensional (3D) visualization techniques that offer users immersive and dynamic displays. However, the impact of these novel image display techniques on user cognitive demands remains largely unexplored. This study aimed to investigate and compare the impact of different lighting conditions, namely photopic and scotopic, on the long-term relative depth judgment task using a volumetric multiplanar display. Electroencephalography (EEG) was employed to record participants' brain activity during the task, and event-related potentials (ERPs) were analyzed using EEGLAB software, specifically focusing on the P3 component. This allowed for a deeper understanding of the neural processes underlying depth perception in varying lighting conditions.

Results showed that, overall, the amplitude of the P3 component was higher in the photopic condition compared to the scotopic condition. Interestingly, the analysis revealed an increase in the amplitude of the P3 component as the task progressed over time, which may indicate the impact of visual fatigue from prolonged use of the volumetric display. The latency of the P3 peak was significantly earlier in the last task under scotopic conditions, while there was no significant difference in the P3 peak latency in the photopic condition between the first and last tasks. This earlier peak in the scotopic condition suggests that using the volumetric multiplanar display in a scotopic environment may improve cognitive processing of depth perception.

Keywords: Electroencephalography (EEG) · event-related potentials (ERPs) · volumetric multiplanar display · depth perception · mental fatigue

1 Introduction

With the ability to create and manipulate virtual objects in a three-dimensional space, this technology has opened new possibilities for a variety of industries such as entertainment, education, healthcare, architecture, design, and oil explorations. In recent years, the advancement of technology has brought about a new era of three-dimensional (3D)

Y. Dekhtyar and I. Saknite (Eds.): NBC 2023, IFMBE Proceedings 89, pp. 172–180, 2023.
https://doi.org/10.1007/978-3-031-37132-5_22

visualization techniques that have enhanced the way we interact with digital media. These techniques offer users immersive and dynamic displays, which are intended to improve their visual experience [1, 2]. Moreover, by developing the image projection, some issues regarding the vergence-accommodation conflict have been relatively resolved. Specifically, by developing the new generation of 3D displays, for instance, in volumetric multiplanar or holographic displays binocular vision is no longer an issue as there is no need for image dissociation to perceive a 3D image on the display [3]. Despite these advancements, the impact of these new image display techniques on human cognitive demands remains largely unexplored. Many previous studies investigated the effect of virtual 3D on the human brain by employing different brain-computer interfaces, for instance, functional magnetic resonance imaging (fMRI) was utilized by Backus [4] to investigate stereo processing in V1 and other areas of the visual cortex. Skrandies [5] examined evoked potential components that are exclusively generated by cortical structures when dynamic random-dot stereograms (dRDS) were presented. Malik [6] evaluated the impact of stereoscopic 3D displays on visual discomfort. Fischmeister [7] investigated slow cortical potentials and source localization to identify the neural correlates of monocular and binocular depth cues. Lastly, Avarvand [8] used EEG, specifically the P1 component of Event-Related Potential (ERP), to study the neurophysiological correlates of vertical disparity in 3D images. However, there are a few studies that have investigated the new display generations and their interaction with the human brain. Pladere [9] studied brain activity during depth perception of stereoscopic and volumetric displays. Naderi [10] studied the user performance objectively by electroencephalography (EEG) analysis when doing a visual task on the volumetric multiplanar display. Furthermore, there is a variety of subjects that scientists are interested in, for example, cognitive load [8], mental fatigue [12, 13], visual attention [14], perception [15], and memory [16] as neural processes that can be affected by using different 3D displays.

To cover this gap, this exploratory study aimed to investigate how different lighting conditions (photopic and scotopic) and the long-term usage of a volumetric multiplanar display affect real 3D image perception. We analyzed EEG signals and extracted ERP waveforms, with a particular focus on the P3 component of ERP, to examine how the human visual system interacts with this new display technology. Based on the physical properties of the volumetric multiplanar display we hypothesized that the cognitive load and perception time was less in scotopic condition with respect to the photopic.

2 Materials and Methods

In this study, we investigate the impact of different lighting conditions on the neural processes underlying relative depth perception in 3D volumetric multiplanar display. The study was approved by the Ethics Committee of the University of Latvia (ZD2019/20807) and was conducted following the Declaration of Helsinki.

2.1 Participants

A total of 18 participants, consisting of 6 males and 12 females, with an average age of 25 ± 6 years, voluntarily took part in the study. Prior to the experiment, we conducted optometric visual tests to ensure their binocular vision function. The participants' inclusion criteria were as follows: normal or corrected-to-normal visual acuity (1.0 or better, decimal units), no binocular dysfunctions, and a stereoscopic acuity of 40 arcsec or better (measured using a Titmus stereo test, Stereo Optical Co., Chicago, IL). Three participants were excluded from the study, with one failing to meet the optometric criteria, one due to noisy brain signals probably electrode failure, and one due to a high rate of mistakes in task performance. Therefore, the final number of participants was 15, consisting of 5 males and 10 females, with an average age of 25 ± 6 years. All participants were unaware of the specific purpose of the study.

2.2 Apparatus

Visual stimuli were presented using a solid-state volumetric display (LightSpace Technologies, model: X1406, 19″ diagonal), which consists of twenty physical image depth planes and temporary image-receiving screens made of short liquid-crystal-based light diffuser elements. These diffusers are synchronized with a high refresh-rate image projection unit and are rapidly switched between highly transparent and light diffusing states to achieve an overall volumetric image refresh rate of 60 Hz. The display has a resolution of 1024 × 768 pixels per image depth plane [17].

2.3 Condition and Procedure

The experiment was conducted under two different lighting conditions. The first condition was a dark room, also known as the scotopic condition, with an illuminance of 1.2 lux. The second condition was a lit room, known as the photopic condition, with an illuminance of 1146 lux.

Each of the independent experiments in this study comprised 160 trials, with the 3D demonstration presented in a pseudo-randomized order in 50% of the trials. The trial began with a fixation cross displayed in the center of the screen for 1 s, followed by the appearance of four rings (with an outer diameter of 0.5° and a line width of 0.1°) located at 1.0° field eccentricity from the center. In the 3D trials, one of the rings (designated as the target) had a different binocular disparity than the other three (distractors), causing it to appear closer to the viewer, as shown in Figs. 1(a) and (b). The participants were instructed to locate the target and indicate its relative position within the display by choosing one of four responses (up, right, down, or left) using the arrow keys on a computer keyboard. After submitting their response (the response time calculated by Windows clock counter), the fixation cross reappeared, and the next trial began. Each experiment was repeated five times, with a short resting period of 1 min between experiments. The total time required to complete the task was approximately 50 min on average, depending on the participants' response times. Participants sat facing the display at a viewing distance of 90 cm.

2.4 EEG/ERP

The Nicolet v32 EEG system was employed to capture the electrical activity of the brain. Twenty-one active electrodes were placed based on the international 10–20 system, and

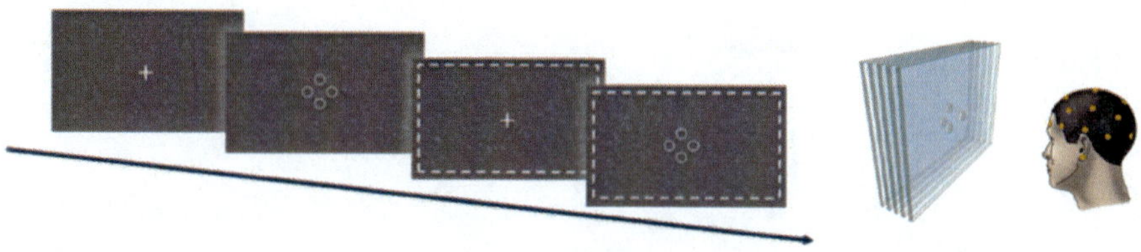

Fig. 1. The experiment design paradigm. Starting with a cross at the center of the display and frequently appearing after each trial for 1 s. 160 trials of rings demonstration. 50% of trials had a 3D effect. The presentation time of rings depended on the response time (a), and schematic illustration of the setup and stimulus on the volumetric multiplanar display (b).

the average of all active electrodes was chosen as a reference. Data collection occurred at a sampling rate of 1024 Hz, with a band-pass filter applied from 1 to 70 Hz. The electrode impedance was kept below 10 KΩ, and data were continuously recorded during the visual search tasks. The open-source toolbox EEGLAB 2022.1.0 connected to MATLAB R2020a (MathWorks Inc., Natick, MA, USA) was used for EEG data analysis. The EEG recordings for each participant were purified using a two-step procedure that involved the built-in software algorithm and visual inspection of variance to reject noise, including baseline, blinking, and muscle activity. The signals were separated into time-locked epochs of 1200 ms duration synchronized with the onset of search arrays, containing 200 ms of prestimulus. Only correct responses were analyzed, and trials with false alarms or misses were excluded. The P3 component's peak amplitude and latency were determined by visually inspecting the ERP waveform within the time frame of 200–600 ms.

3 Results

3.1 Performance Data

Response times (RT) were analyzed to evaluate visual search outcomes. Two-way ANOVA was used to analyze the first and last experiments in photopic and scotopic conditions. A two-way ANOVA revealed that there was not a statistically significant interaction between the effects of lighting conditions and experiment order ($F(1, 26) = 0.016$, $p = 0.9$, $\eta^2_p = 0.0006$). Moreover, there was no statistical difference between the first and last experiments ($F(1, 26) = 1.264$, $p = 0.27$, $\eta^2_p = 0.05$). However, there was a significant difference between photopic and scotopic conditions ($F(1, 26) = 10.25$, $p = 0.003$, $\eta^2_p = 0.28$). Figure 2 shows the average of each group.

3.2 Event-Related Potential (ERP)

We conducted an analysis of the raw EEG signals to extract the peak amplitude and latency of the P3 component of the ERP. The results of ERP analysis are presented in

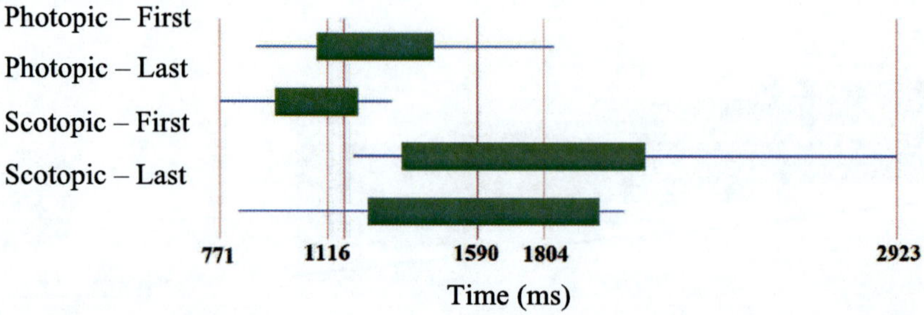

Fig. 2. The average response time of each participant in each group of lighting conditions and experiment order.

Fig. 3, which shows the ERP waveforms average of the occipital and parietal regions (O1, O2, P3, P4, and Pz). As expected, the P3 wave was found to be the dominant wave in these regions.

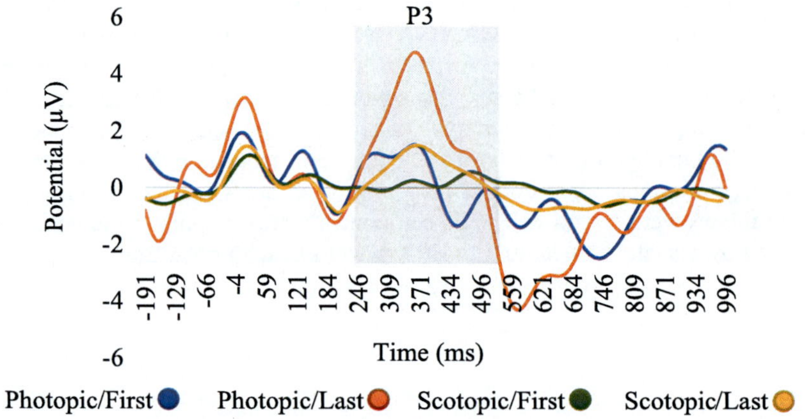

Fig. 3. The average of five occipital and parietal electrodes (O1, O2, P3, P4, and Pz) across all participants. P3 peak highlighted on the waveform.

Besides the waveform, two-way ANOVA showed that there was not a statistically significant interaction between the effects of lighting condition and experiment order (F (1, 36) = 1.571, p = 0.22, $\eta^2_p = 0.042$). However, a statistically significant difference between the first and last experiments (F (1, 36) = 4.42, p = 0.04, $\eta^2_p = 0.11$). Moreover, there was a significant difference between photopic and scotopic conditions (F (1, 36) = 6.23, p = 0.02, $\eta^2_p = 0.15$). Figure 4 shows the peak amplitude of each group.

Moreover, about the peak latency, two-way ANOVA indicated that there was not a statistically significant interaction between the effects of lighting condition and experiment order (F (1, 36) = 2.585, p = 0.12, $\eta^2_p = 0.07$). Furthermore, no significant difference between the first and last experiments (F (1, 36) = 1.17, p = 0.28, $\eta^2_p =$

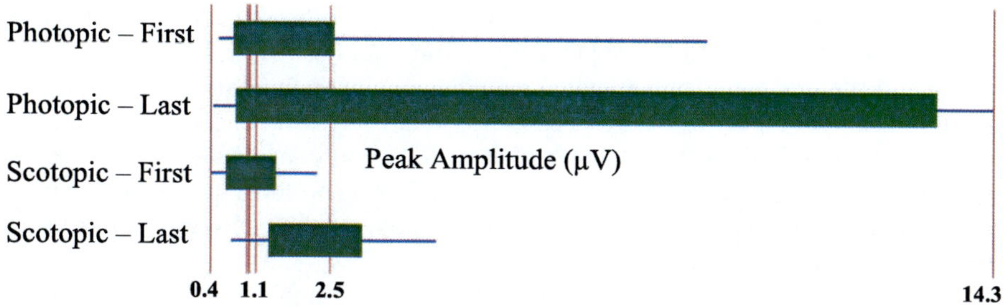

Fig. 4. The peak amplitude average of five occipital and parietal electrodes (O1, O2, P3, P4, and Pz) in different lighting conditions and experiment order.

0.03), and no significant difference between photopic and scotopic conditions (F (1, 36) = 1.310, p = 0.25, $\eta^2{}_p = 0.03$).

However, by paired analysis of each lighting condition between the first and last experiment, the paired-t test indicated that there is a significantly large difference between the First (M = 403.5, SD = 75.4) and Last (M = 358.8, SD = 62.4), t(14) = 3.3, p = .007 experiment in scotopic condition nevertheless, results of the paired-t test in the photopic condition indicated that there is a non-significant small difference between First (M = 346.5, SD = 50.8) and Last (M = 368.4, SD = 40.7), t(14) = 1, p = .367 experiment. Figure 5 shows the peak latency of each group.

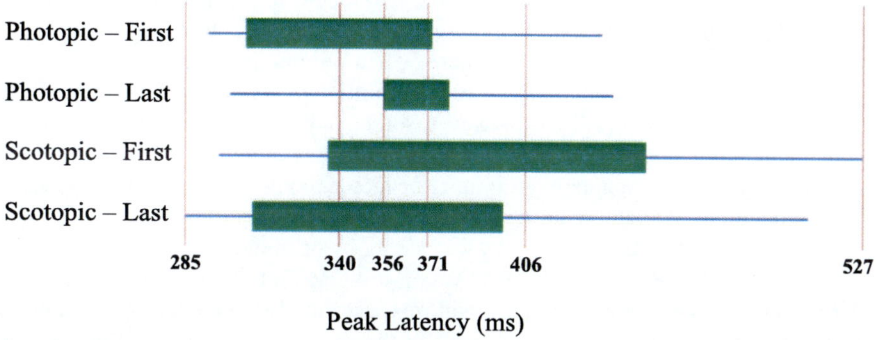

Fig. 5. Peak latency average of five occipital and parietal electrodes (O1, O2, P3, P4, and Pz) in different lighting conditions and experiment order.

4 Discussion

The interaction of the newly developed displays with the human visual system is the point of motivation we conducted a study. The objective of this study was to investigate the effects of different lighting conditions and the long-term usage of a volumetric multiplanar display on human neural processes, specifically by analyzing the amplitude and latency of the P3 component. Generally, lighting condition as a physical factor has a strong effect on human visual perception. Some studies research the illumination on ERP and performance, for instance, Lim [18] concluded that 800 lux illumination excited higher P3 amplitude incorrect response than incorrect. Johannes [19] showed that higher stimulus luminance resulted in higher ERP amplitudes. Moreover, a variety of researchers studied mental fatigue and visual fatigue during prolonged usage of displays. Visual fatigue is caused by the vergence-accommodation conflicts that arise when using conventional 3D displays [20, 21].

In this study, we analyzed the performance level of doing a visual search task on the volumetric display in two different illuminations (photopic and scotopic) conditions. Response time was significantly lower in photopic rather than scotopic, however not significant between the first and last tasks. There could be two points of view; first, by considering each lighting condition separately, there was a shorter response time in the last task than the first in both photopic and scotopic conditions could be regarding the neural facilitation and adaptation to the task due to working memory. However, the results are the opposite of [22] that mentioned prolonged tasks increased the response time. Perhaps motivation is another factor that can affect the result in a positive way. Second, the photopic experiment was performed after the scotopic condition therefore, the significantly shorter response time in the photopic condition could be due to learning the task and enhanced visual working memory [16].

The P3 is an ERP component that appears as a late positive wave at parieto-occipital sites. The identification and assessment of stimuli are frequently associated with the P3 [22, 23]. In this study peak amplitude and latency of P3 were investigated. Significantly higher amplitude in the last task than the first task could be associated with a degree of higher engagement of attention to distinguish the relative depth effect of the visual task [24], but not visual fatigue since there was a shorter response time in the last task. Although many studies showed a lower amplitude of P3 during the repeated task associated with adaptation [22, 23], however, some studies reported that P3 amplitude becomes larger for familiar attended or learned items [25–27].

The latency of components is an important factor to consider in ERP analysis, as higher latency may indicate visual or mental fatigue, or difficulty in decision-making [22]. In this study, while the peak latency was earlier in the photopic condition compared to the scotopic condition, the difference was not significant. This may be due to the working memory, as the photopic experiment was performed after the scotopic condition. Furthermore, there was no difference between the first and last tasks in the photopic condition [16]. However, there was a significantly earlier peak latency in the scotopic condition between the first and last tasks. Lower latency may be associated with neural facilitation in perception, attention, and working memory. Therefore, our results are in line with [16] and suggest that using the volumetric multiplanar display in the scotopic condition can facilitate the learning process.

5 Conclusion

The study aimed to examine the effects of varying lighting conditions and prolonged usage of a volumetric multiplanar display on neural processes. Our results indicate that while response time was faster in photopic conditions, the earlier P3 latency observed in the scotopic condition during the final experiment suggests greater neural facilitation in cognitive processing for detecting depth effects. Additionally, there was no earlier P3 in the photopic condition during the final experiment. Therefore, we conclude that using a volumetric multiplanar display in scotopic conditions is superior to using it in photopic conditions.

Acknowledgment. The work was funded by the Latvian Council of Science ("Development of guidelines for evaluating the visual effects and ergonomics of innovative 3D displays" No. lzp-2021/1-0399).

The authors thank Dr. Roberts Zabels (Lightspace technologies) for creating Fig. 1(b).

References

1. Hong, J., et al.: Three-dimensional display technologies of recent interest: principles, status, and issues. Appl. Opt. (50), 87–115 (2011)
2. Fattal, D., et al.: A multi-directional backlight for a wide-angle, glasses-free three-dimensional display. Nature **495**(7441), 348–351 (2013)
3. Takano, K., Sato, K., Ohki, M.: Improved scattering screen for a multiplanar volumetric holographic display. Opt. Eng. **50**(9), 091315 (2011)
4. Backus, B.T., Fleet, D.J., Parker, A.J., Heeger, D.J.: Human cortical activity correlates with stereoscopic depth perception. Neurophysiology **86**(4), 2054–2068 (2001)
5. Skrandies, W.: The processing of stereoscopic information in human visual cortex: psychophysical and electrophysiological evidence. Clin. EEG Neurosci. **32**(3), 152–159 (2001)
6. Malik, A.S., et al.: EEG based evaluation of stereoscopic 3D displays for viewer discomfort, Biomed. Eng. Online **14**(1), 1–21 (2015)
7. Fischmeister, F.P.S., Bauer, H.: Neural correlates of monocular and binocular depth cues based on natural images: a LORETA analysis. Vision Res. **46**(20), 3373–3380 (2006)
8. Avarvand, F.S., et al.: Objective quality assessment of stereoscopic images with vertical disparity using EEG. J. Neural Eng. **14**(4), 1–22 (2017)
9. Pladere, T., Naderi, M., Zabels, R., Osmanis, K., Krumina, G.: Comparative assessment of brain activity during depth perception of stereoscopic and volumetric images. Light Nat. VIII **11481**, 43–49 (2020)
10. Naderi, M., Pladere, T., Krumina, G.: EEG based assessment of user performance for a volumetric multiplanar display. Proc. SPIE Int. Soc. Opt. Eng. **11350**, 65–71 (2020)
11. Dan, A., Reiner, M.: EEG-based cognitive load of processing events in 3D virtual worlds is lower than processing events in 2D displays. Int. J. Psychophysiol. **122**, 75–84 (2017)
12. Chen, C., Li, K., Wu, Q., Wang, H., Qian, Z., Sudlow, G.: EEG-based detection and evaluation of fatigue caused by watching 3DTV. Displays **34**(2), 81–88 (2013)
13. Cheng, C.-M., Lee, S.-Y., Shu, H.-Y.: Electroencephalographic study of mental fatigue in visual display terminal tasks. J. Med. Biol. Eng. **27**(3), 124–131 (2007)
14. Wróbel, A.: Beta activity: a carrier for visual attention. Acta Neurobiol. Exp. **60**(2), 247–260 (2000)
15. Pegna, A.J., Darque, A., Roberts, M.V., Leek, E.C.: Effects of stereoscopic disparity on early ERP components during classification of three-dimensional objects. Q. J. Exp. Psychol. **71**(6), 1419–1430 (2018)
16. Phillips, S., Takeda, Y.: An EEG/ERP study of efficient versus inefficient visual search. Proc. Ann. Meet. Cogn. Sci. Soc. **31**(31), 383–388 (2009)

17. Osmanis, K., et al.: Advanced multiplanar volumetric 3D display. Emerg. Liquid Crystal Technol. XIII **1055510**, 116–128 (2018)
18. Lim, H.K.: Effects of illuminating condition on ERP and work performance during a counting task. J. Korean Soc. Saf. **15**(1), 167–175 (2000)
19. Johannes, G., Münte, S., Heinze, T.F., Mangun, H.J.: Luminance and spatial attention effects on early visual processing. Cogn. Brain Res. **2**(3), 189–205 (1995)
20. Hoffman, D.M., Girshick, A.R., Banks, M.S.: Vergence – accommodation conflicts hinder visual performance and cause visual fatigue. J. Vis. **8**(3), 33 (2008)
21. Zou, B., Liu, Y., Guo, M., Wang, Y.: EEG-based assessment of stereoscopic 3D visual fatigue caused by vergence-accommodation conflict. J. Display Technol. **11**(12), 1076–1083 (2015)
22. Möckel, T., Beste, C., Wascher, E.: The effects of time on task in response selection - an ERP study of mental fatigue. Sci. Rep. **5**, 1–9 (2015)
23. Haider, A., Fazel-Rezai, R.: Application of P300 event-related potential in brain-computer interface. Event-Related Potentials Evoked Potentials **1**, 19–36 (2017)
24. Heinze, H.J., Luck, S.J., Mangun, G.R., Hillyard, S.A.: Visual event-related potentials index focused attention within bilateral stimulus arrays. I. Evidence for early selection. Electroencephalogr. Clin. Neurophysiol. **75**(6), 511–527 (1990)
25. Noldy, N., Stelmack, R., Campbell, K.: Event-related potentials and recognition memory for pictures and words: the effects of intentional and incidental learning. Psychophysiology **27**, 417–428 (1990)
26. Fabiani, M., Karis, D., Donchin, E.: Effects of mnemonic strategy manipulation in a Von Restorff paradigm. Electroencephalogr. Clin. Neurophysiol. **75**(2), 22–35 (1990)
27. Fabiani, M., Karis, D., Donchin, E.: P300 and recall in an incidental memory paradigm. Psychophysiology **23**(3), 298–308 (1986)

Influence of Acute Mental Stress on the Forehead Photoplethysmographic Signal Waveform

Kristjan Pilt[1(✉)], Deniss Karai[1], Maie Bachmann[1], Marietta Gavriljuk[1], and Ivo Fridolin[1,2]

[1] Tallinn University of Technology, Ehitajate tee 5, 19086 Tallinn, Estonia
kristjan.pilt@taltech.ee

[2] FinEst Center for Smart Cities (Finest Centre), Tallinn University of Technology, Ehitajate tee 5, 19086 Tallinn, Estonia

Abstract. The decreased environment temperature can cause vasoconstriction and drops the skin perfusion, which lowers finger photoplethysmographic (PPG) signal amplitude and signal to noise ratio. The forehead is relatively insulated from external temperature changes due to the presence of the skull and the scalp. Therefore, this site is less affected by temperature compared to other peripheral PPG signal registration sites.

The aim of this study was to characterize the changes in the forehead PPG and second derivative of PPG (SDPPG) waveform parameters related to arterial stiffness for the acute mental stress assessment. The PPG signals were recorded from 42 subjects during eyes open and arithmetic stress test. The signals from 34 subjects (18 females and 16 males) were included to the post-processing and analysis of PPG waveform arterial stiffness related parameters. The changes in the forehead PPG waveform parameters *PPGb*, *PPGAI*, S_{norm}, and SDPPG waveform parameters *b*/*a*, *c*/*a*, *d*/*a*, *e*/*a*, and *AGI* were statistically significantly different ($p < 0.05$) between eyes open and stress test states. The calculated parameters indicated the increase in arterial stiffness due to the induced stress. However, *c*/*a* changed in the opposite direction than expected. In summary, the results indicate that the changes in the forehead PPG and SDPPG parameters could be used similarly to the finger PPG signal for the stress assessment. Nevertheless, further studies are needed.

Keywords: Mental stress · Photoplethysmography · Second derivative PPG signal · Waveform analysis

1 Introduction

Social and medical problems related to mental stress are increasing globally and impact mental health and well-being of people [1, 2]. The acute mental stress assessment can be seen important in everyday life to apply relaxation maneuvers and prevent mental health issues such as anxiety and depression [3].

Physiological signals could provide an objective assessment of mental stress [4]. Brain activates the sympathetic nervous system, which causes an increase in heart rate,

Y. Dekhtyar and I. Saknite (Eds.): NBC 2023, IFMBE Proceedings 89, pp. 181–188, 2023.
https://doi.org/10.1007/978-3-031-37132-5_23

blood pressure, and the release of stress hormones, which can lead to changes in blood flow and the constriction of blood vessels. In addition, it has been found that arterial stiffness increases during the acute mental stress [5]. Therefore, the arterial stiffness related parameters could contribute to the model of mental stress estimation and increase its accuracy.

The photoplethysmographic (PPG) method, which is a relatively simple optical method for measuring relative blood volume changes in the artery and microvascular tissue, has been investigated for the estimation of the acute mental stress using the finger registered pulse waveform indices [6]. It has been found that the induced mental stress affects the arterial stiffness related parameters of finger PPG waveform [7]. However, the decreased environment temperature can cause vasoconstriction and drops the skin perfusion, which lowers finger PPG signal amplitude and signal to noise ratio [8].

Even though the PPG signal is often registered from the finger, as it consists of a large volume of pulsating microvascular tissue, it can be registered from the other sites of the body as well using either transmittance or reflectance mode sensor [9]. The forehead is relatively insulated from external temperature changes due to the presence of the skull and the scalp. Therefore, this site is less affected by temperature compared to other peripheral PPG signal registration sites [10]. However, the morphology of the PPG waveform diverges from the finger recorded signal as the pulse wave propagation path from the heart and the microvascular bed of tissue of the signal registration site have different mechanical and optical properties. Several studies has been carried out on the analysis of forehead PPG signal waveform for the arterial stiffness estimation [11]. It was found that the changes in the forehead vascular bed, which are caused by the stiffening of the blood vessels, can be described with the second derivative of PPG (SDPPG) signal normalized amplitudes b/a and d/a.

Therefore, the aim of this study was to characterize the changes in the forehead PPG signal, its pulsatile waveform, and SDPPG signal parameters related to arterial stiffness for the acute mental stress assessment.

2 Methods

2.1 Experiment Setup

The forehead PPG signal was registered using TT-50 multisignal recorder (Tensiotrace, Estonia) with the synchronous electrocardiographic (ECG) signal registration. The ECG signal was recorded for signal processing purposes to detect the starting points of the cardiac cycles. Both signals were digitized with a sampling rate of 500 Hz. The signals were monitored online during the experiments using a Bluetooth connection.

The forehead PPG sensor was based on the AFE4490 (Texas Instruments, USA) fully integrated analogue front-end chip. The four infrared LEDs with wavelength of 865 nm (DNK1111C, Stanley Electric, Japan) were located circularly around the photodiode BPW34 (Osram, Germany). The distance between the center of each LED and the photodiode was 8 mm. The optical components as well as the AFE4490 with peripheral components were placed in the shielded enclosure.

The PPG sensor was placed inside the cavity of the silicone holder to ensure evenly flat surface attached to the forehead. The silicone sensor holder was attached to the forehead using a wide elastic rubber band.

The blood pressure of the subject was measured using Omron M3 - HEM-7131-E (Omron, Japan) device. The cuff of the device was place to the upper arm of the dominant hand.

2.2 Subjects and Experiment Protocol

The experiments were carried out as a part of the larger project, where the citizens of a city were asked to assess the well-being of the urban environment. Therefore, the health status of the recruited subjects of this experiment was heterogenic. However, the subjects did not have any diagnosed cardiovascular (CV) diseases neither they did not take any CV system related medication. The age of the subjects had to be over 18 years. The study was approved by the Research Ethics Committee of the National Institute for Health Development with the decision no. 1064 (study no. 2351) and the research was conducted in accordance with the Helsinki Declaration.

The experiments took place in a ventilated quiet room, with white walls and natural light from the windows. The subject was first introduced for the experiment and a written informed consent was taken. Thereafter, the subject was instructed for the stress test. In this study, the modified arithmetic stress test (Series seven) was used. The subject was asked to subtract the series of sevens from the given number between 90 and 100 and the result of each subtraction had to be indicated with a pencil on a scale between 0 to 9. In case of a mistake the subject continued with the task. The subtractions started again from the given initial number in case the result was below seven.

The weight of the subject was measured and the ECG electrodes were taped to the chest. Next, the subject was asked to take a sitting position on a comfortable chair and direct the gaze to a white wall at 2 m. After 5 min, the blood pressure of the subject was measured and the PPG sensor was attached to the forehead. In addition, the galvanic skin response electrodes were attached to the ring and middle finger, the electroencephalograph (EEG) electrodes were placed to the forehead to both sides of the PPG sensor, and a temperature sensor was taped to the cheek. All the signals were recorded synchronously, however, in the current publication, only the forehead PPG signal analysis results are reported.

The recording of the signals was started after the attachment of all sensors and the subject was asked firstly to close their eyes for 3 min. Thereafter, the subject was asked to open their eyes and after 4 min the stress test started, which lasted for 2 min.

2.3 Signal Processing and Data Analysis

PPG and ECG signals were post-processed in MATLAB (MathWorks, USA). Firstly, the 1.5 min long segments of the eyes open and stress test were separated from the PPG and ECG signals for further processing. The R-peaks were detected from ECG signal. The Pan-Tompkins algorithm was used for the R-peak detection.

As follows, the arterial stiffness related parameters from the PPG and SDPPG signals were calculated using previously developed and published algorithms [12, 13]. Briefly, the recorded raw PPG signal is inverted and filtered with window method designed (Hamming window function) low- and high-pass filters with the cut-off frequencies of 30 Hz (order of 250) and 0.5 Hz (order of 2000), respectively. Thereafter, the signal is resampled for each heart cycle in a way that the length of the selected heart cycle is 1 s. The signal is filtered with low-pass Parks-McClellan filter with an edge frequency of 6 Hz and transition band of 1 Hz (pass and stop band maximum allowable errors are 0.001), respectively. As a result, the harmonic components of each selected heart cycle is limited to 6 harmonic components. All the resampled and filtered 1 s long PPG waveforms are aligned according to the 50% of the ascending front and the average waveform is calculated (Fig. 1a). Next, the first, second, and fourth derivatives of the averaged PPG waveform are calculated. The distinct wave peaks 'a', 'b', 'c', 'd', and 'e' are detected from the SDPPG waveform using the fourth derivative waveform. The amplitude ratios of the SDPPG waveform peaks *b*/*a*, *c*/*a*, *d*/*a*, and *e*/*a* are calculated. The aging index (AGI) is calculated based on the SDPPG waveform amplitudes:

$$AGI = \frac{b - c - d - e}{a}. \quad (1)$$

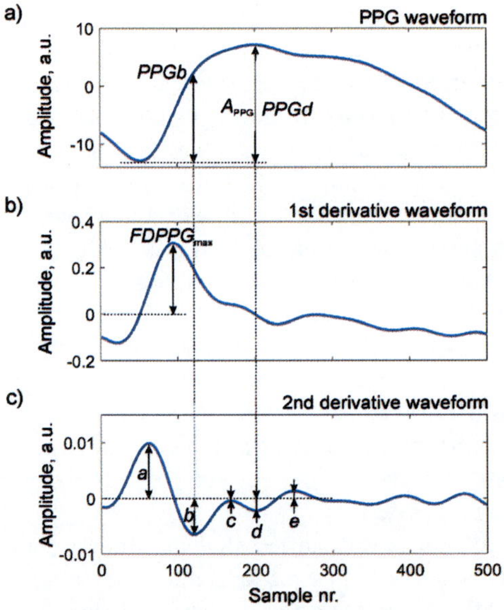

Fig. 1. One period long and averaged a) PPG waveform, b) first derivative PPG waveform, and c) 2nd derivative PPG waveform with the distinct wave peaks and the amplitudes used in calculation of arterial stiffness related parameters.

The amplitudes *PPGb* and *PPGd* are calculated from the PPG signal waveform based on the locations of the wave peaks 'b' and 'd' (Fig. 1a and c). The *PPGAI* is calculated as the amplitude ratio of the amplitudes *PPGb* and *PPGd*:

$$PPGAI = \frac{PPGd}{PPGb}. \quad (2)$$

The slope of the ascending front of the PPG waveform (S_{norm}) is calculated based on the first derivative maximal point ($FDPPG_{max}$) and it is normalized with the amplitude of the PPG signal averaged waveform (A_{PPG}) as follows (see Fig. 1a and b):

$$S_{norm} = \frac{FDPPG_{max}}{0.002 \cdot A_{PPG}}. \tag{3}$$

The 0.002 s corresponds to the time between two samples of the PPG signal.

Firstly, the normal distribution of the data was tested using Anderson-Darling test. Thereafter, the statistical differences were investigated between the parameters from the eyes open and stress signal segments. The nonparametric Wilcoxon signed-rank sum test was used and the $p < 0.05$ was considered as statistically significant difference. The mean and standard deviation were used in case of a normally distributed dataset, otherwise, median and quartiles were calculated.

3 Results

The study was carried out on 42 subjects, however, the signals from 34 subjects were included in the analysis. Some of the signals were excluded due to the poor ECG electrode connections to the hairy chest, a large number of artefacts in the signals during stress test or detected arrhythmia. The general demographic, anthropometric and physiological parameters of the study group are given in Table 1.

The low- and high-pass filtered PPG waveforms during the eyes open and stress test are given in Fig. 2. The heart rate during the stress test is higher as well as the differences between waveforms can be noticed.

Table 1. The basic characteristics of the study group.

	Mean ± SD	Min	Max
Number of subjects	34		
Female	18		
Male	16		
Age, years	41 ± 15.8	18	70
Height, cm	174 ± 8.7	156	192
Weight, kg	80 ± 16.9	52	118
BMI, kg/m^2	26.2 ± 5.41	19.5	39.3
SBP, mmHg	133 ± 16.9	101	172
DBP, mmHg	86 ± 10.0	67	105

Figure 3 shows the box plots with outliers for each calculated parameter during the eyes open and stress test. The *PPGb*, *PPGAI*, *b*/*a*, *c*/*a*, *d*/*a*, *e*/*a*, *AGI*, and S_{norm} were found statistically significantly different between the two states for whole study group ($p < 0.05$).

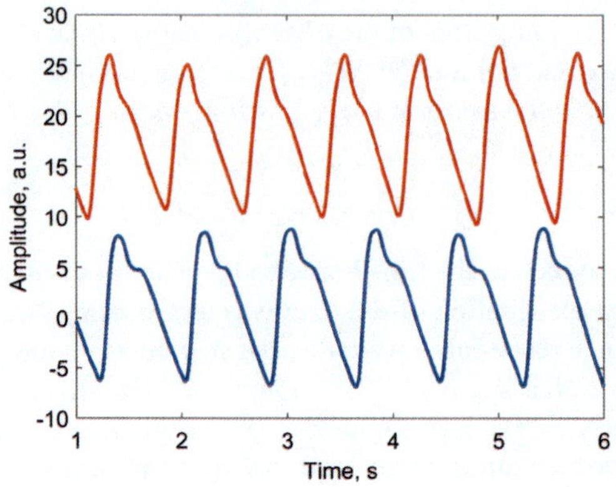

Fig. 2. The low- and high-pass filtered PPG signal segments of 43-year old male subject during eyes open (blue line) and stress test (red line). The PPG signals are offset from each other for illustrative purposes.

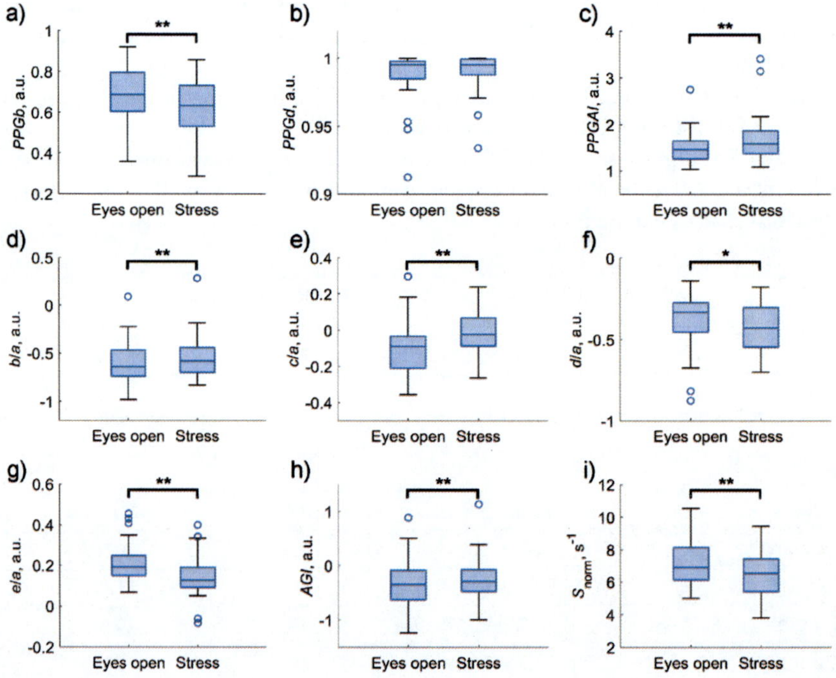

Fig. 3. Box plots with outliers of the arterial stiffness related parameters: a) *PPGb*, b) *PPGd*, c) *PPGAI*, d) *b/a*, e) *c/a*, f) *d/a*, g) *e/a*, h) *AGI*, and i) S_{norm}. The * indicates $p < 0.01$ and ** $p < 0.001$.

4 Discussion

According to the best knowledge of the authors, this is the first attempt to investigate the forehead PPG signal and arterial stiffness related waveform parameters for the stress assessment. The changes in the forehead PPG waveform parameters *PPGb*, *PPGAI*, S_{norm}, and SDPPG signal parameters *b/a*, *c/a*, *d/a*, and *AGI* were found statistically significantly different between the eyes open and the stress test states. The results indicate that the forehead could be considered as an alternative PPG signal registration location for the stress assessment purposes.

It has to be noted that the changes in the parameters due to the induced mental stress were not in the same direction for all subjects. The arithmetic stress test may not induce the stress response equally for the subjects and depends on several factors. The physiology, as well as the previous experience of the subject to handle the stress, are some of the causes.

Compared to the finger, the forehead is a less investigated region for the PPG signal waveform analysis. The forehead measurement location has skull, which is covered with a very thin soft tissue layer of the scalp. Therefore, the visible light rather reflects back from the surface of the skull and the longer wavelengths from the infrared region should penetrate to the deeper layers of the skull and through it. However, in this study the photon propagation at the forehead measurement location was not modelled. Nevertheless, the measurement location for each subject was selected according to the same procedure as described in the methodology section.

The PPG waveform varies between the forehead and finger due to the measurement location anatomical differences. The arterial stiffness related waveform parameters and the parameters calculated from the finger PPG signal behave in the similar manner due to the induced stress [7]. The induced stress increases the arterial stiffness according to the calculated parameters S_{norm}, *PPGb*, *PPGAI*, *b/a*, *d/a*, *e/a*, and *AGI*. From previous studies, it has been found that the aging of the subject decreases the SDPPG parameter *c/a* in case of the finger [14] and forehead [11] PPG waveform, which is expected due to the stiffening of the arteries. However, in this study, it was found that the *c/a* increased due to the stress, which was unexpected and needs further studies.

5 Conclusions

The results indicate that the changes in the forehead PPG and SDPPG parameters could be used similarly to the finger PPG signal for the stress assessment. Therefore, the forehead could be considered as an alternative location for the PPG signal registration for the stress assessment purposes. However, further studies are needed.

Acknowledgements. This work was supported partly by the Estonian Ministry of Research and Education and European Regional Development Fund (grant 2014-2020.4.01.20-0289) and by the Estonian Research Council grant (PSG819).

References

1. Steel, Z., et al.: The global prevalence of common mental disorders: a systematic review and meta-analysis 1980–2013. Int. J. Epidemiol. **43**(2), 476–493 (2014)
2. World Health Organization: WHO Mental health atlas 2014, p. 69. WHO, Geneva (2015)
3. Hamdani, S.U., et al.: Effectiveness of relaxation techniques 'as an active ingredient of psychological interventions' to reduce distress, anxiety and depression in adolescents: a systematic review and meta-analysis. Int. J. Ment. Health Syst. **16**(1), 31 (2022)
4. Arza, A., et al.: Measuring acute stress response through physiological signals: towards a quantitative assessment of stress. Med. Biol. Eng. Compu. **57**(1), 271–287 (2018). https://doi.org/10.1007/s11517-018-1879-z
5. Vlachopoulos, C., Kosmopoulou, F., Alexopoulos, N., Ioakeimidis, N., Siasos, G., Stefanadis, C.: Acute mental stress has a prolonged unfavorable effect on arterial stiffness and wave reflections. Psychosom. Med. **68**(2), 231–237 (2006)
6. Rinkevičius, M., et al.: Photoplethysmogram signal morphology-based stress assessment. In: 2019 Computing in Cardiology (CinC), pp. Page 1–Page 4. IEEE, Singapore (2019)
7. Pilt, K., et al.: Influence of mental stress on the arterial stiffness parameters detected from photoplethysmographic signal waveform. In: 2022 18th Biennial Baltic Electronics Conference (BEC), pp. 1–4. IEEE, Tallinn, Estonia (2022)
8. Fine, J., et al.: Sources of inaccuracy in photoplethysmography for continuous cardiovascular monitoring. Biosensors (Basel) **11**(4), 126 (2021)
9. Allen, J.: Photoplethysmography and its application in clinical physiological measurement. Physiol. Meas. **28**(3), R1-39 (2007)
10. Mannheimer, P.D., O'Neil, M., Konecny, E.: The influence of larger subcutaneous blood vessels on pulse oximetry. J. Clin. Monit. Comput. **18**(3), 179–88 (2004)
11. Pilt, K., Meigas, K., Temitski, K., Viigimaa, M.: Second derivative analysis of forehead photoplethysmographic signal in healthy volunteers and diabetes patients. In: Long, M. (eds.) World Congress on Medical Physics and Biomedical Engineering 2012, IFMBE Proceedings, vol. 39, pp. 410–413. Springer, Heidelberg (2013). https://doi.org/10.1007/978-3-642-29305-4_109
12. Pilt, K., Meigas, K., Ferenets, R., Temitski, K., Viigimaa, M.: Photoplethysmographic signal waveform index for detection of increased arterial stiffness. Physiol. Meas. **35**(10), 2027–2036 (2014)
13. Pilt, K., Meigas, K., Kööts, K., Viigimaa, M.: Photoplethysmographic signal rising front analysis for the discrimination of subjects with increased arterial ageing. Proc. Est. Acad. Sci. **63**(3), 309–314 (2014)
14. Takazawa, K., et al.: Assessment of vasoactive agents and vascular aging by the second derivative of photoplethysmogram waveform. Hypertension **32**(2), 365–370 (1998)

Comparative Oscillometry on Finger Using Pneumatics and Multi-wavelength Photoplethysmography

Jaak Talts[1](✉), Sander Ümarik[2], Jana Kivastik[1], and Kersti Jagomägi[1]

[1] Department of Physiology, University of Tartu, 19 Ravila Street, 50411 Tartu, Estonia
jaak.talts@ut.ee

[2] Tallinn Secondary School of Science, 6 Estonia Avenue, 10148 Tallinn, Estonia

Abstract. Most blood pressure devices capture the oscillometric signal from arm cuff. This location does not allow convenient arterial photoplethysmographic (PPG) signal registration. In contrast, this option is available if using a finger cuff, allowing simultaneous recording of pneumatic and several different PPG signals.

Oscillations from pneumatic, infrared, red, orange, yellow and green PPG channels were recorded in 7 healthy subjects while the cuff pressure was ramped up and down.

The cuff pressure at maximum oscillations was lower during cuff deflation and for longer PPG wavelengths. For pneumatic signal it was closest to the values from yellow and orange PPG. The oscillation envelope was wider during the cuff pressure deflation, with the exception of green PPG.

Parameters for infrared PPG in each experiment were pairwise compared to according values in other channels. The difference found showed the following behaviour of dispersion between experiments: for PPG signals it increased when the difference in wavelength increased, for pneumatic signal it was smaller than for PPG signals during cuff inflation phase.

Keywords: Finger · Oscillometry · Pneumatics · PPG

1 Introduction

Most automatic blood pressure (BP) measurement devices use oscillometric principle. In this technique, the goal is to estimate arterial pressure, hence it would be desirable to rely on a signal which is related only to arteries. However, other blood vessels can also have some role in producing pulsations. Another aspect is that the tissues under edges of the cuff are less pressurized, which means the oscillation envelope does not reflect exactly the ideal condition. Our interest was to quantify the influence of these factors.

Most BP devices capture the oscillometric signal from an arm cuff pneumatically. As the distance between the brachial artery and skin is quite long, this location does not allow convenient arterial photoplethysmographic (PPG) signal registration. Yet, this is possible: Laurent et al. could measure the systolic BP using PPG sensor under arm

Y. Dekhtyar and I. Saknite (Eds.): NBC 2023, IFMBE Proceedings 89, pp. 189–196, 2023.
https://doi.org/10.1007/978-3-031-37132-5_24

cuff [1]. Also, Perpetuini et al. used high sensitivity probes by employing silicon photomultiplier to collect signals from deep vessels [2]. In the finger, arteries are at a depth which is easily accessible by PPG. Moreover, using simultaneous recording of several PPG signals, the contribution of pulsations from vessels at different depths can be investigated. The usage of the multi-wavelength PPG was presented in [3–5]. Questions under in these investigations have been signal-to-noise ratio, the shape and the timing of the pulse. Also, the reaction to external force or pressure, applied to skin, has been investigated. Using that, Lubin et al. compared blood pressure values, estimated from oscillations from infrared, red and green PPG [5].

Raamat et al. compared pneumo- and photoplethysmographic envelopes captured from the wrist [6]. The aim of this study was to compare oscillation envelopes from finger, recorded pneumatically and optically, using five different wavelengths of light. Similarly to [7], we analyzed envelopes during cuff inflation and deflation.

2 Methods

2.1 Equipment

The optical sensor consisted of photodiode (BPW34, Vishay, 430 to 1100 nm) and five LEDs: IR (infrared, TSMF3700, Vishay, λ:870 nm, FWHM:50 nm), red (LR P47F, Osram, λ:625 nm, FWHM:16 nm), orange (LO P47F, Osram, λ:606 nm, FWHM:17 nm), yellow (LY P47F, Osram, λ:590 nm, FWHM:18 nm) and green (LT PWSG, Osram, λ:528 nm, FWHM:33 nm). Those six elements were fixed between two flexible transparent tapes. Distances between centers of adjacent LEDs were 4.5 mm, distance between green LED and photodiode was 10 mm.

The cuff pressure was measured using transducer MPX5050DP (NXP Semiconductors). In addition, pulsating component from the cuff pressure was pneumatically filtered and measured using sensitive pressure transducer NSCDRRN001NDUNV (Honeywell, scale ± 1 inch H_2O).

In one measurement cycle, the current to each LED was sequentially switched on by the microcontroller and signal change from the photodiode was measured. In addition, the cuff pressure and pneumatic pulsation were recorded. The resolution of A/D conversion was 16 bits, the rate of measurement cycles was 100 Hz.

2.2 Experiment

Seven healthy subjects (3 males and 4 females, age from 17 to 79 years) were involved in this study. The subject was asked to sit, be relaxed and breathe normally without gasps. The ribbon with optical elements was wrapped around a proximal phalanx of the middle finger of the left hand, keeping symmetry in relation to finger pulp. Next the ribbon was covered by pneumatic cuff (Fig. 1(a)). During the measurement, the hand rested on the table.

The pressure was raised linearly from 30 mmHg to 160 mmHg and kept at that value for approximately 10 s, after that it was decreased back to 30 mmHg. The rate of pressure change was 3 mmHg/s in both phases. The duration of the experiment was approximately 2 min (Fig. 1(b)).

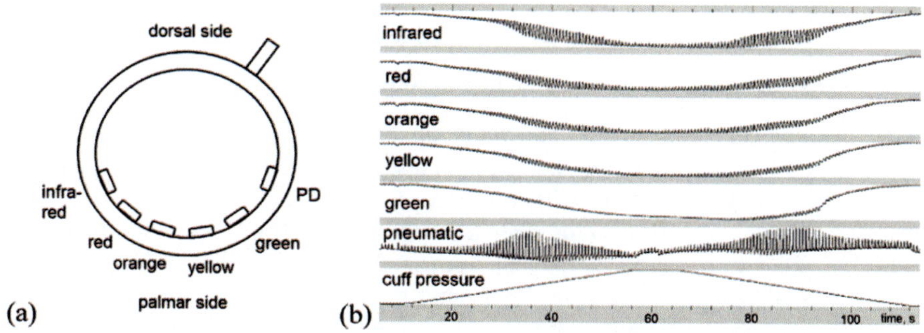

Fig. 1. (a) Arrangement of five LEDs and photodiode (PD) in the finger cuff. (b) Five PPG signals and pneumatic oscillations recorded while the cuff pressure was ramped up and down.

2.3 Data Processing

The signal processing was performed by applying Matlab R2020b (The Mathworks, Inc). The AC components of PPG signals were separated and oscillometric envelopes for all channels were calculated and smoothed. Examples from two experiments are shown in Fig. 2.

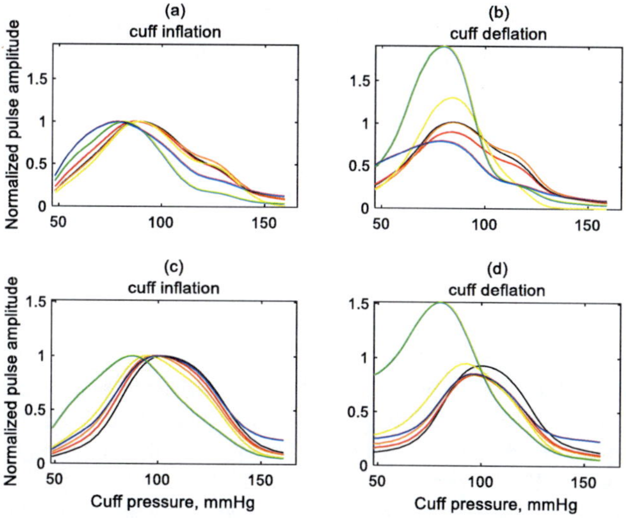

Fig. 2. Examples of smoothed and normalized oscillometric envelopes from two experiments, (a, b) and (c, d) respectively. Blue line shows data from pneumatic, black from infrared channel. Other channels are color-coded. Normalization is based on cuff inflation and is applied to both phases.

To characterize an oscillometric envelope, the maximum oscillation amplitude, cuff pressure at maximum oscillation amplitude and the width at the level of 70% from maximum were calculated.

3 Results

3.1 Cuff Pressure at Maximum Oscillation Amplitude (CP-MO)

Figures 3(a) and (b) show boxplots and means of pressure values when maximum oscillations occurred, for cuff inflation and deflation respectively. For both phases, the longer wavelengths had slightly lower value of CP-MO. This effect was stronger for green light. The mean value of CP-MO for green light over all experiments was approximately 10 mmHg lower than for other optical channels. The mean value for pneumatically recorded CP-MO was closer to yellow or orange channel.

Figure 3(c) shows differences of CP-MO between cuff pressure deflation and inflation. The difference is mostly negative, showing that after being exposed to high cuff pressure blood vessels have some hysteresis and the condition for highest compliance occurs at lower pressure. The mean value of the difference is biggest for green light. Relatively small dispersions in Fig. 3(c) show that the blood pressure for subjects did not change much.

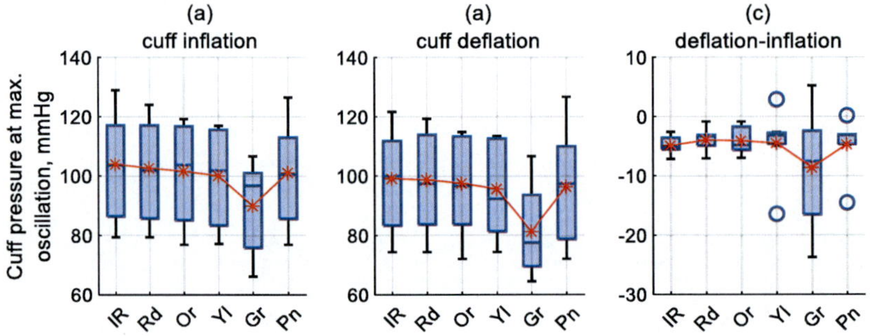

Fig. 3. Cuff pressure at maximum oscillation amplitude. (a) Cuff inflation. (b) Cuff deflation. (c) The cuff pressure difference between deflation and inflation phases.

Pairwise Comparison to IR Channel

Figure 4(a) and (b) show the CP-MO of each channel in relation to according value in IR channel. Note that it has the same tendency as for absolute values of CP-MO on Fig. 3(a) and (b), but showing more compact result, because the variability coming from different subjects is excluded. The dispersion was higher, if the difference in light wavelength (compared to IR) was bigger. Interestingly, for the pneumatic channel, the dispersion is mostly lower than for any optical channel. It may mean that albeit the presence of small bias the pneumatic measurement reflects the true arterial oscillation closely.

3.2 The Width of the Oscillometric Envelope (WOE)

Figure 5(a) and (b) show boxplots and means of the width of the oscillometric envelope at level of 70% of its maximum, for cuff inflation and deflation respectively. In cuff

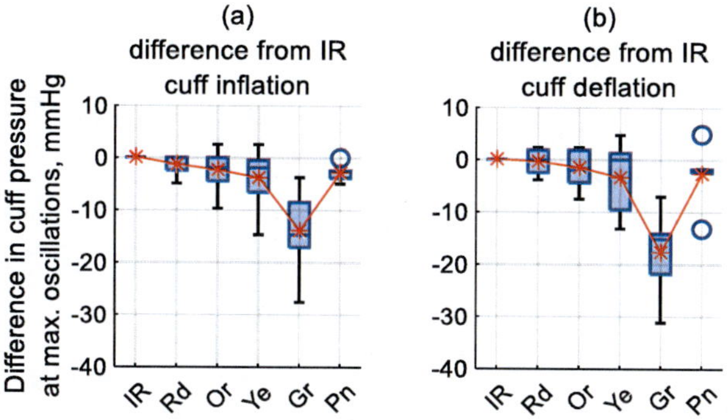

Fig. 4. Cuff pressure at maximum oscillation in relation to according value from infrared PPG in the same experiment. (a) Cuff inflation. (b) Cuff deflation.

inflation phase, the envelope is wider for shorter wavelengths, but in cuff deflation phase this does not hold for green light.

The difference (deflation-inflation) in Fig. 5(c) shows that the oscillometric envelope was wider during cuff deflation, with the exception of green light where the envelope was narrower by approximately 5 mmHg. The dispersion of the difference for WOE was much higher than for CP-MO, likely because the estimation of the width, based only on cuff pressure measurement and without knowing beat-to-beat arterial pressure, can be easily influenced by changes in arterial pressure during cuff pressure scan.

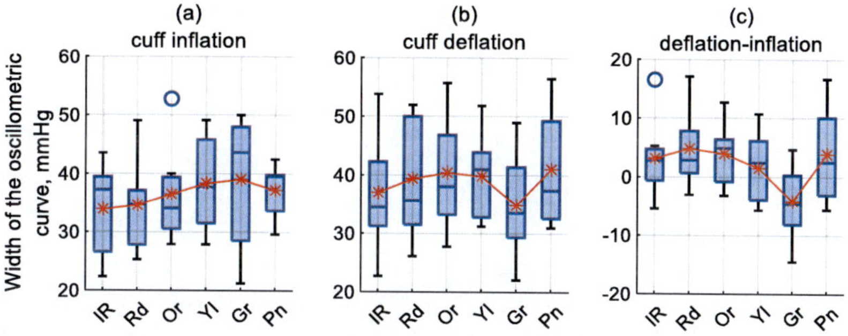

Fig. 5. The width of the oscillometric envelope at level of 70% from maximum. (a) Cuff inflation. (b) Cuff deflation. (c) The difference between width in cuff deflation and inflation phases.

Pairwise Comparison to IR Channel

Figures 6(a) and (b) show the differences of the WOE in relation to according value in IR channel. Similarly to pairwise comparison of CP-MO, the dependency of the difference follows the same character as dependency of absolute values. The dispersion is not so low as for CP-MO and the values of the pneumatic channel are not so close to according values of the IR channel. Like in the analogous comparison for CP-MO, dispersion is mostly higher, if the difference in light wavelength (compared to IR) was bigger.

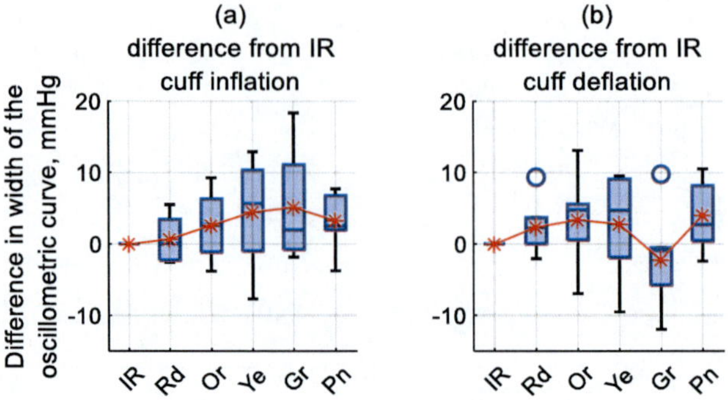

Fig. 6. The width of the oscillometric envelope in relation to according value for infrared channel in the same experiment. (a) Cuff inflation. (b) Cuff deflation.

3.3 Amplitude Characteristics

Figure 7(a) shows how different maximum oscillation amplitudes were during cuff inflation and deflation. Note the remarkably higher amplitudes in green light channel during cuff deflation. This change is also well visible in Fig. 2, and it is a sign of hyperemia of smaller vessels after occlusion by the high cuff pressure at the end of the cuff inflation phase.

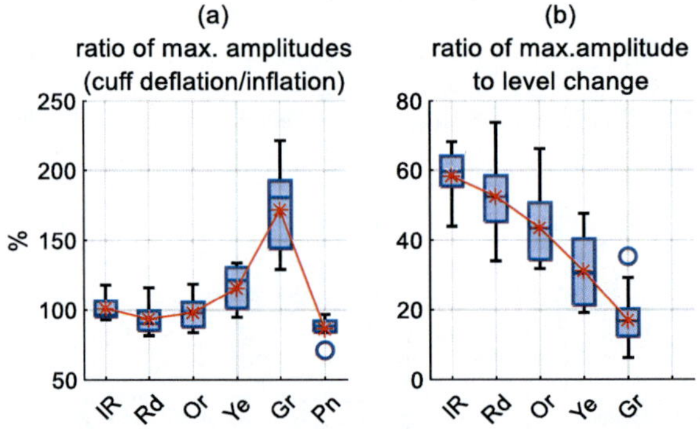

Fig. 7. (a) Ratio showing the change of maximum amplitude during cuff deflation in relation to value obtained from previously conducted inflation phase. (b) Ratio showing the percentage of maximum oscillation amplitude to full signal span.

The ratio of oscillation amplitude to overall signal level change in PPG channels is shown on Fig. 7(b). For IR, the majority of level change is coming from the oscillations. Contrarily, for green PPG the oscillations are very small compared to level changes.

4 Discussion

In this study, the oscillometric envelope was examined using cuff pressure scan and assuming the mean BP level remains constant. Although this principle is used widely in BP measurement devices, more accurate results could be achieved when beat-to-beat BP level is simultaneously measured. In this case the pulse amplitudes can be related more correctly to transmural pressure, not to cuff pressure. The simplification of using only the cuff pressure may lead to estimation of wider WOE when during scan the BP changes in the same direction as the cuff pressure does; and the estimated WOE would be narrower, when the BP change is in the opposite direction.

The LEDs in our experiments were located in consideration of possible light path length through the tissues: green is closest to photodiode and IR is farthest. So actually, the PPG system is in reflectance mode for green, and gradually turns into more and more transmittance mode for every longer wavelength. To avoid this additional influencing factor, some authors irradiated light from several sources into common fiber light guide, so the contact distance between the source and detector is constant for every wavelength [3] or placed several LEDs very compactly onto one common base [4]. However, in this case the common ligth path for all LEDs should be short, because shorter lightwaves can not reach longer distances and the ability of longer lightwaves to collect signal over larger region would be lost. So we tried to use the potential of all PPG channels, but it should be kept in mind that the wavelength was not the single factor.

Main variation in parameters of oscillometric envelope came from different blood pressures in subjects. For subjects with higher mean arterial pressure the CP-MO is higher, for subjects with higher pulse pressure the WOE is larger. We did not notice isolated age-related differences.

Some variation of results is caused by different size of fingers of subjects. It can be expected that in persons with thicker fingers the oscillometric envelopes at neighbouring wavelengths differ more.

The information source of most BP devices is pneumatic, i.e., pressure pulses. Because of that it would be useful to know factors affecting the pneumatic oscillometric envelope. To estimate arterial pressure, the best information source would be related to arteries, but not to smaller vessels. Infrared PPG has the best ability to reach deeper tissues and therefore, has the best potential to reflect true arterial volume changes. As the finger is fully surrounded by the cuff, the pneumatic signal should reflect arterial volume changes well. But our experiment showed that for pneumatic signal the CP-MO was mostly closer to yellow or orange channel, which can lack capturing some portion of arterial signal. One explanation is that the CP-MO for pneumatic signal differs from infrared because under ends of the cuff the pressure transmission to deeper tissues is weaker and part of the signal is collected over region under different transmural pressure. Hence one can say, the cuff side effect for pneumatic channel was approximately equal to inability to capture whole arterial volume change by yellow or orange light PPG.

5 Conclusion

In this methodological study the finger was used as a model for any PPG-based oscillometric measurement. The simultaneous registration of pneumatic and several PPG signals showed differences in oscillation envelopes, reflecting that each signal is collected over specific region of tissues. Every other measurement site has its own properties; but the same tendencies we found in fingers, describing how the results depend on wavelength of PPG, are present for all sites.

One concern related to continous finger BP measurement by volume clamp method is the change of vascular tone which alters the level of infrared PPG channel [8]. Because of that the measurement should be stopped frequently to recalibrate the setpoint. Using hardware described in this study, it would be possible to investigate whether information from other PPG channels could help compensate the level shifts in infrared PPG channel.

References

1. Laurent, C., Jönsson, B., Vegfors, M., Lindberg, L.-G., Jönsson, B.: Non-invasive measurement of systolic blood pressure on the arm utilising photoplethysmography: development of the methodology. Med. Biol. Eng. Compu. **43**(1), 131–135 (2005)
2. Perpetuini, D., et al.: Multi-site photoplethysmographic and electrocardiographic system for arterial stiffness and cardiovascular status assessment. Sensors **19**(24), 5570 (2019)
3. Spigulis, J., Gailite, L., Lihachev, A., Erts, R.: Simultaneous recording of skin blood pulsations at different vascular depths by multiwavelength photoplethysmography. Appl. Opt. **46**(10), 1754–1759 (2007)
4. Liu, J., Yan, B.P.-Y., Dai, W.-X., Ding, X.-R., Zhang, Y.-T., Zhao, N.: Multi-wavelength photoplethysmography method for skin arterial pulse extraction. Biomed. Opt. Express **7**(10), 4313–4326 (2016)
5. Lubin, M., Vray, D., Bonnet, S.: Blood pressure measurement by coupling an external pressure and photo-plethysmographic signals. In: 42nd Annual International Conference of the IEEE 2020, Montreal, Canada, pp. 4996–4999. IEEE (2020)
6. Raamat, R., Jagomägi, K., Talts, J., Kivastik, J.: Simultaneous pneumo- and photoplethysmographic recording of oscillometric envelopes applying a local pad-type cuff on the radial artery. In: Pamidis, P.D., Pallikarakis, N. (eds.) The XII MEDICON 2010, pp. 144–147. Springer, Cham (2010). https://doi.org/10.1007/978-3-642-13039-7_36
7. Liu, C., Zheng, D., Griffiths, C., Murray, A.: Oscillometric waveform difference between cuff inflation and deflation during blood pressure measurement. In: Computing in Cardiology Conference 2014, Cambridge, USA, pp. 849–852, CCAL (2014)
8. Wesseling, K.H., de Wit, B., van Der Hoeven, G.M.A., van Godoever, J., Settels, J.J.: Physiocal, calibrating finger vascular physiology for Finapres. Homeostasis **36**, 67–82 (1995)

Accuracy and Precision of the HRV Measurement by ECG, PPG and Mobile App

Matti Huotari[1](✉), Erkki Vihriälä[2], Kari Määttä[3], Teemu Myllylä[2], and Juha Röning[1]

[1] BISG, University of Oulu, Oulu, Finland
matti.huotari7@gmail.com
[2] OPEM, University of Oulu, Oulu, Finland
[3] CAS, University of Oulu, Oulu, Finland

Abstract. Accuracy and precision determination of the heart rate variability (HRV) or especially the variability of HRV require high fidelity measurement of inter beat intervals (IBI) in the ECG, inter pulse interval in the PPG signals, and in CorSense sensor for mobile Elite HRV application. These signals are collinear which happen when PPG, ECG, and Elite HRV data contain accurate and precise variables in a parallel measurement.

The ECG, PGG, and Elite HRV values are taken into consideration for opportunity to realize a practical, cheap, and easy health condition detection system for cardiovascular diagnosis. PPG could be applied for measuring the physiological state of individuals along their ages in daily life and it could connect healthcare applications seamlessly with other biosensors. This work is collecting each signal of the ECG, PPG, and mobile HRV sensor parallel and simultaneously of 35 persons (11 female). The results visualization is based on the precise and accurate data on the merged Poincaré plot analysis so that extra or lagging beats were cancelled. Then the data is highly correlated with each other based on the correlation coefficient (R^2), and the variance inflation factor (VIF) which is also very informative factor. Elite HRV may be a preferred method of HRV data collection for young person due to its practicality. However, older people needs an accurate device to get reliable result to get rid many erratic heart beats such as premature heart beats. Accordingly, few studies exist that asses the accuracy of combined ECG, PPG & Elite HRV.

Keywords: Photoplethysmogram (PPG) · Electrocardiogram (ECG) · heart rate variability (HRV) · mobile Elite HRV app · CorSense sensor · Poincaré plot

1 Introduction

Health and well-being depend very much on your own. Deep knowing your heart and especially your heart rate (HR) can save you. That's why development of accurate automatic HR measurement systems using a minimum number of bio-signals, e.g., photoplethysmographic devices (PPG), have become a challenging issue for personalized medicine. According to the report on the global PPG Sensors market size, only the local PPG sensor market in the USA was valued USD 874.93 million during last year and

Y. Dekhtyar and I. Saknite (Eds.): NBC 2023, IFMBE Proceedings 89, pp. 197–204, 2023.
https://doi.org/10.1007/978-3-031-37132-5_25

is expected to expand at a CAGR of 42.1% during the forecast period, reaching USD 7204.82 million by the year 2028. The report can predict about many upcoming custom trends and it is believed to change in future consumer behavior in PPG markets using HR and especially heart rate variability (HRV) documentation. Here we designed both ECG and PPG amplifier for the accurate and precise comparison measurements with the CorSense and the Elite HRV mobile app with Samsung phone.

2 Methodology of the Heart Rate Variability Measurement

Heart Rate Variability (HRV) means the difference in time between heartbeats (R-R intervals, or inter beat intervals (IBI)), and variability of HRV likewise means the difference in time between HRV measured in milliseconds [ms]. For the accurate HRV readings, it is necessary to make sure that the PPG measured location from the fingertip is placed over the sensor window and lighted correctly, and not moved or talked during the PPG measurement. The body should be in the same position during all readings, while the heart rate and breathing should be stable.

Our PPG sensor construction is based on the compact reflective phototransistor in which the reflective photomicrosensor incorporates an emitter and a detector, and they are combined. The fingertip is located in the sensing area of the photomicrosensor, where the blood reflects and scatters the optical beam of the emitter and changing the amount of optical energy reaching the detector as a function of synchronously beating capillary blood flow. In the PPG sensor the photomicrosensor's emitter and detector are on the same side of the finger as reflective mode whereas in the CorSense sensor of Elite HRV application in the studied PPG sensor the photomicrosensor's emitter and detector are as transmission mode on the opposite side of the measured fingertip. In the mobile PPG, a heart pulse curve is captured continuously producing the heart rate data, and HRV which is shown continuously on the mobile display. Later it is possible to calculate also the variability of HRV. The high resolution ECG data was collected with means of a high impedance ECG amplifier with the sampling rate of 1 kHz. The ECG electrodes were the monitoring disposable INTCO sensors. These sensors measure one channel high resolution ECG with three electrodes with the amplitude resolution of 16 bit ADC. The PPG data collection was based on the photomicrosensor's emitter and detector (component: Omron ee-sy113) which are on the same side (reflective mode) of the fingertip with the sampling rate 1 kHz. The Elite HRV CorSense PPG sensor is working on the other way in transmission mode with sampling rate 500 Hz. Before the measurement we used skin warm tap water, and the comfortable skin temperature was received for improving the quality and accuracy of the PPG signals, but it did not cause any effect on the quality or precision of ECG signal. However, the room area, devices and device construction, and personal factors can affect the variability in the measured signals and can cause artefacts especially in optical signal which can cause errors within the analysis of the signal.

Detailed information on the key regulatory HRV and variability of HRV parameters will create novel opportunities both for health technology intervention and sickness conditions where anemia, ischemia, inflammation, diabetes type II, heart diseases, COVID-19 virus, or lack of oxygen uptake could be present [1–3].

Both ECG and PPG sensor provide acceptable measurement results, and the heart rate variability sensor CorSense provide also an acceptable agreement for the measurement of Poincaré plot parameters (SD1, SD2 in ms) when compared with ECG and PPG records. HRV parameters could predict heart disease at early stages, what's why precision and accuracy in achieved results are important. Because Poincaré plot analysis is almost immune to artefacts these artefacts are easily detected. Detailed information on the key regulatory HRV parameters will create novel opportunities both for health technology intervention and sickness conditions where anemia, ischemia, inflammation, heart diseases and lack of oxygen uptake could be present.

In the Figs. 1 and 2 it is shown the ECG amplifier with its circuit components and the PPG amplifier with its circuit components, respectively.

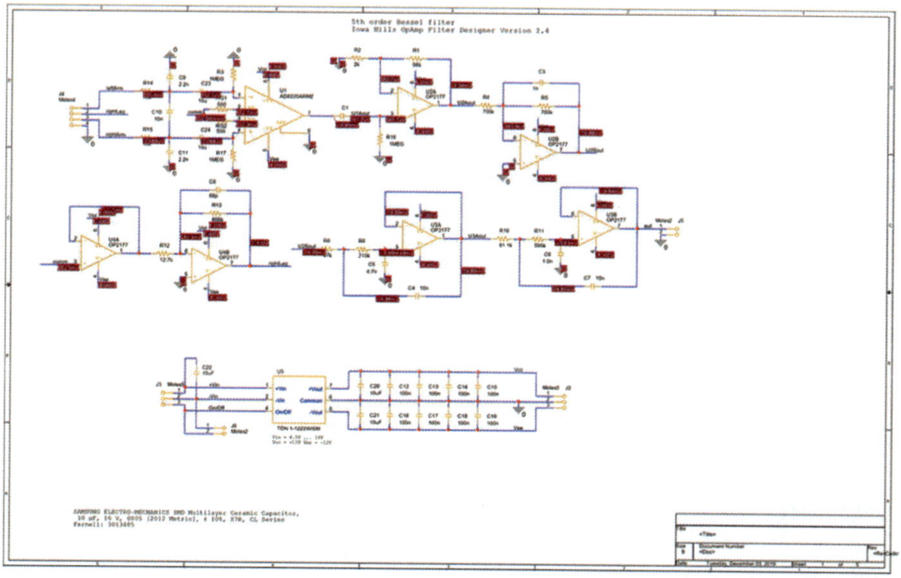

Fig. 1. The ECG amplifier for the ECG application with battery (1.5 V) powered, designed by K. Määttä.

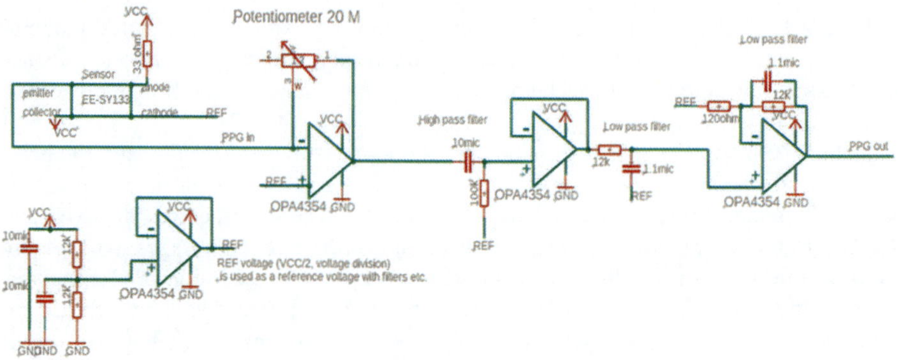

Fig. 2. The OPA4354 devices are the photodiode transimpedance amplifiers for the Photoplethysmographic application with battery (9 V) powered, designed by E. Vihriälä.

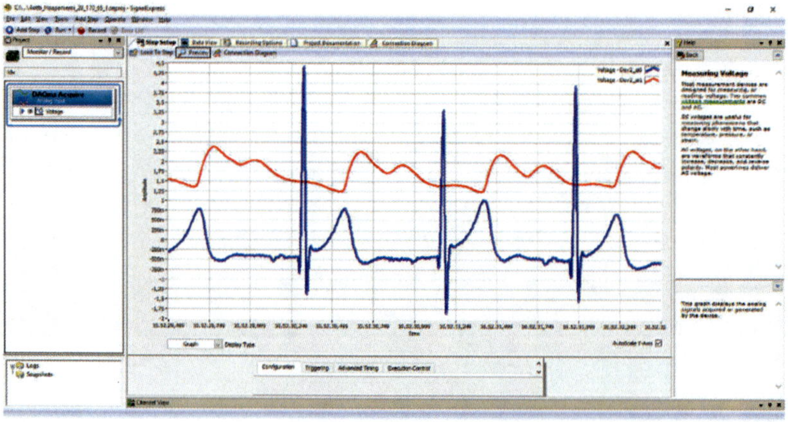

Fig. 3. The measured PPG (red) and ECG (blue) in an arbitrary voltage scale with means of the NI Signal Express software.

3 Results

In Fig. 3 it is shown the measurement signal record for 3 s of a 28 y male, ECG (blue) and PPG (red). In Fig. 4, the IBI data measured with ECG (black), with PPG (red), and with mobile CorSense for Elite HRV app (blue) clearly overlap on the whole record time shown as a function beat number (Fig. 4 (A)). On the Fig. 4 B the IBI points overlap showing good correlation coefficient, R^2 and VIF value. The Poincaré plot shows accurately the measured signal triadic (ECG, PPG, Elite HRV) on the IBI cases (Fig. 4 (C)). The same kind of data is shown in Fig. 5 for an older male test persons. In each case the ellipse of the Poincaré plot for ECG has the smallest number of SD1 and SD2 compared with the ellipses of the PPG or Elite HRV data. The IBI data do not overlap so clearly as in Fig. 4 but the trend and range are in good relation in Fig. 5. Each applications (ECG, PPG, and CorSense with Elite HRV) can be implemented to monitor

HRV using 5 min length measurements for the young but carefully for the elderly cases. In the elderly cases there is much scattered data in each analysis step, especially the correlation coefficient (R^2) and VIF value. The R^2 is a statistical measure of the strength of a linear relationship between IBI of ECG vs. IBI of PPG, or IBI of ECG vs. IBI of Elite HRV.

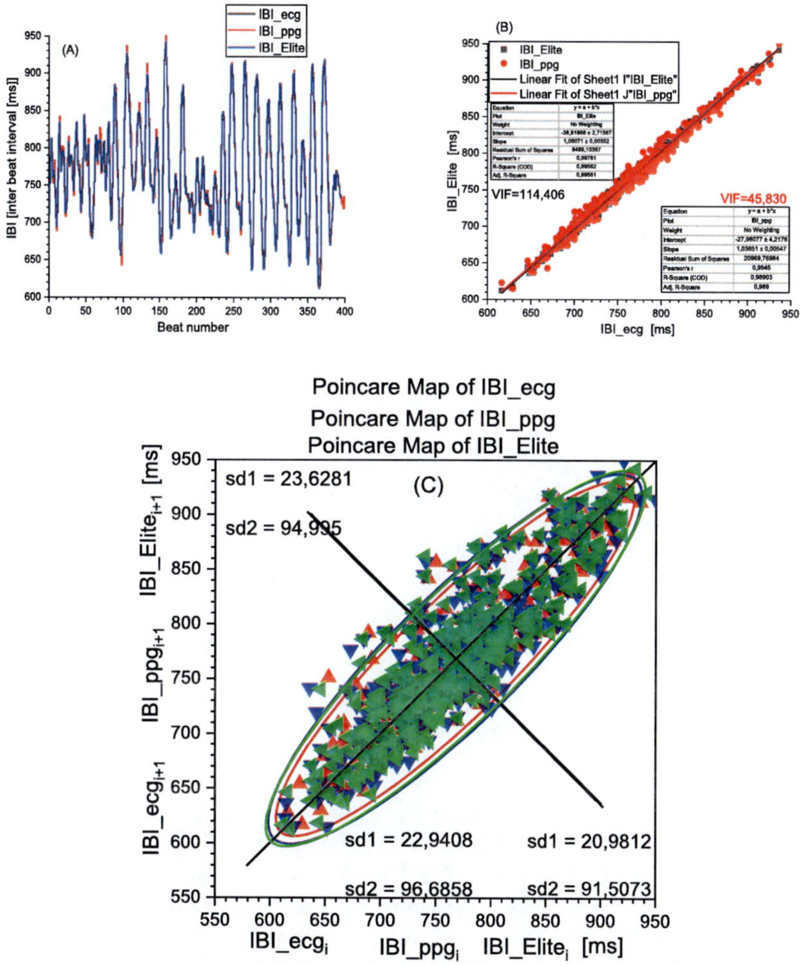

Fig. 4. (A) The IBI measured with ECG (black), PPG (red), and mobile CorSense with Elite app (blue) in the overlapped absolute scale, (B) the correlation between IBI of ECG vs IBI of PPG (red), and IBI of Elite HRV (black, overlap) with the VIF (114 for ECG vs Elite HRV, and 45 for ECG vs PPG), and (C) the corresponding merged Poincaré plots with the parameters where ECG(red), PPG (blue); and Elite (green) for a 28 y healthy male. The SD1 value 20,9812 for ECG, 23,6281 for PPG, and 22,9408 for Elite HRV, and the SD2 value, 91,5073, 94,995, and 96,6858, respectively (in Fig, B & C axis in [ms]). (Color figure online)

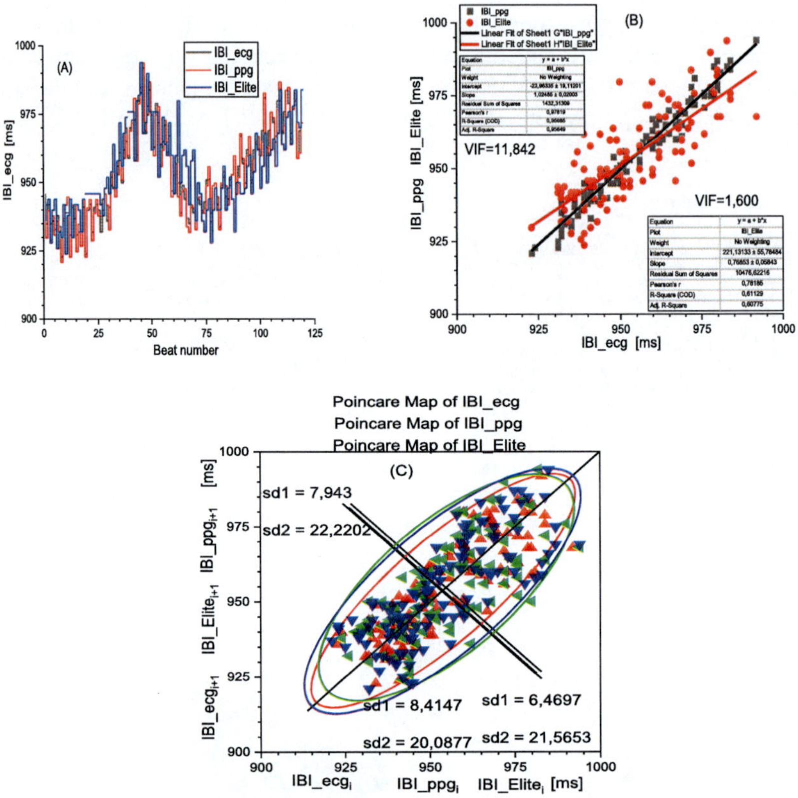

Fig. 5. (A) The IBI measured with ECG (black), PPG (red), and mobile CorSense with Elite app (blue) in the overlapped absolute scale, (B) the correlation between IBI of ECG vs IBI of PPG (red), and IBI of Elite HRV (black) with the VIF, and (C) the corresponding merged Poincaré plots with the parameters where ECG (red), PPG(blue); and Elite (green) for a 71 y healthy male (in Fig. 5 B & C axis in [ms]). (Color figure online)

In the Table 1 it is shown the measurement results of 35 test persons (11 females) for the ECG and Elite HRV calculations with gender, correlation coefficient (R^2), age, SD1, and SD2 for both ECG and Elite HRV. SD1 is under 10 over the 60 y people, and their SD2 is under 25 with a few exceptions. We conclude that healthy elderly people present decreased HRV compared with healthy young people indicating that they have decreased autonomic nervous modulation when compared to young people. In the young persons the both values are invariably larger than older people. The older people could be prone to various diseases, especially cardiovascular diseases, if their HRV values are low. It is important to research the function between standard measuring devices and new mobile instruments before they could be adapted to use in public over HR variability and other parameters. A poor performance of HRV estimation with PPG devices was found in uncontrolled conditions highlighting the need for improvements in the biomedical technology and biological signal analysis methods [1, 2].

Table 1. Table on the results of 35 test persons with, gender, age, correlation coefficients (R^2), SD1, and SD2 for the ECG and Elite HRV

gender	age	R^2	SD1 ECG	SD2 ECG	SD1 Helite HRV	SD2 Helite HRV
M	**42**	**0,995**	**34,6**	**118,8**	**38,7**	**129,2**
M	**28**	**0,996**	**30,99**	**127,7**	**33**	**129,6**
M	**69**	**0,932**	**6,25**	**11,1**	**9,21**	**12,15**
M	**74**	**0,877**	**20,45**	**27,384**	**18,825**	**27,1557**
M	**63**	**0,92**	**8,555**	**35,65**	**16,589**	**37,555**
M	**85**	**0,8316**	**6,7**	**30,1**	**10,284**	**31,147**
M	**34**	**0,992**	**20,611**	**43,881**	**22,872**	**44,628**
M	**65**	**0,994**	**10,688**	**40,519**	**11,894**	**40,808**
M	**25**	**0,998**	**17,211**	**54,612**	**17,587**	**56,546**
M	**54**	**0,985**	**9,415**	**22,689**	**10,397**	**24,026**
F	**75**	**0,96**	**7,1**	**14,689**	**9,818**	**16,386**
M	**30**	**0,991**	**5,07**	**22,207**	**5,658**	**22,815**
F	**27**	**0,992**	**11,35**	**31,537**	**11,214**	**31,8444**
F	**26**	**0,984**	**21,495**	**44,549**	**22,259**	**45,817**
M	**26**	**0,9967**	**97,775**	**145,316**	**98,889**	**151,161**
F	**39**	**0,993**	**25,27**	**68**	**26,194**	**68,75**
M	**43**	**0,992**	**10,133**	**58,111**	**12,21**	**60,072**
M	**50**	**0,991**	**8,6855**	**25,31**	**8,631**	**25,59**
F	**68**	**0,991**	**7,899**	**23,316**	**8,995**	**24,662**
F	**79**	**0,982**	**9,636**	**11,299**	**9,6587**	**11,415**
M	**34**	**0,999**	**85,996**	**178**	**87,307**	**181,512**
M	**60**	**0,956**	**5,874**	**23,,827**	**8,0358**	**24,5**
M	**49**	**0,997**	**28,39**	**67,127**	**29,071**	**67,602**
M	**35**	**0,989**	**37,707**	**55,173**	**38,116**	**56,336**
F	**35**	**0,993**	**42,138**	**58,638**	**44,056**	**59,14**
M	**27**	**0,997**	**26,614**	**91,342**	**28,381**	**94,85**
F	**34**	**0,995**	**24,172**	**42,257**	**22,818**	**41,756**
F	**27**	**0,998**	**67,821**	**87,596**	**69,293**	**90,54**
M	**75**	**0,978**	**8,739**	**16,581**	**10,043**	**16,881**
M	**24**	**0,997**	**24,657**	**59,748**	**26,449**	**63,161**
M	**62**	**0,992**	**12,789**	**24,984**	**13,766**	**26,08**

(*continued*)

Table 1. (*continued*)

gender	age	R^2	SD1	SD2	SD1	SD2
			ECG	ECG	Helite HRV	Helite HRV
M	**23**	**0,997**	**24,833**	**67,792**	**25,345**	**69,505**
M	**36**	**0,988**	**14,484**	**23,188**	**15,933**	**24,071**
M	**21**	**0,997**	**12,793**	**55,828**	**13,176**	**57,258**
M	**25**	**0,986**	**24,736**	**65,584**	**27,658**	**70,678**

References

1. Hoog, A.C., Mai, Y., Peltokangas, M., Leonhardt, S., Oksala, N., Vehkaoja, A.: Accuracy of heart rate variability estimated with reflective wrist-PPG in elderly vascular patients. Sci. Rep. **11**(1), 8123 (2021)
2. Graham, S.A.: Associations between heart rate variability measured with a wrist-worn sensor and older adults' physical function: observational study. JMIR Mhealth Uhealth. **7**(10), e13757 (2019)
3. Sahroni, A., Hassya, I.A., Rifaldi, R., Ul Jannah, N., Irawan, A.F., Rahayu, A.W.: HRV assessment using finger-tip photoplethysmography (pulserate) as compared to ECG on healthy subjects during different postures and fixed breathing pattern. Procedia Comput. Sci. **161**, 535–543 (2019)

Approach for Calculating Hemodialysis Machines for a Specific Target Population

Fabiola M. Martinez-Licona[1](✉), Raul E. Molina-Salazar[2], and Alma E. Martinez-Licona[1]

[1] Electrical Engineering Department, Universidad Autonoma Metropolitana Iztapalapa, Mexico City, Mexico
fmml@xanum.uam.mx

[2] Economy Department, Universidad Autonoma Metropolitana Iztapalapa, Mexico City, Mexico

Abstract. Chronic Kidney Disease (CKD) is one of the non-communicable diseases that is constantly growing. In Mexico, about 10% of deaths are due to CKD, and it also represents one of the most frequent comorbidities in the population. Treatment options include organ replacement, which is clinically and financially complex, and hemodialysis. There is insufficient information on the number and status of patients with CKD in Mexico, so determining the need to care for them must be approached through an approximation. This paper presents a proposal to determine the technological resource (number of hemodialysis machines) to serve a previously identified population profile through contextualizing the available clinical, socioeconomic, and technological information and adapting a resource allocation method. Infrastructure, clinical, social, and economic profile indicators were proposed and incorporated into an adapted version of the Resource Allocation Working Party (RAWP) focus formula from population to equipment to calculate the State Weighted hemodialysis machines. Using the 2020 data published by the Mexican Ministry of Health and The National Institute of Statistics and Geography, the developed approach found that the requirement of more than 20 000 machines is estimated to meet the needs for the female population between 24–49 years old with low levels of schooling, informal economic activities and still living in underdeveloped central areas. The approach has significant limitations due to the availability and reliability of the information and the exclusion of private services. CKD care and financing through hemodialysis must be planned considering the different contexts in which the population is found.

Keywords: Hemodialysis units · Health resource allocation · CKD

1 Introduction

Chronic Kidney Disease (CKD) is a non-communicable disease that continues to rise in prevalence. Mexico ranks sixth in the world for premature death rates related to CKD and presents a country's age-standardized CKD mortality rate of 58.1 deaths per 100 thousand inhabitants, doubling the figure in 17 years [1]. The significant rise in the mortality rate can be attributed to the increasing burden of risk factors such as diabetes,

Y. Dekhtyar and I. Saknite (Eds.): NBC 2023, IFMBE Proceedings 89, pp. 205–214, 2023.
https://doi.org/10.1007/978-3-031-37132-5_26

hypertension, and obesity, an inadequate access to preventive measures and resources facilitating early disease management, and the limited availability of renal replacement therapy (RRT), particularly among the low-income demographic [2]. CKD treatment options include organ replacement, which is both clinically and financially challenging [3], and hemodialysis. Peritoneal dialysis is another option that has been shown to be cost-effective with respect to hemodialysis [4, 5]; other findings show that timely referral and proper education about dialysis options can help identify a substantial group of patients who can be effectively treated with home dialysis [6].

Studies on access to these treatments to identify equity gaps have underpinned the relevance of acting in low and middle-income countries [7], addressing the needs of minority groups [8] and comorbidities [9]. Access to treatments in Mexico poses a situation of inequity because the cost and dedication in time demanded by the patient cannot always be covered, the number of establishments that offer these treatments is not completely displayed in the reported records, and there is no certainty about the regulatory compliance to which they must submit to offer a safe and effective service. The health system is fragmented; the economically active population receives health insurance according to their job description; economic instability causes more significant movements in jobs, and this, in turn, causes greater pressure that is reflected in the neglect of chronic diseases, in addition to the financial weight that treatments imply and that frequently cannot be absorbed. Due to insufficient information on the number and condition of CKD patients in Mexico, determining the appropriate care for them requires an approximate approach.

The disease analysis approach to identify lags in access to treatment has shown essential differences when approached from the gender perspective. Women have more predialysis chronic kidney disease (CKD) than men. This difference can be attributed to the longer life expectancy of women and the possible overdiagnosis of CKD when using estimated glomerular filtration rate equations. Men generally have a better perceived health-related quality of life than women on renal replacement therapy (RRT), as per studies. On the other hand, women report a higher symptom burden and greater symptom severity than men [10].

A recent study carried out in the most significant public health systems in Mexico indicates that the cost of the first year of hemodialysis per capita is around 39 thousand dollars [11], an amount that most of the population cannot fully cover. The economic and social backwardness in which women in rural areas live, where they dedicate themselves to low-paid or unpaid work, and therefore without social security, places them in a vulnerable situation to face the burden of this disease. Even though this population profile is not the only vulnerable one (children and older people), it does represent a problem of a social and economic nature since women in the productive age range represent the primary breadwinner in 33% of Mexican households [12].

Based on the results of a study on equity in CKD for the female population in Mexico, this paper presents a proposal to determine the number of hemodialysis machines required to meet the target population requirements. Taking as a premise the probability of mortality from CKD for a woman, social, economic, and demographic characteristics that determine this condition were identified, which were used to build a model that used principles of resource allocation.

2 Methods

2.1 Determinants

The results obtained from [13] claim that the highest incidence rate for chronic kidney disease in the female population corresponds to the adult woman of intermediate age who works in the tertiary sector (informal activities that do not have a fixed remuneration such as salary and are excluded from institutional health services), with low education and who live in rural areas of the central zone of the country. The intermediate age of adulthood was set between 24 and 49 years since this interval precedes mature women (50+) and follows young adults (18–24), although some sources place the maximum up to 44 years [14]. Rural areas are those whose development is behind urban areas; the interval considered in the reference source was taken, while the states that make up the country's central zone were selected considering their geographical location. Based on this description, the information for this profile was obtained from the National Institute of Statistics and Geography (INEGI) [15] and the Mexican Ministry of Health [16]. The results section specifies how we obtained the variables and their corresponding values for each state.

2.2 Model

The model aimed to determine the necessary number of hemodialysis machines in each state, considering the target population specifications, infrastructure, and epidemiological information. The multiple criteria decision-making process, known as the analytic hierarchy process (AHP), was utilized to define and evaluate the variables used in the model. The hierarchical analysis identified the variables with the highest weight to propose a resource allocation model based on the classic principles of population-weighted allocation. These principles consider factors such as age structure, health needs, and service delivery costs. A combination of these factors is used to calculate the weighted population (WP) of an authority. A similar approach was made to determine the number of mammography machines for screening [17]. The model uses the following equation:

$$WP = POP * (1 + a) * (1 + n) * (1 + c) \tag{1}$$

The unweighted population (*POP*) uses age (a), needs (n), and cost (c) adjustments [18], which were used to determine the weight for each characteristic set, and the model followed a customized approach for hemodialysis resource allocation for the rural areas of the central States.

State's weighted hemodialysis machines (SW_{HdM}): the objective is to determine the required number of machines for the target population. We changed the emphasis in the original model from the population to the rural State's technological infrastructure capacity.

Adjustments: Adjustments were made to reflect the target population's characteristics, infrastructure, and epidemiological data. The formula for the model is expressed as

$$SW_{HdM} = NSW_{HdM} * (1 + FP) * (1 + Prof_{SocEc}) * (1 + Inf) \tag{2}$$

NSW_{HdM} is the unweighted State's hemodialysis machines calculated under different operational capacity assumptions, and *FP*, $Prof_{SocEc}$, and *Inf* are objective adjustments for female target population, Socio economic profile and infrastructure information.

3 Results

3.1 Model Variables

The selection of variables for the estimation were female population, schooling level, residence area, economic activity, number of hemodialysis machines and CKD prevalence in target population. As mentioned, we obtained these variables by analyzing CKD incidence determinants for the target population [13]. Table 1 shows the weights of each variable and the category to which they were assigned. The weights were calculated by taking the average of the heuristically assigned values for each variable's impact on the others, ensuring their consistency in each instance, and analyzing them concerning the potential solutions to the problem at hand, which in this scenario were the categories in which they were classified.

Table 1. Variable weights form AHP

Variable	Weight	Category
Female population (24–49)	0.092	*Target population*
Schooling level	0.149	*Socio economic*
Residence area	0.160	*Target population*
Economic activity	0.053	*Socio economic*
Number of hemodialysis machines	0.321	*Equipment*
CDK prevalence in target population	0.224	*Epidem. Indicators*

Then, we proceeded with defining the measurement of each variable and its integration into the model.

3.2 Target Population

Based on the profile of the target population, intermediate adult women living in rural areas in the center of the country, we identified the following variables:

- Women between 24–49 years old living in rural areas (FP_{Rural}). The proportion of this population concerning the total target population in the state ($FP_{RuralSt}$) was calculated.
- Rural areas.- Populations between 10,000 and 20,999 inhabitants.
- Central part of Mexico.- Mexico City, Hidalgo, State of Mexico, Morelos, Tlaxcala, and Querétaro.

We added the number of municipalities where the rural areas where the target population lives (Mun_{FP_R}) and the total number of municipalities (Mun_{Tot}) to obtain the proportion. In addition, information on the municipalities was added since, administratively, the rural areas are designated to them. Table 2 shows the values found to determine this variable and its determination for each rural region.

Table 2. Target population in the States rural areas

State	FP_{Rural}	$FP_{RuralSt}$	***Prop. $FP_{Rural}/FP_{Rural\ St}$***	Mun_{FP_R}	Mun_{Tot}	***Prop. Mun FP_R/Tot***
Mexico City	35 805	1 833 770	**0.0195**	3	17	**0.1765**
Hidalgo	94 213	580 578	**0.1623**	19	84	**0.2262**
Mexico State	375 741	3 262 149	**0.1152**	55	125	**0.4400**
Morelos	64 381	363 833	**0.1770**	16	36	**0.4444**
Tlaxcala	53 912	247 317	**0.2180**	18	60	**0.3000**
Queretaro	64 764	468 056	**0.1384**	7	18	**0.3889**

3.3 Socioeconomic Profile

We generated the socioeconomic profile from the information on schooling and economic activity. The determinant of low schooling was obtained by counting the number of women in the target population (24–49 in rural areas of central Mexico) who were illiterate, without schooling, or with incomplete primary school (Sch_{Rural}), and we obtained the proportion concerning the state (*Prop.* $Sch_{Rural}/Sch_{Rural\ St}$). For the economic variable, the determinant indicates tertiary activities not covered by any health insurance scheme. As the information cannot be obtained directly, we proceeded with the calculation of the proportion of the uninsured target population (*Prop. NHE* $_{FP\_R/FPTot}$) based on the percentages of the uninsured population by state ($\%NEH_{St}$) and the calculation of the target population (FP_{NEH_St}). The results are shown int Table 3.

Table 3. Target population socio economic profile

State	Sch_{Rural}	$Sch_{RuralSt}$	***Prop.*** $\boldsymbol{Sch_{Rural}/Sch_{Rural\ St}}$	$\%NEH_{St}$	FP_{NEH_St}	***Prop. NHE FP_R/FPTot***
Mexico City	8 590	341 683	**0.0251**	27.4	9 810	**0.0195**
Hidalgo	21 645	276 422	**0.0783**	30.3	28 546	**0.1623**
Mexico State	98 181	971 752	**0.1010**	33.7	126 624	**0.1152**
Morelos	26 755	140 112	**0.1910**	28.1	18 091	**0.1770**
Tlaxcala	16 175	74 561	**0.2169**	28.2	15 203	**0.2180**
Queretaro	12 409	132 578	**0.0936**	20.9	13 535	**0.1384**

3.4 Infrastructure

This category includes the number of hemodialysis machines *(#HdM)*. We formed the variable by considering the presence of hemodialysis machines in rural areas concerning the number of municipalities in the state. The relationship between the number of municipalities and rural areas was calculated as an intermediate variable ($V = \#Mun/Rural$) later to calculate the proportion of hemodialysis teams in rural areas (*Prop. #HdU/Rural*), as shown in Table 4.

Table 4. Infrastructure

State	*#HdM*	$V = \#Mun/Rural$	$V/\,Mun_{Tot}$	***Prop. #HdM/Rural***
Mexico City	182	0.3000	0.0176	**3.2118**
Hidalgo	5	0.7600	0.09000	**0.0452**
Mexico State	179	0.4583	0.0037	**0.6563**
Morelos	12	0.7273	0.0202	**0.2424**
Tlaxcala	18	0.9000	0.0150	**0.2700**
Queretaro	1	0.3684	0.0205	**0.0205**

3.5 Number of Hemodialysis Machines

To determine the number of hemodialysis machines per State, we calculated the average of the variables for each category (bold values from Tables 2 to Table 4) and then included them in Eq. 2. The unweighted State's number of hemodialysis machines ($NSW_{HdU\text{-}6}$) corresponds to the number of machines for every six patients based on considering three 6-h shifts with 3 h per treatment. Calculations were also made for the case of four 6-h

Table 5. Number of hemodialysis machines per State for the target population

State	*FP*	$Prof_{SocEc}$	*Inf*	$NSW_{HdM\text{-}6}$	$SW_{HdM\text{-}6}$	$^*SW_{HdM\text{-}8}$
Mexico City	0.0980	0.0223	3.2118	5 967.50	**28 213.07**	**21 159.80**
Hidalgo	0.1942	0.1203	0.0452	15 702.17	**21 958.06**	**16 468.54**
Mexico State	0.2776	0.1081	0.6563	62 623.50	**146 845.02**	**110 133.77**
Morelos	0.3107	0.1840	0.2424	10 730.17	**20 687.77**	**15 515.83**
Tlaxcala	0.2590	0.2175	0.2700	8 985.33	**17 491.09**	**13 118.32**
Queretaro	0.2636	0.1160	0.0205	10 794.00	**15 533.12**	**11 649.84**

* *The* $NSW_{HdM\text{-}8}$ *results are not shown*

shifts with the same treatment time, giving eight patients attended per day ($NSW_{HdU\text{-}8}$). The results are shown in Table 5.

The data obtained correspond to the number of units for the established target population. We incorporated the epidemiological variable by considering that in Mexico, the prevalence of CKD was 12.2% by 2022 [19] was taken as a basis to redo the calculations of the target population variable. Since a hemodialysis machine requires facilities to operate correctly, the number of hemodialysis units (*HdU*) was calculated based on a model of 16 machines [20]. The results obtained for the number of units under these conditions are shown in Table 6.

Table 6. Number of hemodialysis units per State for the target population considering CKD prevalence

State	$SW_{HdM\text{-}6}$	$HdU_{HdM\text{-}6}$	$SW_{HdM\text{-}8}$	$HdU_{HdM\text{-}8}$
Mexico City	**3 415.12**	213.44	**2 561.34**	160.08
Hidalgo	**2 519.08**	157.44	**1 889.31**	118.08
Mexico State	**17 206.04**	1 075.37	**12 904.53**	806.53
Morelos	**2 374.32**	148.39	**1 780.74**	111.29
Tlaxcala	**1 971.71**	123.23	**1 478.79**	92.42
Queretaro	**1803.94**	112.74	**1 352.96**	84.55

4 Discussion

Calculating the number of equipment to meet a health need depends on many factors, such as the mode of use, the target population, or even the infrastructure required to make them work. In the case of the hemodialysis machines, the situation increases the degree of complexity since the machine by itself will not operate correctly: specialized facilities are required to manage the quality of the water that will be used, distribution of equipment to attend to patients comfortably, human resources that care for the patient and supervise the treatments, etc. The segmentation of the target population with specific and contextualized characteristics allows dimensioning the challenge that involves meeting their needs with technological resources and the financial burden it implies. The results reaffirm prompt action to maintain control over the disease burden in this sector, which is vulnerable by nature.

The degree of complexity involved in specifying the resources in health interventions has led to the creation of theories and methods to meet the needs that arise from this problem. Proposals have been developed to address health resource allocation problems through simulation, optimization, and forecasting techniques [21] or to analyze equity and efficiency at the primary level [22]. In this work, a first approximation was developed to determine the size of the technology park from a theoretical framework directed to defining the health resources by identifying the determinants that describe it (population, need, and cost). Our adaptation was based on the factors specifying the target population found in [13] to clarify the specific determinants for the problem (population, socioeconomic profile, and infrastructure).

The central zone of Mexico has states with economic and social attractions for the population, such as the states' capital cities: Mexico City even as the national capital or Queretaro as a highly industrial city. The State of Mexico has the largest population in the country, and its geographical condition and demography show a distribution of important urban and rural areas. In the case of Mexico City, the rural areas identified in the population censuses meet the criteria mentioned in the target population and correspond to bordering areas of the state where there are lags in the provision of healthcare services, so the population has to move to the regions with the highest concentration in the state where the hospitals that can provide the service are located.

Although a high number of the required equipment is obtained, some aspects that limit its interpretation should be considered. The information comes from official sources; although the demographic and socioeconomic data have a high degree of reliability, the same does not happen with those related to the hemodialysis machine and the epidemiological part. This situation is more notable when Table 1 indicates that these factors have greater weight than the rest. The list of hemodialysis machines used belongs to public institutions; private services are not included. There is no record of the treatments that patients take; since hemodialysis is a recurrent treatment, the quality of life is altered when it is not followed up, and the lack of records does not allow it to be adequately tracked. The number of machines was calculated based on continuous shifts and with a minimum time considered per patient, although the average time is higher [23]; it should be taken into account that this time must cover the preparation of the patient before and after treatment, hemodialysis, cleaning and adaptation of the station for the next patient, and that patients do not present additional special requirements as a consequence of comorbidities. These factors will alter the results, causing an increase in the number of machines. The model tried to calculate by integrating contextualized information to appreciate the target population's situation better. Regardless of whether the values obtained are high or not, the need to address this health problem with programs aimed at planning care areas that are physically closer to people, that promote its prevention and control its progress, is evident. as it has been exposed in other studies as in [24].

5 Conclusions

We proposed a resource allocation-based model to determine the required hemodialysis machines for a specific target population in Mexico. The female population between 24–49 years old with low levels of schooling, informal economic activities, and living in underdeveloped central areas need more than 20 thousand machines to meet their needs. Further analysis of the determinants of access to the service and the equipment proposals is addressed, particularly modifications to the model parameter calculation, alternative model proposals and approaching the lack of data.

References

1. Bikbov, B., et al.: Global, regional, and national burden of chronic kidney disease, 1990–2017: a systematic analysis for the global burden of disease study 2017. Lancet **395**(10225), 709–733 (2020). https://doi.org/10.1016/s0140-6736(20)30045-3
2. Gonzalez-Bedat, M., Rosa-Diez, G., Pecoits-Filho, R., et al.: Burden of disease: prevalence and incidence of ESRD in Latin America. Clin Nephrol **83**(7 Suppl 1), 3–6 (2015)
3. Garcia-Garcia, G., Tonelli, M., Ibarra-Hernandez, M., et al.: Access to kidney transplantation in Mexico, 2007–2019: a call to end disparities in transplant care. BMC Nephrol **22**, 99 (2021). https://doi.org/10.1186/s12882-021-02294-1
4. Klarenbach, S.W., Tonelli, M., Chui, B., Manns, B.J.: Economic evaluation of dialysis therapies. Nat. Rev. Nephrol. **10**(11), 644–652 (2014). https://doi.org/10.1038/nrneph.2014.145
5. Assanatham, M., Pattanaprateep, O., Chuasuwan, A., et al.: Economic evaluation of peritoneal dialysis and hemodialysis in Thai population with end-stage kidney disease. BMC Health Serv. Res. **22**, 1384 (2022). https://doi.org/10.1186/s12913-022-08827-0
6. Mendelssohn, D.C., Mujais, S.K., Soroka, S.D., Brouillette, J., et al.: A prospective evaluation of renal replacement therapy modality eligibility. Nephrol. Dial. Transplant. **24**(2), 555–561 (2009). https://doi.org/10.1093/ndt/gfn484
7. Van Biesen, W., Jha, V., Abu-Alfa, A.K., et al.: Considerations on equity in management of end-stage kidney disease in low-and middle-income countries. Kidney Int. Suppl. **10**(1), E63–E71 (2020). https://doi.org/10.1016/j.kisu.2019.11.004
8. Roberts, B.T., Rodgers, G.P.: NIDDK initiatives addressing health disparities in chronic diseases. J. Clin. Investig. **130**(10), 5036–5038 (2020). https://doi.org/10.1172/JCI141563
9. Mohamed, N.E., et al.: Association between chronic kidney disease and COVID-19-related mortality in New York. World J. Urol. **39**(8), 2987–2993 (2021). https://doi.org/10.1007/s00345-020-03567-4
10. Carrero, J., Hecking, M., Chesnaye, N., et al.: Sex and gender disparities in the epidemiology and outcomes of chronic kidney disease. Nat. Rev. Nephrol **14**, 151–164 (2018). https://doi.org/10.1038/nrneph.2017.181
11. Sánchez-Cedillo, A., Cruz-Santiago, J., Mariño-Rojas, F.B., Hernández-Estrada, S., García-Ramírez, C.: Disease burden: renal failure, dialysis-hemodialysis and renal transplantation in Mexico. Cost of Disease (In Spanish). Rev. Mex. Traspl **9**(1), 15–25 (2020). https://doi.org/10.35366/94025

12. National Institute of Statistics and Geography (INEGI). https://cuentame.inegi.org.mx/poblacion/hogares.aspx
13. Aguilar-Bustamante, F., Molina-Salazar, R.E., Amozurrutia-Jimenez, J.A., Martinez-Licona, F.: Healthcare equity and chronic renal failure in Mexico: an analysis for the case of women. Medwave **19**(3), e7619–e7619 (2019). https://doi.org/10.5867/medwave.2019.03.7619
14. National population council (CONAPO) at: http://www.conapo.gob.mx/work/models/CONAPO/Resource/1342/1/images/02introduccion.pdf. 28 Mar 2023
15. National Institute of Statistics and Geography (INEGI). https://www.inegi.org.mx/programas/ccpv/2010/default.html#Datos_abiertos
16. Ministry of Health at https://datos.gob.mx/busca/dataset/recursos-en-salud-nivel-central. 3 Feb 2023
17. Martinez-Licona, F.M., Martinez-Vazquez, C.M.: A proposed model for calculate the number of mammography machines for the south-west area of Mexico. In: IX Latin American Congress on Biomedical Engineering Proceedings, Florianopolis Brazil (in press)
18. Peacock, S., Smith, P.: The resource allocation consequences of the new NHS needs formula. York Centre of Health Economics, University of York (1995)
19. Kidney disease in Mexico: prevention, promotion, care, and follow-up. Mexico City Government at. https://www.gob.mx/salud/prensa/119-enfermedad-renal-en-mexico-prevencion-promocion-atencion-y-seguimiento?idiom=es. Accessed 5 Apr 2023
20. Ávila-Palomares, P., López-Cervantes, M., Durán-Arenas, L.: Calculating the optimum size of a hemodialysis unit based on infrastructure potential. Salud Pública de México (52), 315–323 (2010). https://www.scielo.org.mx/pdf/spm/v52n4/v52n4a07.pdf
21. Ordu, M., Demir, E., Tofallis, C., Gunal, M.: A novel healthcare resource allocation decision support tool: a forecasting-simulation-optimization approach. J. Oper. Res. Soc. **72**(3), 485–500 (2021). https://doi.org/10.1080/01605682.2019.1700186
22. Zhang, Y., Wang, Q., Jiang, T., Wang, J.: Equity and efficiency of primary health care resource allocation in mainland China. Int. J. Equity Health **17**, 1–12 (2018). https://doi.org/10.1186/s12939-018-0851-8
23. National Institute of Diabetes and Digestive and Kidney Diseases. Hemodialysis at, 14 April 2023. https://www.niddk.nih.gov/health-information/kidney-disease/kidney-failure/hemodialysis
24. Agudelo-Botero, M., González-Robledo, M.C., Reyes-Morales, H., et al.: Health care trajectories and barriers to treatment for patients with end-stage renal disease without health insurance in Mexico: a mixed methods approach. Int. J. Equity Health **19**, 90 (2020). https://doi.org/10.1186/s12939-020-01205-4

Patient and Occupational Dosimetry Aspects of Holmium-166 Radioembolization: Three Case Studies from Vilnius University Hospital Santaros Klinikos

Kirill Skovorodko[1,2](✉), Marius Kurminas[1,3], Inga Andriulevičiūtė[1], and Renata Komiagienė[1,3]

[1] Vilnius University Hospital Santaros Klinikos, Santariškiu 2, Vilnius, Lithuania
kirill.skov@gmail.com

[2] State Research Institute Center for Physical Sciences and Technology (FTMC), Savanorių Ave. 231, Vilnius, Lithuania

[3] Department of Radiology, Nuclear Medicine and Medical Physics, Faculty of Medicine, M. K. Čiurlionio st. 21/27, Vilnius, Lithuania

Abstract. Radioembolization with Holmium-166 microspheres is a new treatment method which was launched for the clinical use about ten years ago. Radioembolization with Ho-166 microspheres is a minimally invasive procedure in which radioactive microspheres are delivered directly to liver tumors via an interventional radiological procedure. This procedure was launched in Santaros Klinikos in 2022, with two patients suffering from hepatocellular carcinomas and one patient with metastatic cholangiocarcinoma being treated. This study investigated occupational exposure levels under working conditions and patient dosimetry aspects.

For evaluation of occupational exposure, passive and active dosimeters were used. Patient dosimetry calculation was done with dosimetry software Q-Suite 2.1 software (Quirem BV). The planar and SPECT/CT images were acquired with a GE Infinia Hawkeye SPECT/CT system. Interventional surgical procedures were performed using cath lab angiography systems.

The administered activities for treatment planning were 132, 155 and 170 MBq, whilst the treatment activities were 9.30, 9.78 and 9.84 GBq respectively.

Exposure rates immediately after the surgical procedure and immediately before discharge 18 h later were measured at different positions around the patient. Contamination measurements were also carried out. After the treatment procedures mean residual activity in the re-usable administration box, vial and the lines were assessed.

Keywords: Radionuclide Therapy · Radioembolization · Occupational Exposure · Patient Dosimetry · Personalized Medicine

Y. Dekhtyar and I. Saknite (Eds.): NBC 2023, IFMBE Proceedings 89, pp. 215–220, 2023.
https://doi.org/10.1007/978-3-031-37132-5_27

1 Introduction

Radioembolization is a minimally invasive treatment procedure for patients suffering from inoperable liver cancer [1]. For this procedure two kinds of radionuclides loaded into microspheres are used: Yttrium-90 isotope microspheres (SIR-Spheres resin microspheres or glass microspheres) and Holmium-166 microspheres (poly l-lactic acid), other microspheres were not authorized for commercial use in Europe [2]. These two radionuclides emit high-energy beta particles, however, Ho-166 also emits a low energy and low intensity gamma photons, that allows Ho-166 location with a gamma camera. These characteristics of the isotope make it possible to perform personalized dosimetry-based treatment planning with Ho-166 scout dose [3]. For the procedure to be successful, the prescribed activity for the treatment dose must be evaluated by dosimetry-based treatment planning procedure. The scout procedure may be also carried out using Tc-99m MAA (macroaggregated albumin) but localisation is inaccurate as the biodistribution would be different from that of poly 1-lactic acid [4].

The use of ionizing radiation for human imaging or treatment purposes is strictly regulated by various standards, guidelines and legislation [5–7]. Administered activity should be measured and must not deviate from the prescribed administered activity by more than 5% [6]. Absorbed doses to target volumes should be individually planned [7]. Various sources of uncertainty arise from volume-of-interest, activity meter and SPECT measurements and dosimetry software calculations [8].

Another concern is associated with personnel ionizing radiation exposure control and the patient discharge from the hospital. Since the legislation and recommendations are mostly based on I-131 treatment, there is a lack of recommendations for other therapeutic radionuclides. The aim of the current study was to evaluate occupational and patient doses during radioembolization with Ho-166 microspheres procedure.

2 Materials and Methods

2.1 General Information

Radioembolization with Holmium-166 microspheres was launched in Vilnius University Hospital Santaros Klinikos in 2022. In 9-month period (during 2022–2023) three patients were treated.

2.2 Microspheres Properties

The physical and chemical properties of Ho-166 microspheres allow the effective use of this radiopharmaceutical for therapeutic procedures. Microspheres are made of poly-L-lactic acid (PLLA), which contains Ho-166 isotope. The Ho-165 is embedded in the PLLA matrix an is activated to Ho-166 by neutron activation in a nuclear reactor. Ho-166 emits high-energy beta particles (1774.32 keV; yield 48.8% and 1854.9 keV; yield 49.9%) and gamma rays (80.57 keV; yield 6.6% and 1379.40 keV; yield 0.9%, half-life of 26.8 h [9]. Microspheres diameter are in the range of 15–60 μm, density 1.4 g/cm^3, administered through a catheter to a hepatic artery. The paramagnetic properties of Ho make possible its localisation using MRI.

2.3 Planning Procedures (Dosimetry and Treatments Process)

Patient dosimetry calculation was done with Q-Suite 2.1 software (Quirem BV). The same application was used for treatment planning and for treatment evaluation. Treatment planning was done with Ho-166 scout dose. The treatment dose of Ho-166 was administered approximately 2–3 weeks after the scout dose.

The scout and the treatments dose were injected under interventional angiography procedure by the physician accompanied by nuclear medicine radiology technologist.

2.4 Activity Measurements

The readings of activity meter (Veenstra VDC-404 and VDC-405) used for checking of activity (<5% deviation) were verified by the secondary standard chamber Capintec CRC-15R, No. 158488 (4π γ ionization chamber) brought to hospitals by the Ionizing Radiation Metrology Laboratory of the FTMC (Center for Physical Sciences and Technology) that is the National Metrology Institute (NMI) in Lithuania.

2.5 Imaging Process

The planar and SPECT/CT images were acquired with GE Infinia Hawkeye SPECT/CT system. The data were acquired using protocol parameters: low dose CT, zoom 1, matrix size 128 $\times$ 128, rotation mode, time per projection 30 s, energy peak 81 keV $\pm$ 15%, low-energy high-resolution collimator, H mode. Interventional surgical procedures were performed using Philips Azurion 7 cath lab angiography system.

2.6 Dose Rate and Contamination Monitoring

For the discharge process and occupation dose rate measurements, ATOMTEX AT1121 ambient dosimeter was used. For contamination measurements, NuviaTech Healthcare CoMo 170 contamination monitor was used. Medical personnel radiation exposure data was measured with POLIMASTER PM1610B-01 and The Thermo ScientificEPD Tru-Dose. The equipment was metrologically verified by the Vilnius Metrology Centre. The personnel constantly wear passive dosimeters (thermoluminescent TLD and Optically Stimulated Luminescence OSL) in order to measure personal dose equivalent Hp(10) (chest area) and Hp(0.07) (finger) for individual monitoring.

3 Results and Discussion

3.1 Treatment Planning and Treatment Procedure

The administered activities for treatment planning were 132, 155 and 170 MBq, whilst the treatment activities were 9.30, 9.78 and 9.84 GBq respectively. The quantitative assessment after the scout dose showed no radiation exposure risk of lung shunting (calculated lung dose was lower than 25 Gy).

The average duration of the interventional procedure for the administration of the scout dose of Ho-166 was 90 min (the median radiation exposure time was 750 s) and for the treatment dose 76 min (the median radiation exposure time was 482 s). The average value of kerma-area product for the scout dose administration was 78.37 Gy · cm^2, for the treatment dose injection was 36.29 Gy · cm^2.

The residual activity in the system after the treatment procedure was in the range of 11.8–60.1 MBq (lower than 2%).

3.2 Radiation Safety

Since ALARA (as low as reasonably achievable) principles are applicable for any procedure that involves the use of radioactive material, in order to properly evaluate risk associated with the Ho-166 procedure, ionizing radiation dose rate was measured during different procedure steps: during Ho-166 transportation package unpacking, activity measurement and loading to the re-usable administration box (Fig. 1), during Ho-166 administration process near the patient, during the patient's stay on the ward and during the discharge process. The results of passive dosimetry measurements did not show an excess of dose levels compared to the previous measurement period, thus, an electronic personal dosimeter was used for a detailed analysis of radiation exposure. Since the interventional radiologist received the radiation exposure dose mostly from the scattered X-ray radiation from patient [10, 11], a detailed assessment of exposure is required to evaluate the exposure of the whole body and extremities.

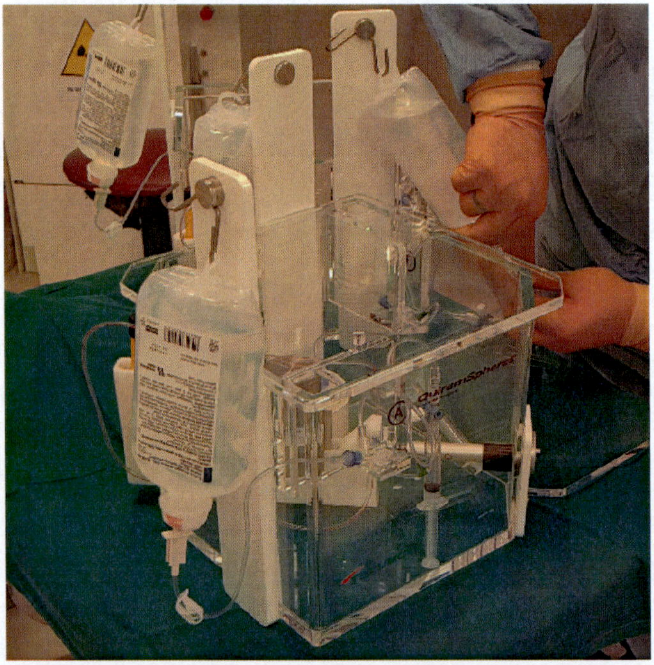

Fig. 1. Re-usable shielded administration box for administration of Ho-166 microspheres (angiosuite – with Customer Kit and Delivery Set).

3.3 Dose Rate During Different Measurement Steps and Contamination Results

The dose rate during the unpacking of the treatment dose was up to 0.18 mSv/h (for ~10 GBq) at 1 m from the surface of the packaging.

The average dose received by the interventional radiologist for the whole body (the electronic dosimeter was worn on waist level) was 50 ± 11 μSv per procedure. This radiation exposure dose was obtained during an interventional radiology procedure with Ho-166 using an angiography system. Immediately after the administration of the therapeutic dose, the occupational dose rate at one meter from the right side (liver side) of the patient was up to 80 μSv/h, the dose rate results from the different positions are shown in Table 1. The dose rate at a distance of 1 m from the patient before discharge (18 h following therapeutic Holmium administration) was in the range 24 to 32 μSv/h.

The contamination measurement showed that after the scout and treatment procedures, the surface in the operational theatre room was free from contamination.

Table. 1. Dose rate results from the different positions and places.

Source description and conditions	Distance from the source (cm)	Dose rate (μSv/h)
Ho-166 shielded package with ~150 MBq	10	15–20
Ho-166 shielded package with ~10 GBq	100	140–180
Exactly after ~9.5 GBq injection, interventional radiologist position	5–10	180–210
After ~9.5 GBq injection, near the patient	100	65–80
After ~9.5 GBq injection, near the patient	200	18–30
After ~9.5 GBq injection, near the patient	300	2–5

4 Conclusions

The prescribed activity for a treatment dose was evaluated using a dosimetry-based treatment planning procedure with Ho-166 scout dose. No radioactive surfaces contamination was found after the procedures.

References

1. van Roekel, C., et al.: Evaluation of the safety and feasibility of same-day Holmium-166 -Radioembolization simulation and treatment of hepatic metastases. J. Vasc. Interv. Radiol. **31**(10), 1593–1599 (2020). https://doi.org/10.1016/j.jvir.2020.01.032

2. d'Abadie, P., Hesse, M., Louppe, A., Lhommel, R., Walrand, S., Jamar, F.: Microspheres used in liver radioembolization: from conception to clinical effects. Molecules **26**, 3966 (2021). https://doi.org/10.3390/molecules26133966
3. Smits, M.L.J., et al.: The superior predictive value of 166Ho-scout compared with 99mTc-macroaggregated albumin prior to 166Ho-microspheres radioembolization in patients with liver metastases. Eur. J. Nucl. Med. Mol. Imaging **47**(4), 798–806 (2019). https://doi.org/10.1007/s00259-019-04460-y
4. Klaassen, N.J.M., Arntz, M.J., Gil Arranja, A., Roosen, J., Nijsen, J.F.W.: The various therapeutic applications of the medical isotope holmium-166: a narrative review. EJNMMI Radiopharm. Chem. **4**(1), 1–26 (2019). https://doi.org/10.1186/s41181-019-0066-3
5. International Atomic Energy Agency: Radiation Protection and Safety of Radiation Sources: International Basic. Safety Standards. Safety Standards, No. GSR Part 3. Vienna (2014)
6. International Atomic Energy Agency: Radiation Protection and Safety in Medical Uses of Ionizing Radiation. Specific Safety Guides, No. SSG-46. Vienna (2018)
7. Council Directive 2013/59/Euratom of 5 December 2013 laying down basic safety standards for protection against the dangers arising from exposure to ionising radiation, and repealing Directives 89/618/Euratom, 90/641/Euratom, 96/29/Euratom, 97/43/Euratom and 2003/122/Euratom (2013)
8. Gear, J.I., et al.: EANM practical guidance on uncertainty analysis for molecular radiotherapy absorbed dose calculations. Eur. J. Nucl. Med. Mol. Imaging **45**(13), 2456–2474 (2018). https://doi.org/10.1007/s00259-018-4136-7
9. Nijsen, J.F., Krijger, G.C., van Het Schip, A.D.: The bright future of radionuclides for cancer therapy. Anti Cancer Agents Med. Chem. **7**(3), 271–290 (2007). https://doi.org/10.2174/187152007780618207
10. Stella, M., Braat, A.J.A.T., van Rooij, R., et al.: Holmium-166 radioembolization: current status and future prospective. Cardiovasc. Intervent. Radiol. **45**, 1634–1645 (2022). https://doi.org/10.1007/s00270-022-03187-y
11. Braat, A.J.A.T., Prince, J.F., van Rooij, R., Bruijnen, R.C.G., van den Bosch, M.A.A.J., Lam, M.G.E.H.: Safety analysis of holmium-166 microsphere scout dose imaging during radioembolisation work-up: a cohort study. Eur. Radiol. **28**(3), 920–928 (2017). https://doi.org/10.1007/s00330-017-4998-2

Electroencephalography as an Objective Indicator of Stress

Marietta Gavriljuk[1], Tuuli Uudeberg[1], Kristjan Pilt[1], Deniss Karai[1], Ivo Fridolin[1,2], and Maie Bachmann[1](✉)

[1] Tallinn University of Technology, Ehitajate tee 5, 19086 Tallinn, Estonia
maie.bachmann@taltech.ee

[2] FinEst Center for Smart Cities (Finest Centre), Tallinn University of Technology, Ehitajate tee 5, 19086 Tallinn, Estonia

Abstract. Mental stress can lead to different health problems or may increase the risk of accidents, especially in cases of high personal responsibility (pilots, policemen, military specialists etc.). Therefore, there is a need for an objective assessment of mental stress in everyday life to prevent serious mental disorders and accidents. The aim of this study was to find out whether, with the help of adding the secondary stress caused by the serial sevens test, the frontal EEG theta band power can differentiate the change in primary stress causing fatigue. The results demonstrate that in the case of increased stress, the added cognitive load does not get as many resources as at baseline. The results suggest that frontal EEG theta band power combined with the cognitive test could be a potential tool to determine the prevalence of mental stress.

Keywords: Mental stress · electroencephalography · theta band power · serial sevens test

1 Introduction

Stress is a serious public health concern that can force significant psychological or biological change upon an organism. It results from physical or psychosocial disequilibrium where the body tries to preserve homeostasis or ensure survival [1]. Research has indicated that stress has a direct effect on our health through autonomic and neuroendocrine responses. Besides that, the indirect influence is expressed through changes in health behaviors [2]. Unlike physical strain which is often quantified, the definition of psychological stress is a more difficult concept [1]. The assessment of mental stress is often based on subjective self-report questionnaires [3] which raises the need for objective evaluation of mental stress in everyday life to prevent serious mental disorders and accidents. Nowadays, most of the stressors we encounter are psychological rather than physical [4].

Numerous research studies have focused on the noninvasive detection and classification of stress levels through biological signals that are expected to express changes in the

Y. Dekhtyar and I. Saknite (Eds.): NBC 2023, IFMBE Proceedings 89, pp. 221–226, 2023.
https://doi.org/10.1007/978-3-031-37132-5_28

normal functioning of the body. Signals such as electroencephalogram (EEG), electrocardiogram (ECG), electromyogram (EMG), galvanic skin response (GSR), blood pressure (BP), skin temperature (SKT), photoplethysmogram (PPG) and others can extract specific information and be used as markers of stress [5]. Since brain is the key organ, that activates many neuronal circuits when a situation is perceived as stressful [6], the utilization of EEG can unravel details of acute stress responses [7]. Compared to other neuroimaging techniques, EEG modality has several advantages such as low cost, high temporal resolution and ease of use.

Different protocols are implemented to induce stress in laboratory settings, such as the Stroop test (Stroop), cold pressor test (CPT), Trier Social Stress Test (TSST), Serial Sevens Test (SST), and others. In order to obtain the EEG signal during the subject exposure to the stress conditions, the selected protocol should be completed in a steady position to avoid any unwanted electrical input induced by the body's physical movements. Arithmetical tests can be completed with minimized movements thus SST is a suitable stressor to measure brain activity and monitor mental workload changes. Mental load is the aspect of cognitive load originating from the interaction between task and subject characteristics [8].

Numerical cognition is associated with a frontoparietal brain network and studies have indicated that theta band power provides evidence for the specific neural activation in this region [9]. Additionally, changes in frontal theta band power were associated with the development of mental fatigue [9]. While the primary stress and related mental fatigue is hidden in the normal variability of the brain activity, the cognitive task-related secondary stress as a theta band power increase can be measured. This increase, whether higher or lower, can indicate the level of mental fatigue.

In the present study we aim to investigate whether with the help of adding the secondary stress caused by SST, the frontal EEG theta band power can differentiate the baseline and follow-up condition, in which a subject is expected to have different primary stress levels.

2 Methods

2.1 Subjects and Recording Protocol

The EEG data were obtained from a group of 42 subjects consisting of 21 females and 21 males with a mean age of 43 years, and a standard deviation (SD) of 13 years. The subjects were recruited among the residents of one Estonian town, with a population of 53 424 inhabitants, who did not have any known diagnosis of cardiovascular disease, mental disorder, head trauma, and prescribed medications. This study was approved by the Research Ethics Committee of the National Institute for Health Development with the decision no. 1064 (study no. 2351) and research was conducted in accordance with the Declaration of Helsinki. All subjects signed written informed consent.

Initially, the test subject was introduced to the study procedure, answered questions about demographic and anthropometric parameters, and underwent weight and blood pressure measurements. Electrocardiography (ECG), photoplethysmography (PPG), electroencephalography (EEG), galvanic skin response (GSR), finger and cheek temperature signals were simultaneously recorded during the experiment. Only the EEG signal analysis is reported in the current publication.

Thereafter, the subject was comfortably sitting on a chair facing the white wall at a 2 m distance. Next, the baseline signals were recorded in a sitting position at rest, starting from the 3 min of eyes closed recording, followed by 4 min of eyes open (EO) recording, and concluded by the 2 min of cognitive task where the test subject performed arithmetic serial sevens task. Next, the subject spent about two hours in different outdoor urban environments while periodically completing the arithmetic task. Upon the subject's return to the room, the same recording protocol was applied resulting in the follow-up EEG signals recorded at higher fatigue condition.

2.2 Data Acquisition

The EEG signals were recorded using Enobio 8 (Neuroelectrics Barcelona SLU) device using adhesive electrodes. The electrodes were placed on the forehead (Fp1 and Fp2) according to the international 10/20 system, while two electrodes for reference and ground were placed behind the right ear. Raw EEG signals were recorded at a sampling frequency of 500 Hz.

The recorded EEG signals were digitally filtered at the cut-off frequencies of 3 Hz and 47 Hz. The signals were visually inspected and signal segments with artefacts were removed. Therefore, further analysis was performed on 110 s EO resting and SST EEG signal segments from 39 subjects, where EO resting and SST segments were obtained as the last 110 s of the corresponding segment. MATLAB (The MathWorks, Inc.) software was used for EEG signal processing.

2.3 EEG Analysis

First, each 110 s EEG signal segment was cut into 10 s epochs. Power spectral density (PSD) was estimated by means of the Welch averaged periodogram method. Epochs were divided into overlapping sections (50%), with a length of 1024 points. Thereafter, theta band power was calculated at 4–7 Hz for each epoch. After that, epoch-based median theta power was obtained for resting and SST segments. The normalized theta power difference of rest and SST conditions was calculated as:

$$TP_{diff} = \frac{\theta\_P_{SST} - \theta\_P_{Rest}}{\theta\_P_{SST} + \theta\_P_{Rest}} \qquad (1)$$

The normalized theta power difference represents the difference between the theta band power in the EO cognitive arithmetic task condition (θ_P_{SST}) and the theta band power in the EO resting condition (θ_P_{Rest}), where the normalization is performed relative to their sum.

The paired sample Wilcoxon signed rank test was applied for the statistical comparison of the normalized theta power difference between baseline and follow-up recordings. The confidence level of 0.05 was considered statistically significant to the Wilcoxon p values.

3 Results and Discussion

Figure 1 presents the theta power values for eyes open resting and SST condition. There is a statistically significant increase in theta power in SST condition ($p = 0.0011$). The results are in agreement with the review performed by Chikhi et al. [10] concluding that increased cognitive load is often indicated in increased frontal theta band power.

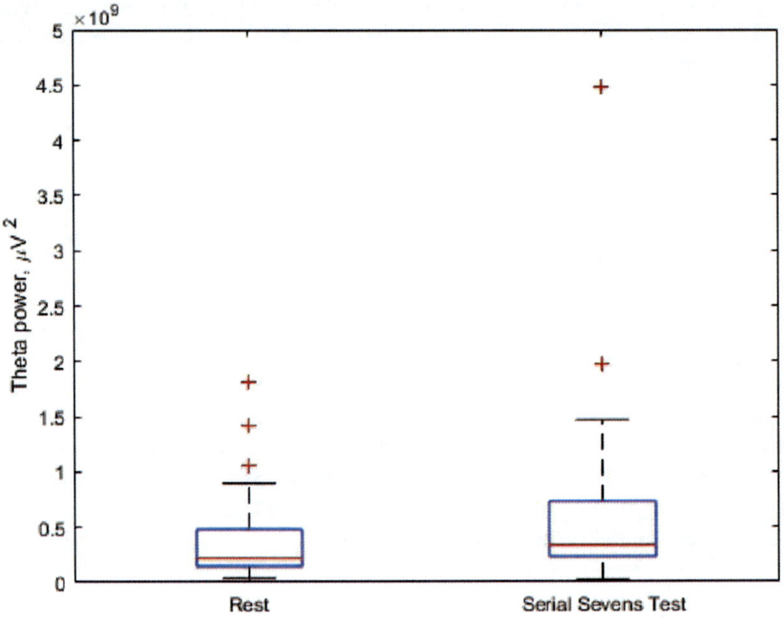

Fig. 1. A statistically significant difference in theta power for eyes open resting and Serial Sevens Test ($p = 0.0011$). Indicated are median (horizontal line), 25th and 75th percentiles (box), minimum and maximum values (whiskers) and outliers (+).

Due to the high theta power variability in the group (interindividual differences), the normalized theta power difference between rest and SST was compared to analyze the changes for baseline and follow-up recordings at the group level.

Figure 2 presents the normalized theta power differences for baseline and follow-up recordings. The positive median values for baseline and follow-up conditions indicate the increased theta power in STT in both conditions. Still, the difference is statistically significantly higher for baseline recordings compared to follow-up recordings indicating higher brain activity on calculations during baseline, while at follow-up the cognitive workload on calculations is lower. As the location where the test was performed was the same for both conditions, the difference cannot be explained by the recording environment. It is presumed that during follow-up the subjects had higher fatigue levels, so the amount of available cognitive resources for STT was lower [11].

As the EEG reflects the brain state which in turn is sensitive to recording conditions, it is suggested to perform the EEG studies always at the same time of day, while the

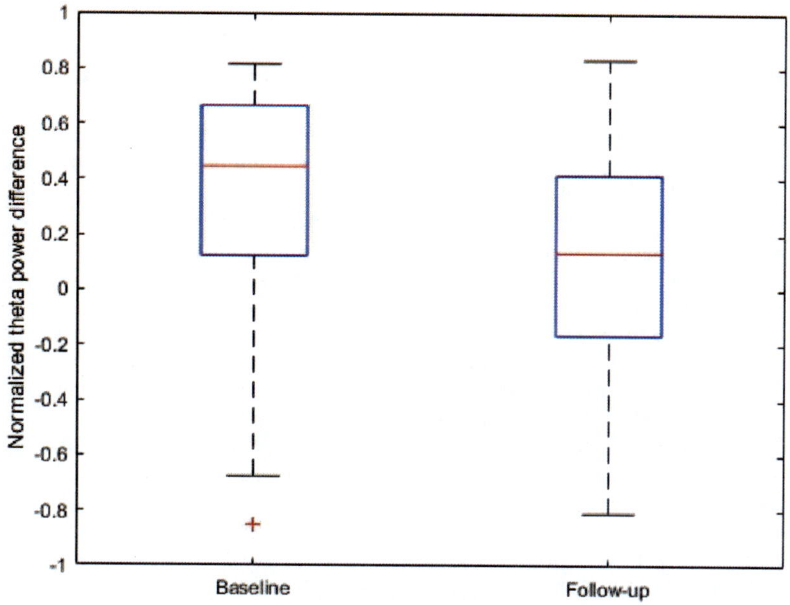

Fig. 2. Statistically significant increase in normalized theta power difference for baseline condition compared to follow up ($p = 0.0076$). Indicated are median (horizontal line), 25th and 75th percentiles (box), minimum and maximum values (whiskers) and an outlier (+).

morning is preferred as the brain is not yet loaded with different everyday assignments. On the other hand, the sensitivity to recording conditions could also be put to work as an advantage – the theta band power together with a cognitive test could be used to evaluate the stress caused by specific environments etc.

The current study has some limitations to consider. As the group's overall variability of the normalized theta power difference was high, considering the age could decrease the overall variability. It is presumed that SST has age limitations and there is an age group for whom this method works better than others. The present study did not consider males and females separately, while it has been found that the EEG power at different conditions varies for males and females [12].

4 Conclusion

The results of the performed study confirmed that frontal EEG theta band power increases statistically significantly due to cognitive load. Employing the SST allowed to differentiate the condition in which the brain was lacking the resources due to increased fatigue. The study suggests that theta band power combined with a cognitive test could be a potential tool to determine the prevalence of mental stress.

Acknowledgements. This work was supported partly by the Estonian Ministry of Research and Education and European Regional Development Fund (grant 2014-2020.4.01.20-0289).

References

1. Thoits, P.A.: Stress and health: major findings and policy implications. J. Health Soc. Behav. **51**, 41–53 (2010)
2. Tomiyama, A.J.: Stress and obesity. Annu. Rev. Psychol. **70**, 703–718 (2019)
3. Cohen, S., Kamarck, T., Mermelstein, R.: A global measure of perceived stress. J. Health Soc. Behav. **24**(4), 385–396 (1983)
4. Björntorp, P.: Do stress reactions cause abdominal obesity and comorbidities? Obes. Rev. **2**(2), 73–86 (2001)
5. Ladakis, I., Chouvarda, I.: Overview of biosignal analysis methods for the assessment of stress. Emerging Sci. J. **5**(2), 233–244 (2021)
6. de Kloet, E.R., Joëls, M., Holsboer, F.: Stress and the brain: from adaptation to disease. Nat. Rev. Neurosci. **6**(6), 463–475 (2005)
7. Aspiotis, V., et al.: Assessing electroencephalography as a stress indicator: a VR high-altitude scenario monitored through EEG and ECG. Sensors (Basel) **22**(15), 5792 (2022)
8. Spüler, M., Walter, C., Rosenstiel, W., Gerjets, P., Moeller, K., Klein, E.: EEG-based prediction of cognitive workload induced by arithmetic: a step towards online adaptation in numerical learning. ZDM Math. Educ. **48**(3), 267–278 (2016). https://doi.org/10.1007/s11858-015-0754-8
9. Trejo, L., Kubitz, K., Rosipal, R., Kochavi, R., Montgomery, L.: EEG-based estimation and classification of mental fatigue. Psychology **6**, 572–589 (2015)
10. Chikhi, S., Matton, N., Blanchet, S.: EEG power spectral measures of cognitive workload: a meta-analysis. Psychophysiology **59**(6), 14009 (2022)
11. Tran, Y., Craig, A., Craig, R., Chai, R., Nguyen, H.: The influence of mental fatigue on brain activity: evidence from a systematic review with meta-analyses. Psychophysiology **257**(5), e13554 (2020). https://doi.org/10.1111/psyp.13554
12. Cantillo-Negrete, J., Carino-Escobar, R.I., Carrillo-Mora, P., Flores-Rodríguez, T.B., Elías-Vinas, D., Gutiérrez-Martínez, J.: Gender differences in quantitative electroencephalogram during a simple hand movement task in young adults. Rev. Invest. Clin. **68**(5), 245–255 (2016)

Semi-automatic Approach to Estimate the Degree of Non-alcoholic Fatty Liver Disease (NAFLD) from Ultrasound Images

Simone Kresevic[1(✉)], Milos Ajcevic[1], Mauro Giuffrè[2], Pierpaolo Pupa[3], Lory Saveria Crocè[2], and Agostino Accardo[1]

[1] Department of Engineering and Architecture, University of Trieste, Trieste, Italy
simone.kresevic@phd.units.it

[2] Department of Medicine, Surgery and Health Sciences, University of Trieste, Trieste, Italy

[3] Prodigys Technology Technology s.r.l., Trieste, Italy

Abstract. The early diagnosis of Non-Alcoholic Fatty Liver Disease (NAFLD) is crucial to prevent fibrosis progression or the onset of advanced chronic liver disease. Among the non-invasive methods, ultrasound (US) B-mode imaging is recommended for population screening and follow-up. Hamaguchi's score was proposed to improve the evaluation of the fatty liver from US images.

In our study, we aimed to assess objectively the Hamaguchi score through an advanced semi-automatic analysis of US images.

The study encompassed a dataset of 325 bariatric patients with NAFLD diagnosed by liver biopsy who underwent ultrasound assessment at the Liver Clinic at Trieste University Hospital. The classification models for the estimation of the three Hamaguchi sub-scores were produced by semiautomatic US image analysis based on clustering and Convolutional Neural Network (CNN) with transfer learning techniques. The results showed that the produced models were able to estimate the three sub-scores with high classification accuracy.

The predictive models produced for the estimation of liver brightness hepatorenal echo contrast, the diaphragm deep attenuation, and the vessel blurring sub-scores presented a classification accuracy of 92.6%, 84.8%, and 90.9%, respectively.

In conclusion, in this preliminary study, the results assessed the possibility to produce the NAFLD computer-aided diagnostic models based on analysis of US images.

Keywords: Non-Alcoholic Fatty Liver Disease · Hamaguchi's score · Artificial Intelligence · Ultrasound images

1 Introduction

Non-Alcoholic Fatty Liver Disease (NAFLD) represents a spectrum of diseases that ranges from simple steatosis (i.e., Non-Alcoholic Fatty Liver) to Non-Alcoholic Steatohepatitis (NASH). NAFLD is a common liver disease and is the major cause of liver

Y. Dekhtyar and I. Saknite (Eds.): NBC 2023, IFMBE Proceedings 89, pp. 227–235, 2023.
https://doi.org/10.1007/978-3-031-37132-5_29

failure worldwide. NAFLD can progress to cirrhosis, liver failure, or hepatocellular carcinoma. It is estimated that around one-third of the adult population in developed countries is affected by this disease [1, 2].

NAFLD is caused by excessive fat deposition in the liver and is almost asymptomatic. Thus, early evaluation and monitoring of liver functions are important to prevent advanced fibrosis or hepatocellular carcinoma [3]. The gold standard for direct liver steatosis and inflammation quantification is liver biopsy. However, performing liver biopsies is invasive and expensive, making it unsuitable for broad screening at the population level [4].

Among the non-invasive liver imaging methods, Ultrasound-Based (US-B) techniques are the most suitable for the quantification of liver steatosis in individuals of all ages [3]. Indeed, the clinical practice guidelines have recommended US-B modalities as the preferred first-line diagnostic procedure for imaging in adults [5]. In daily practice, physicians use specific features of a US image to evaluate hepatic steatosis. These features include analyzing the echogenicity of the liver, examining the visibility of intrahepatic vessels, and assessing the degree of diaphragm blurring as reported by Hamaguchi *et al.* [6]. However, different studies report that this diagnostic method is limited by its low intra- and inter-observer reproducibility. In particular, Strauss et al. showed in a study composed of 168 US examinations an intra- and inter-observer agreement of 54.7%–67.9% and 47.0%–63.7%, respectively [7, 8]. Furthermore, the diagnostic accuracy of conventional B-mode ultrasonography is lower in bariatric patients [9]. To improve objectivity, Hamaguchi et al. introduced a semi-quantitative scoring method based on US imaging [10].

Hamaguchi's scoring system involves comparing the echogenicity of the liver to the echogenicity of the kidney, evaluating liver brightness (scored from 0 to 3), assessing the degree of diaphragmatic contour attenuation by the liver (scored from 0 to 2), and analyzing liver vessel blurring (scored from 0 to 1). The total of these sub-scores yields Hamaguchi's score, which ranges from 0 (representing a healthy liver) to 6 (indicating a fatty liver). However, despite improving the standardization of hepatic lipid accumulation in the liver, Hamaguchi's score is still a subjective measure that strongly relies on the physician's expertise evaluating the US image.

In the last decade, Artificial Intelligence (AI) has been used in various medical fields to detect and forecast patterns or correlations within large datasets. These AI-based methods have been demonstrated to be helpful in the diagnostic process. A recent meta-analysis focused on using AI-based methods for liver diagnosis revealed high accuracy in diagnosing and staging NAFLD based on diagnostic imaging [11]. Although AI has been extensively studied in the diagnostic field of NAFLD, there has been no prior application of AI-based methods in the analysis for the estimation of Hamaguchi's score (from 0 to 6) in bariatric patients. Thus, this study aims to create algorithms, using clustering and Convolutional Neural Networks (CNN) with transfer learning techniques, to estimate different sub-items that compose Hamaguchi's score by analysis of bariatric patients' US images.

2 Materials and Methods

2.1 Data Acquisition and Dataset Definition

The clinical and radiological data of 325 bariatric patients (mean age 50.6 ± 14.1 years, mean Body Mass Index (BMI) 39.5 ± 8.6 kg/m^2, 34.5% males and 65.5% females) who underwent US assessment at the Liver Clinic at Trieste University Hospital were analyzed. The inclusion criteria were liver biopsy-based NAFLD diagnosis, and US assessment characterized by the visibility of liver and renal parenchyma, visibility of intrahepatic vessels, and visibility of the diaphragm.

To calculate the sub-scores composing Hamaguchi's score, three parameters need to be evaluated in the US image. Therefore, three separate sub-score specific datasets were created: the Hepatorenal dataset, the Diaphragm dataset, and the Vessel dataset.

However, it may not always be possible to assess all three parameters on a single US scan due to variations in the physical conformation of bariatric patients. Some patients may have all three parameters visible in a single US image, while others may have the liver and the kidney in one image and the diaphragm and intrahepatic vessels in another. Additionally, some patients may have all three parameters visible in three separate US images. Therefore, for each patient, from one to three images were included and labeled to create three aforementioned datasets.

Each of the three datasets underwent a thorough evaluation and labeling process, where four physicians independently assigned the sub-scores constituting Hamaguchi's score. In cases where the ratings were inconsistent, the scores were reviewed and determined through consensus. After labeling the images, pre-processing techniques were employed to locate and center the US cone containing the relevant information.

2.2 Framework Proposed for Semi-automatic Evaluation of Hamaguchi's Score

The proposed approach for automatic estimation of Hamaguchi's score is based on separate estimation of sub-scores by semiautomatic analysis of US images. Since each sub-score assessment relies on identifying different features, three sub-score-related algorithms were developed. A semi-automatic evaluation based on clustering was adopted to assess the liver brightness hepatorenal echo contrast score (sub-score 1). However, for evaluating the diaphragm deep attenuation score (sub-score 2) and the vessel blurring score (sub-score 3), a semi-automatic methodology was implemented using two CNNs with transfer learning techniques. Due to the variability of the US images and moderate sample size, additional pre-processing was necessary to produce effective classification algorithms. This involved manually delineating the regions of interest (ROIs) to define the Diaphragm and Vessel datasets.

Estimation of Liver Brightness Hepatorenal Echo Contrast Sub-score. An excessive fat component in the liver causes it to appear brighter compared to the kidney (typically isoechogenic [5]). With the support of physicians, two regions of interest (ROIs) were manually identified: one in the lighter region of the liver parenchyma and the other in the darker area of the cortical region of the kidney. A Python script was used to select two ROIs (each of size 5x5 pixels) and process them to extract the features of interest. This allowed to obtain features related to the echogenicity of the liver for each

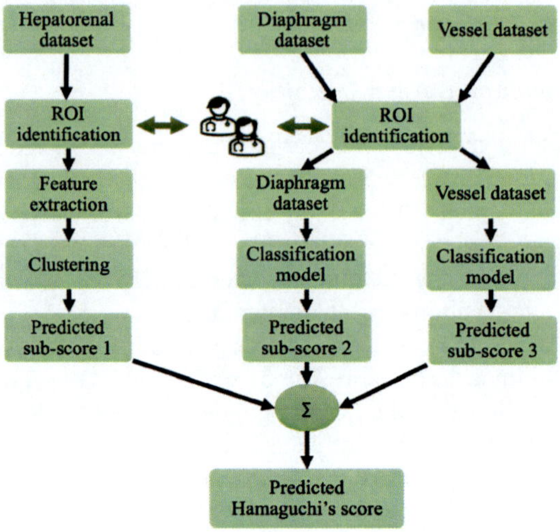

Fig. 1. The proposed framework

US image sample. All previously saved features are later analyzed with K-means by setting the number of classes to four (liver brightness hepatorenal echo contrast score could be 0, 1, 2, or 3). Based on this clustering, a classification model was constructed to assign a score ranging from 0 to 3 to the feature extracted from a US image of the liver. Subsequently, it was assessed whether the four clusters correctly predicted the sub-score by comparing the prediction with the physicians' assessment.

Estimation of Diaphragm Deep Attenuation Sub-score and Liver Vessel Blurring Sub-score. The diaphragm deep attenuation sub-score is evaluated according to how well the diaphragm is visible: the greater the presence of fat in the liver, the greater the absorption of ultrasound signal and thus the lower the visibility of the diaphragm. Similarly, the liver vessel blurring sub-score is also affected by the presence of fat [5]. To evaluate these sub-scores, CNNs and transfer learning techniques were used.

Before implementing and training the CNNs, each image dataset was divided into three sub-sets: training set, validation set, and test set. The ratio of partitioning was 75%, 15%, and 10%, respectively. The test set was solely used after the model had undergone full training. To give correct assessments, the test set must be well maintained and contain a variety of images covering the various classes of each sub-score.

To achieve high classification accuracy in the CNNs, despite using the transfer learning technique, a substantial number of images are still necessary in the training set, and it is crucial to balance the sample case histories. Therefore, data augmentation was employed. This technique involves applying transformations to the US images to expand the size of the datasets and balance them, thus preventing overfitting. The original US images were subjected to rotational, shear distortion, zoom, and horizontal flip operations randomly to create new US images with different variations.

To select the most appropriate and effective network, for both the sub-scores, ten CNNs were implemented using variations on the base model imported or the number of neurons used in the top layers (Table 1). The networks' development was implemented with Python scripts by importing the TensorFlow and Keras libraries. In particular, the pre-trained networks VGG-16 and VGG-19 from Keras are imported as the base model. A block of convolutional layers with trainable parameters is added to the output of the VGG-Net to extract additional features. These are followed by a flatten layer that converts the previous layer's output into a one-dimensional array, which is passed to the top layer. The top layer architecture differs between the CNN models developed depending on the number of convolutional layers and nodes per layer. However, all top layers use the same activation function, the Rectified Linear Unit (ReLU). Dropout layers are placed between dense layers of all models to prevent overfitting. The models' output layer differs between the two sub-scores: the diaphragm deep attenuation sub-score could be 0, 1, or 2, so it should be able to predict three classes. Instead, the vessel blurring score should only involve two classes.

The diaphragm deep attenuation score is a multi-class problem, so the CNN output is an array of probabilities associated with each score value. The output is determined by the index in the array with the highest probability. The activation function is a SoftMax, and the loss function is categorical cross-entropy. Since the vessel blurring score is a binary problem, the CNN output layer needs only one neuron that reports the predicted parameter value and the probability of the prediction. The activation function is a sigmoid, and the loss function calculation is binary cross entropy.

The performances of models were evaluated by classification accuracy, loss, Area Under the ROC Curve (AUC), and precision. Subsequently, the model with the best performances on the validation set was assessed on the test set.

Table 1. Architectures of the ten CNNs implemented to estimate the diaphragm deep attenuation sub-score and liver vessel blurring sub-score

#	Base model	Additional convolutional block	Classification block
1	VGG-16	-	Flatten layer Dense layer (512 nodes) Dense layer (256 nodes) Output layer
2	VGG-16	Conv2D layer (512 nodes) Conv2D layer (512 nodes) MaxPooling2D layer	Flatten layer Dense layer (512 nodes) Dense layer (256 nodes) Output layer
3	VGG-16	Conv2D layer (512 nodes) Conv2D layer (256 nodes) MaxPooling2D layer	Flatten layer Dense layer (4096 nodes) Dense layer (512 nodes) Output layer

(continued)

Table 1. *(continued)*

#	Base model	Additional convolutional block	Classification block
4	VGG-16	Conv2D layer (512 nodes) Conv2D layer (256 nodes) MaxPooling2D layer	Flatten layer Dense layer (4096 nodes) Dense layer (4096 nodes) Output layer
5	VGG-16	Conv2D layer (512 nodes) Conv2D layer (256 nodes) MaxPooling2D layer	Flatten layer Dense layer (512 nodes) Dense layer (256 nodes) Dense layer (128 nodes) Output layer
6	VGG-16	Conv2D layer (512 nodes) Conv2D layer (256 nodes) Conv2D layer (128 nodes) MaxPooling2D layer	Flatten layer Dense layer (1024 nodes) Dense layer (512 nodes) Dense layer (256 nodes) Output layer
7	VGG-16	Conv2D layer (512 nodes) Conv2D layer (256 nodes) Conv2D layer (128 nodes) MaxPooling2D layer	Flatten layer Dense layer (1024 nodes) Dense layer (512 nodes) Output layer
8	VGG-19	-	Flatten layer Dense layer (512 nodes) Dense layer (256 nodes) Output layer
9	VGG-19	Conv2D layer (512 nodes) Conv2D layer (256 nodes) MaxPooling2D layer	Flatten layer Dense layer (512 nodes) Dense layer (256 nodes) Output layer
10	VGG-19	Conv2D layer (512 nodes) Conv2D layer (256 nodes) Conv2D layer (128 nodes) MaxPooling2D layer	Flatten layer Dense layer (1024 nodes) Dense layer (512 nodes) Dense layer (256 nodes) Output layer

3 Results

The model developed, by using the clustering approach, for the estimation of liver brightness hepatorenal echo contrast score showed a high classification accuracy of 92.6%. In addition, the maximum error for this sub-score was one point in the misclassified cases.

For the models designed to estimate the diaphragm deep attenuation and vessel blurring sub-scores, architecture #10 and architecture #2 from Table 1 demonstrated the best classification performance on the validation dataset, respectively.

The classification accuracy, loss, AUC, and precision values obtained for these models are presented in Table 2.

Table 2. The performance parameters of the models which showed the best classification performance for diaphragm deep attenuation and vessel blurring sub-scores.

	Accuracy	Loss	AUC	Precision
Diaphragm deep attenuation	80.69%	0.49	0.93	81.0%
Vessel blurring	82.90%	0.47	0.87	82.0%

The identified models were subsequently evaluated on the test dataset. The confusion matrixes for diaphragm deep attenuation and vessel blurring sub-score models are reported in Table 3-A and 3-B, respectively.

Table 3. Comparison of the predicted sub-score and the sub-score assigned by the physicians (true label) on the test set represented by a normalized confusion matrix

A. Diaphragm deep attenuation sub-score				
True label	SCORE 0	0.73	0.27	0
	SCORE 1	0	1.00	0
	SCORE 2	0	0.18	0.82
		SCORE 0	SCORE 1	SCORE 2
			Predicted label	

B. Vessel blurring sub-score			
True label	SCORE 0	0.82	0.18
	SCORE 1	0.09	0.91
		SCORE 0	SCORE 1
		Predicted label	

The algorithm for the estimation of diaphragm deep attenuation showed an accuracy of 84.8% on the test set. The misclassified scores were under- or over-estimated by no more than one point. Regarding the vessel blurring sub-score model, the identified model presented a classification accuracy of 90.9% on the test dataset. In both cases, the models confirmed the performance obtained during the validation process showing a good generalization performance.

4 Discussion

In our study conducted on bariatric patients, we aim to assess the feasibility of estimating Hamaguchi's score on US images using advanced image analysis techniques based on AI methods. The results obtained in this study demonstrated that AI-based methods can accurately estimate the three sub-scores required to determine Hamaguchi's score. The models developed showed high classification accuracy for all three sub-scores (92.6% on the liver brightness hepatorenal echo contrast sub-score, 84.8% on the diaphragm deep attenuation sub-score and 90.9% on the liver vessel blurring sub-score), indicating their clinical relevance. The findings suggest that decision support systems based on these models could potentially aid in diagnosing liver disease, reducing errors associated with intra- and inter-operator assessments.

There is growing interest in AI-based methods for the diagnosis of liver disease. Recently, different studies reported promising results by models proposed for diagnosis and staging assessment of NAFLD based on diagnostic imaging [11]. In a recent study of 240 non-bariatric participants, a model based on deep learning techniques was developed to classify NAFLD from US images [12]. The aforementioned study showed high performance on the interpretation of the presence of NAFLD (AUC = 0.933), but it presents a poor diagnostic ability in distinguishing intermediate classes (AUC < 0.70) [12] and it classifies the patients as: "normal", "mild", "moderate" and "severe".

However, to the best of our knowledge, no prior studies have aimed at estimating the Hamaguchi score in bariatric patients based on machine learning semi-automatic US image analysis. The main limitation of the proposed approach is that the operator selects the ROI on which to perform the automatic assessment of each sub-score. An automatic tool for selecting regions of interest (ROIs) could be developed to further improve the proposed approach. The findings of this pilot study need to be validated and possibly enhanced by using a larger dataset that is more balanced in terms of clinical characteristics. Also, it can be applied to evaluate US images of bariatric patients and general NAFLD patients' US images.

In conclusion, this study evaluated the feasibility of developing computer-aided diagnostic models for NAFLD based on the analysis of US images.

References

1. Manne, V., Handa, P., Kowdley, K.V.: Pathophysiology of nonalcoholic fatty liver disease/nonalcoholic Steatohepatitis. Clin. Liver Disorder **22**(1), 23–37 (2018)
2. Huang, D.Q., El-, H.B., Loomba, R.: Global epidemiology of NAFLD-related HCC: trends, predictions, risk factors and prevention. Nat. Rev. Gastroenterol Hepatol. **18**(4), 223–238 (2021)
3. Ibacahe, C., et al.: Accuracy of a semi-quantitative ultrasound method to determine liver fat infiltration in early adulthood. Diagnostics (Basel) **10**(6), 431 (2020)
4. Ting Soon, G.S., Wee, A.: Liver biopsy in the quantitative assessment of liver fibrosis in nonalcoholic fatty liver disease. Indian J. Pathol. Microbiol. **64**(Supplement), S104–S111 (2021)
5. Ferraioli, G., Soares Monteiro, L.B.: Ultrasound-based techniques for the diagnosis of liver steatosis. World J. Gastroenterol. **25**(40), 6053–6062 (2019)
6. Targher, G.: What's past is prologue: history of nonalcoholic fatty liver disease. Metabolites **10**(10), 397 (2020)
7. Strauss, S., Gavish, E., Gottlieb, P., Katsnelson, L.: Interobserver and intraobserver variability in the sonographic assessment of fatty liver. AJR Am. J. Roentgenol. **189**(6), W320–W323 (2007)
8. Cengiz, M., Sentürk, S., Cetin, B., Bayrak, A.H., Bilek, S.U.: Sonographic assessment of fatty liver: intraobserver and interobserver variability. Int. J. Clin. Exp. Med. **7**(12), 5453–5460 (2014)
9. Tamaki, N., Ajmera, V., Loomba, R.: Non-invasive methods for imaging hepatic steatosis and their clinical importance in NAFLD. Nat. Rev. Endocrinol. **18**(1), 55–66 (2022)

10. Hamaguchi, M., Kojima, T., Itoh, Y., et al.: The severity of ultrasonographic findings in nonalcoholic fatty liver disease reflects the metabolic syndrome and visceral fat accumulation. Am. J. Gastroenterol. **102**(12), 2708–2715 (2007)
11. Decharatanachart, P., Chaiteerakij, R., Tiyarattanachai, T., Treeprasertsuk, S.: Application of artificial intelligence in chronic liver diseases: a systematic review and meta-analysis. BMC Gastroenterol. **21**(1), 10 (2021)
12. Cao, W., An, X., Cong, L., Lyu, C., Zhou, Q., Guo, R.: Application of deep learning in quantitative analysis of 2-dimensional ultrasound imaging of nonalcoholic fatty liver disease. J. Ultrasound Med. **39**(1), 51–59 (2020)

Ocular Following Response Measurement: Comparing Infrared Eye-Tracking and High-Resolution Video-Oculography

Aleksandar Miladinović[1](✉), Christian Quaia[2], Miloš Ajčević[3], Simone Kresevic[3], Francesco Bassi[3], Stefano Pensiero[1], and Agostino Accardo[3]

[1] Institute for Maternal, Child Health-IRCCS Burlo Garofolo, Trieste, Italy
aleksandar.miladinovic@burlo.trieste.it

[2] Laboratory of Sensorimotor Research, Department of Health and Human Services, National Eye Institute, National Institutes of Health, Bethesda, MD, USA

[3] Department of Engineering and Architecture, University of Trieste, Trieste, Italy

Abstract. Ocular Following Responses (OFRs) are reflexive eye movements that help process visual motion and stabilize gaze. They are sensitive to various stimulus properties and can potentially be used for visual processing in humans, as well as a diagnostic tool for identifying stereodeficencies. However, recording OFRs can be challenging as they are typically quantified over a short time window, and measuring them noninvasively using infrared eye-tracking systems can be limited by their precision and spatial resolution. Recently, a new technique called high-resolution video-oculography (HR-VOG) has emerged, which claims to provide more accurate measurements of OFRs. In this study, we aimed to compare the measurement of OFRs using HR-VOG and infrared eye-tracking techniques. Our findings suggest that HR-VOG may provide more accurate and reliable measurements of OFRs than standard infrared eye-tracking. While both techniques were able to register eye-displacements induced by vertically drifting images during the open-loop period, only HR-VOG was able to detect significant differences between experimental conditions, indicating its potential for more precise measurements. The limitations of infrared eye-tracking, such as the need for precise positioning and calibration, make it unsuitable for clinical applications, especially in children.

This study highlights the potential benefits of HR-VOG as a more reliable tool for studying OFRs in both research and clinical settings. However, future studies with larger sample sizes and a broader range of stimuli should be conducted to further investigate the potential benefits and limitations of HR-VOG and infrared eye-tracking techniques.

Keywords: Ocular following response · OFR · Eye-tracking · Video-oculography · VOG

Mr. Christian Quaia is a US state employee, particularly of the Laboratory of Sensorimotor Research, National Eye Institute, of the National Institutes of Health (NIH)

Y. Dekhtyar and I. Saknite (Eds.): NBC 2023, IFMBE Proceedings 89, pp. 236–245, 2023.
https://doi.org/10.1007/978-3-031-37132-5_30

1 Introduction

Ocular Following Responses (OFRs), reflexive eye movements triggered by the motion of textured patterns in the visual field, have been studied extensively in both humans and non-human primates due to their important role in visual motion processing and gaze stabilization [1–4]. These eye movements, which occur at ultra-short latencies of around 70 ms in humans, are the initial component of the optokinetic nystagmus response that supports the translational vestibulo-ocular reflex system in the stabilization of gaze [5]. Recent studies have shown that OFRs can also be induced, albeit at longer latencies, by relatively small stimuli [6, 7], suggesting that they may also contribute to the initial phase of smooth pursuit eye movements. The strong dependency of OFRs on various stimulus properties, such as size, contrast, and spatial and temporal frequency content, has made them a powerful tool for investigating the visual motion processing in humans. Recently, it has been inferred that, since OFRs are sensitive to interocular correlations [8], they must be mediated by disparity-tuned cortical neurons, and in principle can thus be used as a diagnostic tool for identifying stereodeficiencies [9, 10]. Several clinical tests are currently available to assess binocular depth perception and visual processing, but most of them require patient cooperation, and so are poorly suited for infants and young children [11]. An objective test that can study visual processing without dependence on patient cooperation would be highly beneficial to develop.

However, one of the problems in recording Ocular Following Responses (OFRs) is that they are typically quantified over a very short time window, the so-called open-loop period. The typical ocular following latency in humans corresponds to 70–80 ms, while the 160 ms marks the end of the open-loop period of the movement [7, 8, 12, 13]. During this period, the eye movements are largely driven by visual motion and not yet influenced by feedback from the pursuit system.

The magnitude of OFRs during this period is small, making them difficult to record accurately. Until recently, the only way to measure them in humans was through the use of invasive scleral coils. This technique is not suitable for clinical applications as it requires sedation or anesthesia, and it is uncomfortable and inconvenient for the subject.

Measuring OFRs noninvasively using most common infrared eye-tracking has been shown to be difficult due to the limitations of the system [14, 15]. Additionally, some studies have reported that infrared limbal eye-tracking may suffer from drift and have limited spatial resolution, which may result in inaccuracies when measuring eye movements [14, 15]. These limitations have led to the development of new techniques that claim to provide more accurate and precise measurements of eye movements. One such technique is high-resolution video-oculography (HR-VOG), which has emerged in recent years. HR-VOG claims to provide more accurate measurements of OFRs. To the best of our knowledge, no studies offer a comparison between HR-VOG and the standard infrared eye-tracking technique in the measurement of OFRs in humans. Therefore, the current study aims to compare these two approaches and investigate whether the HR-VOG provides more accurate measurements of OFRs.

2 Methods

2.1 Study Population

This study enrolled five adult subjects (4 males and 1 female) with an average age of 26 ± 3.1 years. All subjects had normal or corrected-to-normal vision and stereopsis measured by the digital stereoacuity test (<100 arc sec) [16]. The study adhered to the tenants of declaration of Helsinki. All participants provided written informed consent before participating in any of the experiment.

2.2 Behavioral Paradigm

In the experimental setup, the participants were seated in a dimly lit room and their heads were stabilized using padded chin and helmet. They were positioned facing a monitor (ASUS VG248QE) that was placed at a distance of 50 cm from the corneal vertex. The monitor had a resolution of 1920 × 1080 pixels and a vertical refresh rate of 144 Hz. The height of the chair was adjusted to align the subject's eyes with the center of the screen. Two lenses, one low pass (blue) and one high-pass (red), with cut-off wavelength of 550 nm, are positioned in front of the participants' eyes. At the start of each trial, the screen was filled with a mid-luminance (7.0 cd/m^2) blank background using only red and blue color channels (Fig. 1b). Stimuli were presented within a square aperture (28° side) centered on the screen and had a mean luminance of 7.0 cd/m^2. Each stimulus was generated by randomly assigning a high or low luminance value (symmetric around the mean luminance) to consecutive pairs of pixel rows (0.06°), and the resulting stimulus was then filtered with low-pass spatial filter. The gain of the filter was zero above 0.75 cycles per degree (cpd) and one below 0.375 cpd, with the transition following a raised-cosine function. The root mean square contrast of the stimulus was set to 30%, which kept the Michelson contrast below 100% to pre-vent saturations. The study utilized a motion of the 1D pattern, where the images were shifted either upwards or downwards at a drift speed of approximately 50°/s. Two conditions were considered: correlated and anti-correlated. In the correlated condition, both monocular images drifted in the same direction with zero disparity and interocular correlation of 1.0 (Fig. 1a, upper image). In the anti-correlated condition, one monocular image was obtained by contrast-reversing the other, resulting in binocular anti-correlated stimulus with zero disparity and interocular correlation of –1.0 (Fig. 1a, lower image). The comparison between responses to correlated (corr) and anticorrelated (anti-corr) stimuli is particularly interesting as they have identical spatial and temporal frequency content and contrast but differ only in interocular correlation.

The high-resolution video-oculography (HR-VOG) system recorded the movements of the right eye, whereas the movements of the left eye were recorded using the infrared eye-tracker. During the experiment, a total of 120 trials, consisting of 30 blocks of stimuli for each condition, were presented to the subject.

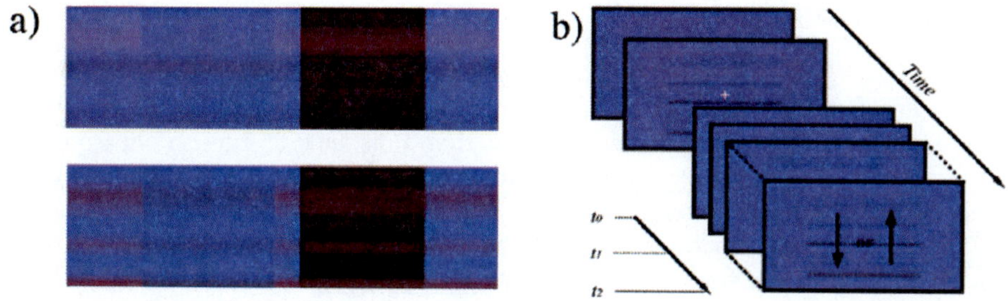

Fig. 1. a) Presented correlated (upper) and anticorrelated stimuli (lower) and how their appear to the participant after filter lenses; b) stimuli fast drifting stimuli (50°/s) upward or downward for a duration of 200 ms.

2.3 High-Resolution Videooculography (HR-VOG)

The HR-VOG system includes a high-resolution camera (FLIR Grasshop-per 3 GS3-U3-51S5M-C) with a resolution of 2448 × 2048 pixels, fitted with a C-Mount 50v mm f/2.8 lens, which records in the near-infrared spectral range. For comparison the average pupil size recorded with the camera is from 60 to 120 pixels. To block the visible spectrum and capture images only in the near-infrared range, a Hoya R72 IR filter with a cut-off wavelength of 720 nm was placed in front of the lens. To ensure adequate lighting conditions, we utilized three custom-built infrared (IR) LED illuminators (one positioned on each side of the subject, and one placed in front and below the subject) as shown in Fig. 2. For each trial, we acquired three frames: t0 = 0 ms (fixation cross offset/motion onset), t1 = 80 ms, which corresponds to the typical ocular following latency in humans to stimuli of similar size and contrast used in previous studies [6, 7], and t2 = 160 ms, which marks the end of the open-loop period of the movement. The exact timing was achieved through the utilization of a specially designed hardware trigger box, which detects the initial frame of the stimulus and generates trigger impulses for t0, t1, and t2.

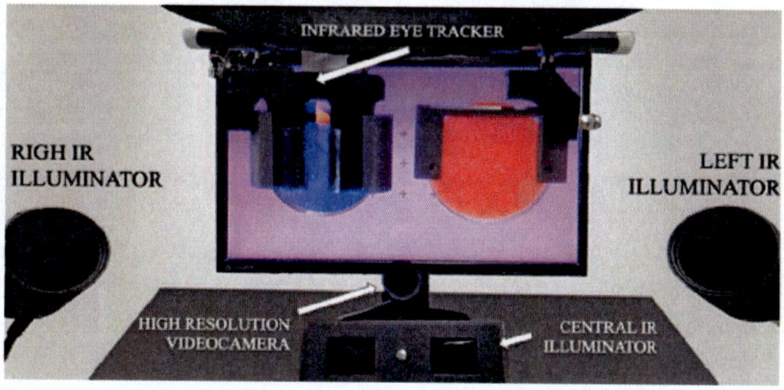

Fig. 2. Recording apparatus

2.4 Infrared (IR) Eye-Tracking

OFRs were recorded on left eye using an IRIS model 6500 (Skalar Medical, Delft, Netherlands) device based on the infrared limbal reflection method (declared resolution of 0.03°). The emitter-detector pair is mounted at a distance of 1.5–2.5 cm from the eye. An iterative process of sensor positioning, and calibration is used, which involves a calibration pattern on the screen to ensure that the emitter-detector pair is accurately positioned. The signal is acquired along with the signal from the hardware trigger box, sampled at 500 Hz, and saved for further analysis.

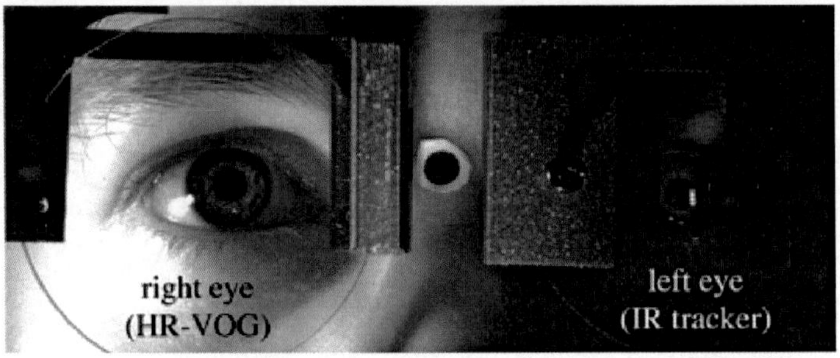

Fig. 3. Image sample taken by high-resolution infrared videocamera

2.5 Data Processing and Statistical Analysis

After obtaining a reference frame from HR-VOG at time t0, our algorithm extracted the displacement of the head marker and pupil center at frames t1 and t2 with sub-pixel resolution, using the reference frame as a guide. The eye displacement during the fixation epoch (frames t1 vs t0) and movement epoch (frames t2 vs t1) was calculated.

The signal obtained from the infrared tracker was epoched from –25 ms to 180 ms. To avoid characteristics drift, the –25 ms to 0 were taken as baseline and subtracted from the epoch. For comparison with the HR-VOG, the data were sampled at the correspondence of t0, t1, and t2. A procedure for both approaches was applied to automatically exclude trials from the analysis in which fixation was poor (for example, because of the presence of saccades, microsaccades, or large head movements). The algorithm determined that neither the head nor the eyes were moving during the fixation period, while only the eyes were moving during the movement period. Accordingly, the displacement of the head (in space) and the eye (in space and relative to the head) during the fixation period, and the displacement of the head (in space) during the movement period were assessed.

Any trials in which any of these measures were identified as outliers were excluded. After that, the values of each measure were sorted in ascending order across all trials, and identified outliers as those with a value larger than three times the 68th percentile value. On average, this procedure resulted in excluding 15% of the trials (ranging from 5% to 25%). For each system, we then computed the difference (mean and SD) of the vertical displacements induced by upward and downward correlated and anticorrelated motion. The vertical eye displacements induced by correlated and anticorrelated upward and downward moving stimuli were compared for each subject using a paired t-test with a 0.05 significance level. The relative differences between eye-displacement induced by correlated stimuli (UP corr - DW corr) and anticorrelated (UP anti corr - DW anti-corr) were compared using a bootstrapping method. Levene's test is used to test if two system have have equal variances for each condition. All statistical analyses were performed in Python using the statsmodels and scipy packages.

3 Results

Table 1. And Table 2. Show the responses to vertically drifting images measured by HR-VOG and Infrared eye-tracker respectively. The results showed that there were significant differences in the eye displacements induced by upward and downward moving stimuli during the open-loop period (t2-t1) for both correlated and anticorrelated experimental conditions recorded with both HR-VOG system and infrared eye-tracking system.

Table 1. Responses to vertically drifting images measured by HR-VOG (Δy: Vertical eye displacement in 80–160 ms time window (deg); Sub: subject; N: number of correct trials; p: unpaired t-test, ΔCR, ΔACR: difference between UP and DW eye-displacement induced by correlated and anticorrelated pattern motion respectively)

	Correlated pattern induced motion			Anticorrelated pattern induced motion			Correlated vs Anticorrelated		
Sub	$\Delta y \pm$ SD (N) UP	$\Delta y \pm$ SD (N) DW	p	$\Delta y \pm$ SD (N) UP	$\Delta y \pm$ SD (N) DW	p	ΔCR $\pm$ SD	ΔACR $\pm$ SD	p
1	0.211 $\pm$ 0.096 (19)	-0.006 $\pm$ 0.109 (23)	< 0.001	0.162 $\pm$ 0.126 (25)	0.069 $\pm$ 0.146 (23)	0.025	0.217 $\pm$ 0.145	0.093 $\pm$ 0.193	0.008
2	0.211 $\pm$ 0.101 (17)	0.002 $\pm$ 0.117 (18)	< 0.001	0.166 $\pm$ 0.129 (22)	0.072 $\pm$ 0.119 (16)	0.033	0.209 $\pm$ 0.154	0.094 $\pm$ 0.176	0.032
3	0.236 $\pm$ 0.123 (24)	-0.125 $\pm$ 0.084 (24)	< 0.001	0.158 $\pm$ 0.086 (25)	-0.100 $\pm$ 0.072 (25)	< 0.001	0.361 $\pm$ 0.149	0.258 $\pm$ 0.112	0.014
4	0.181 $\pm$ 0.095 (23)	-0.084 $\pm$ 0.080 (20)	< 0.001	0.136 $\pm$ 0.108 (23)	-0.054 $\pm$ 0.071 (21)	< 0.001	0.265 $\pm$ 0.124	0.19 $\pm$ 0.129	0.028
5	0.158 $\pm$ 0.062 (8)	-0.095 $\pm$ 0.097 (10)	< 0.001	0.060 $\pm$ 0.099 (7)	-0.060 $\pm$ 0.108 (15)	0.028	0.253 $\pm$ 0.115	0.12 $\pm$ 0.146	0.032
Avg	0.199 $\pm$ 0.095 (18)	-0.062 $\pm$ 0.097 (19)		0.136 $\pm$ 0.110 (20)	-0.015 $\pm$ 0.103 (20)		0.261 $\pm$ 0.137	0.151 $\pm$ 0.151	

When comparing the eye displacement induced by the two conditions (correlated ΔCR and anticorrelated ΔACR), significant differences were observed only in data recorded by HR-VOG. The bar plot in Fig. 4 with standard error mean (SEM) whiskers shows the results of comparing the eye displacement induced by the two conditions (correlated and anticorrelated).

Table 3 shows the results of the Levene's test, which aimed to compare the variances in the eye-displacement (UP, DW) and stimulus condition (correlated, anticorrelated) data recorded by two different eye-tracking systems: HR-VOG and infrared (IR) eye-tracker. A p-value of less than 0.05 indicates that the two systems have different variances.

Figure 5 offers the representation of the eye-displacements and velocity traces recorded by infrared eye-tracking induced by correlated and anticorrelated motion patterns for the subject 3. We can observe that the standard deviation of the traces of upward and downward motions induced by correlated and anticorrelated overlapped.

Table 2. Responses to vertically drifting images measured by Infrared limbal eye-tracker (Δy: Vertical eye displacement in 80-160ms time window (deg); Sub: subject; N: number of correct trials; p: unpaired t-test, ΔCR, ΔACR: difference between UP and DW eye-displacement induced by correlated and anticorrelated pattern motion respectively)

	Correlated pattern induced motion			Anticorrelated pattern induced motion			Correlated vs Anticorrelated		
Sub	$\Delta y \pm$ SD (N) UP	$\Delta y \pm$ SD (N) DW	p	$\Delta y \pm$ SD (N) UP	$\Delta y \pm$ SD (N) DW	p	ΔCR $\pm$ SD	ΔACR $\pm$ SD	p
1	0.134 ± 0.116 (29)	– 0.026 ± 0.096 (22)	<0.001	0.112 ± 0.111 (28)	0.023 ± 0.071 (20)	0.003	0.160 ± 0.150	0.089 ± 0.132	0.104
2	0.097 ± 0.04 (26)	– 0.097 ± 0.06 (28)	<0.001	0.099 ± 0.192 (28)	– 0.055 ± 0.054 (23)	0.001	0.194 ± 0.072	0.154 ± 0.200	0.324
3	0.129 ± 0.208 (22)	– 0.16 ± 0.18 (23)	<0.001	0.112 ± 0.215 (29)	– 0.127 ± 0.207 (22)	<0.001	0.289 ± 0.275	0.239 ± 0.299	0.548
4	0.078 ± 0.074 (26)	– 0.056 ± 0.1 (28)	<0.001	0.048 ± 0.091 (29)	– 0.057 ± 0.091 (29)	<0.001	0.134 ± 0.124	0.105 ± 0.128	0.434
5	0.058 ± 0.111 (25)	– 0.111 ± 0.135 (28)	<0.001	0.092 ± 0.124 (28)	– 0.023 ± 0.154 (27)	0.004	0.169 ± 0.175	0.115 ± 0.198	0.284
Avg	0.099 ± 0.11 (26)	– 0.09 ± 0.114 (26)		0.093 ± 0.147 (28)	– 0.048 ± 0.116 (24)		0.189 ± 0.159	0.141 ± 0.191	

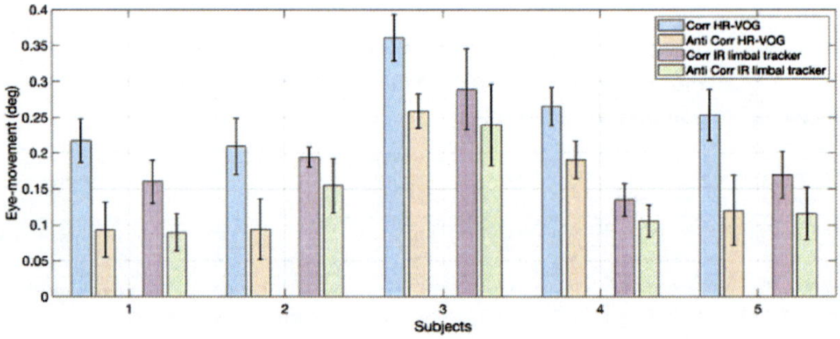

Fig. 4. Bar plots with standard error of the mean whiskers eye-movements of two conditions recorded by HR-VOG and infrared eye-tracker.

Table 3. The outcomes of the Levene's test that examines whether the variances in each eye-displacement direcotion (UP, DW) and stimulus condition (correlated, anticorrelated) recorded by the HR-VOG and infrared (IR) eye-tracking system are equal.

	correlated		anticorrelated	
Sub	Δy UP$_{\text{HR-VOG vs IR eye-tracker}}$	Δy DW$_{\text{HR-VOG vs IR eye-tracker}}$	Δy UP$_{\text{HR-VOG vs IR eye-tracker}}$	Δy DW$_{\text{HR-VOG vs IR eye-tracker}}$
1	0.210	0.657	0.946	0.015*
2	0.005*	0.113	0.001*	0.002*
3	0.035*	0.115	0.009*	<0.001*
4	0.536	0.103	0.552	0.445
5	0.002*	0.910	0.214	0.042*

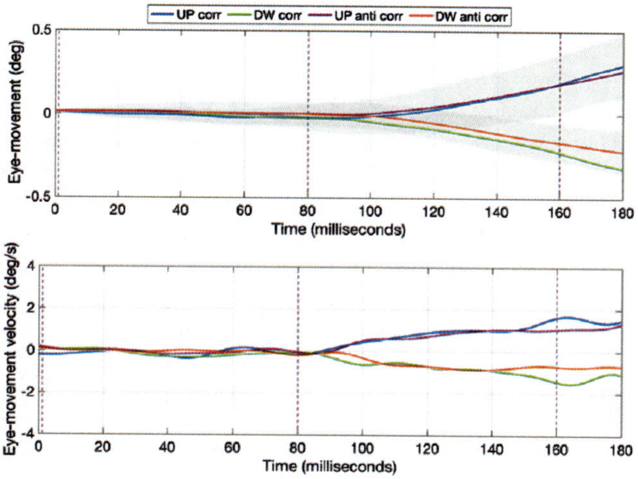

Fig. 5. OFRs position and velocity traces recorded by infrared eye-tracking induced by correlated and anticorrelated motion pattern.

4 Discussion

The present study aimed to compare the measurement of OFRs using HR-VOG and infrared eye-tracking techniques. Our results showed that both techniques were able to accurately register eye-displacements induced by vertically drifting images during the open-loop period. Indeed, the eye-displacement and eye-velocity reflexes depicted in Fig. 5 are similar to those recorded by scleral coils [7], indicating that the infrared eye-tracking measurements reflect OFRs rather than noise. Similarly, the average amplitude of OFRs in the correlated condition (0.261 ± 0.137) is similar to the average amplitude of OFRs produced by vertically drifting stimuli reported in [17]. Despite that, the precision of the infrared eye-tracking system did not allow finding differences between the two experimental conditions. In contrast, the HR-VOG technique was able to register significant differences in eye-displacement induced by the two experimental conditions (correlated and anticorrelated). This suggests that HR-VOG may provide more accurate and reliable measurements of OFRs than the standard infrared eye-tracking technique. Our findings are in line with previous studies that have highlighted the limitations of the infrared eye-tracking system for measuring eye movements [14, 15]. These studies have reported that infrared limbal eye-tracking may suffer from drift and have limited spatial resolution, leading to inaccuracies in measuring eye movements. In contrast, HR-VOG claims to provide accurate and precise measurements of eye movements [17]. In our study, we found that HR-VOG was able to measure significant differences between the two experimental conditions, indicating that it may be a more reliable tool for measuring OFRs. Although the infrared eye-tracking system has a higher temporal resolution, allowing the study of eye-displacements and velocity traces during time, it has some limitations that make it unsuitable for clinical applications, especially in children where OFRs can be potentially used to diagnose stereodeficiencies. One of the main limitations is the need for tedious and accurate positioning of the sensors, which requires calibration for accurate recordings [15].

According to the results of Levene's test, there was no significant variance difference between the two eye-tracking systems for the correlated downward condition. However, for the anticorrelated condition, there was usually a significant difference in variance (excluding subject 4). This can be attributed to the fact that in the down-drifting anticorrelated condition, the eye-lids may cover the limbus (the border between the iris and sclera) [18], making it more difficult for a used infrared eye-tracker to accurately detect smaller eye-movements compared to the down-drifting correlated condition. These results suggest that the HR-VOG and IR eye-tracking systems may not always produce comparable results, depending on the subject and condition being tested.

In addition, it is important to note that our study was conducted on a small sample size and involved only vertical motion stimuli. Another possible concern is that two separate eyes were recorded during the experiment. Although in [9] it stated that the choice of using the dominant eye was arbitrary and considered not influential in healthy subjects, this has not been experimentally validated and may introduce bias. On the other hand, recording OFRs in two separate sessions can lead to other issues when comparing the two eye-tracking systems. These may include variations in the number of artifacts, subject fatigue, and slight changes in room lighting conditions.

Future studies with larger sample sizes and a horizontal moving stimulus should be conducted to further investigate the potential benefits and limitations of HR-VOG and infrared eye-tracking techniques.

To our knowledge, this is the first study that directly compared the measurement of OFRs using HR-VOG and infrared eye-tracking techniques. Our results suggest that HR-VOG may provide more accurate measurements of OFRs than the standard infrared eye-tracking technique making the HR-VOG be a more suitable method for studying OFRs in both research and clinical settings.

Acknowledgments. This study was funded by the Italian Ministry of Health, Grant/Award Number: 2764554 Ricerca Corrente 2021 and partially supported by the master's programme in Clinical Engineering of the University of Trieste.

References

1. Masson, G.S., Perrinet, L.U.: The behavioral receptive field underlying motion integration for primate tracking eye movements. Neurosci. Biobehav. Rev. **36**, 1–25 (2012)
2. Masson, G.S.: From 1D to 2D via 3D: dynamics of surface motion segmentation for ocular tracking in primates. J. Physiol. **98**, 35–52 (2004)
3. Kawano, K.: Ocular tracking: behavior and neurophysiology. Curr. Opin. Neurobiol. **9**, 467–473 (1999)
4. Miles, F.A., Busettini, C., Masson, G.S., Yang, D.S.: Short-latency eye movements: evidence for rapid, parallel processing of optic flow. In: Vaina, L.M., Beardsley, S.A., Rushton, S.K. (eds.) Optic Flow and Beyond. Synthese Library, vol 324. Springer, Dordrecht (2004). https://doi.org/10.1007/978-1-4020-2092-6_5
5. Miles, F.A., Kawano, K., Optican, L.M.: Short-latency ocular following responses of monkey. I. Dependence on temporospatial properties of visual input. J. Neurophysiol. **56**, 1321–1354 (1986)

6. Quaia, C., Sheliga, B,M., Fitzgibbon, E,J., Optican, L,M.: Ocular following in humans: spatial properties. J. Vis. (2012)
7. Sheliga, B.M., Quaia, C., Cumming, B.G., Fitzgibbon, E.J.: Spatial summation properties of the human ocular following response (OFR): dependence upon the spatial frequency of the stimulus. Vis. Res. **68**, 1–13 (2012)
8. Quaia, C., Optican, L.M., Cumming, B.G.: Binocular summation for reflexive eye movements. J. Vis. **18**, 7 (2018)
9. Quaia, C., FitzGibbon, E.J., Optican, L.M., Cumming, B.G.: Binocular summation for reflexive eye movements: a potential diagnostic tool for stereodeficiencies. Invest. Ophthalmol, Vis. Sci. **59**, 5816–5822 (2018)
10. Sheliga, B.M., Quaia, C., FitzGibbon, E.J., Cumming, B.G.: Ocular-following responses to white noise stimuli in humans reveal a novel nonlinearity that results from temporal sampling. J. Vis. **16**, 8 (2016)
11. Jonas, D.E., Amick, H.R., Wallace, I.F., et al.: Vision screening in children aged 6 months to 5 years. JAMA **318**, 845 (2017)
12. Barthélemy, F.V., Vanzetta, I., Masson, G.S.: Behavioral receptive field for ocular following in humans: dynamics of spatial summation and center-surround interactions. J. Neurophysiol. **95**, 3712–3726 (2006)
13. Gellman, R.S., Carl, J.R., Miles, F.A.: Short latency ocular-following responses in man. Vis. Neurosci. **5**, 107–122 (1990)
14. Hutton, S.B.: Eye tracking methodology. In: Klein, C., Ettinger, U. (eds.) Eye Movement Research. SNPBE, pp. 277–308. Springer, Cham (2019). https://doi.org/10.1007/978-3-030-20085-5_8
15. Thomson, D.: Eye tracking and its clinical application in optometry. Optician **2017**, 6045–6051 (2017)
16. Bonfanti, S., Gargantini, A., Esposito, G., Facchin, A., Maffioletti, M., Maffioletti, S.: Evaluation of stereoacuity with a digital mobile application. Graefes Arch. Clin. Exp. Ophthalmol. **259**(9), 2843–2848 (2021). https://doi.org/10.1007/s00417-021-05195-z
17. Miladinović, A., Quaia, C., Ajčević, M., et al.: Ocular-following responses in school-age children. PLoS ONE **17**, e0277443 (2022)
18. Holmqvist, K., Örbom, S.L., Hooge, I.T.C., et al.: Eye tracking: empirical foundations for a minimal reporting guideline. Behav. Res. Methods **55**, 364–416 (2022)

Check for updates

Assessment of Children Eye Movement Performance: An Eye-Tracker Approach

Evita Serpa(✉), Madara Alecka, Asnate Berzina, Viktorija Goliskina, Evita Kassaliete, Anete Klavinska, Marija Koleda, Rita Mikelsone, Elizabete Ozola, Tomass Ruza, Aiga Svede, Daniela Toloka, Sofija Vasiljeva, Liva Volberga, Ilze Ceple, and Gunta Krumina

Faculty of Physics, Mathematics and Optometry, Department of Optometry and Vision Science, University of Latvia, Jelgavas str. 1, Riga, Latvia
evita.serpa@lu.lv

Abstract. Eye movement disorders can have various impacts on reading difficulties, such as tracking difficulties, unstable fixation, issues related to visual processing and attention. A comprehensive understanding of patient's visual functions and reading ability may require a precise evaluation of their eye movements by vision or speech specialists. Eye tracking is a widely applied method for assessing eye movement parameters during reading and other visual tasks. By using eye tracking, it is possible to track eye movements across words and sentences without requiring any overt verbal or motor response from the child. The aim of our study was to develop an objective method for the assessing eye movement performance in children using eye-tracking technology. We tested this method on 53 s-grade school-aged children (7 and 8 years old) using special reading tasks displayed on a computer screen and eye movement recording with a Tobii Pro Fusion eye tracking device (250 Hz). Speech therapists assessed the children's reading skills using the Acadience Reading test. Our results indicated a correlation between the children's reading performance and the number of eye fixations, average fixation duration, and total reading time. Based on our results, we conclude that the developed method based on eye-tracking works well both as a screening method and as a diagnostic method for assessing eye movements during reading. This method will be particularly useful for optometrists, speech therapists, and other specialists involved in children's vision, health, and academic achievements.

Keywords: Acadience Reading test · Eye Tracking · Fixation Duration · Fixations · Reading Speed

1 Introduction

Research has shown that reading difficulties are a common problem among children, with approximately 5–12% of children experiencing significant difficulties with reading [1]. Additionally, children with reading difficulties are frequently diagnosed with various visual function anomalies. Detecting and preventing visual function problems is crucial to avoid their further impact on educational development [2]. This emphasizes the importance of objectively assessing reading ability.

Y. Dekhtyar and I. Saknite (Eds.): NBC 2023, IFMBE Proceedings 89, pp. 246–250, 2023.
https://doi.org/10.1007/978-3-031-37132-5_31

Precise control and coordination of eye movements with cognitive processes are crucial during reading [3]. Previous studies suggest that eye-tracking may be useful in identifying children with abnormal reading development [4] and may detect the earliest stages of cognitive decline [5]. Eye-tracking studies have demonstrated that children with reading difficulties have poorer control of vertical and horizontal eye movements [6] and they demonstrate more and longer fixations during reading [7]. The time spent fixating on a text is often associated with a measure of reading time [8].

The Acadience Reading test is designed to assess reading skills in children from preschool age (6 years) to sixth grade. Depending on the age of the child, the test evaluates reading accuracy, retell quality, and other measures. However, the reading composite score, which combines multiple sub-test scores, provides the best overall estimate of the student's reading abilities [9]. Eye-tracking is a widely applied for determining eye movement parameters during reading and other visual tasks, and eye trackers are now accessible not only to scientific laboratories but also to primary vision care specialists. Therefore, the aim of our study is to develop a method for registering and assessing eye movement parameters during reading and to determine a correlation between the results obtained using this method and those obtained from the widely used Acadience Reading test.

2 Method

Our developed method for assessing eye movement performance in children during reading is based on an eye-tracking approach. We recorded eye movements using Tobii Pro Fusion eye tracker (250 Hz) and Titta Master toolbox [10]. A special reading task was designed to be displayed on a computer screen at a distance of 65 cm. The reading task consisted of an age-appropriate text of 24 words, with a letter size of 0.66 degrees and a space of 1.17 degrees between lines. The text width was 14.9 degrees. We used the I2MC algorithm [11] to detect the average fixation duration, the total reading time, and the number of eye fixations.

To evaluate our developed method, we tested it on 53 s-grade school-aged children (7 and 8 years old). The text used in the reading task was adapted to the age of the children. The children's reading skills were assessed by speech therapists using the Acadience Reading test, which revealed that 28% of the children had low reading scores. We analysed the data using SPSS 22.0 (SPSS Inc., Chicago, IL, USA).

3 Results

Dividing children into two groups based on their reading skills, we found that those with reading difficulties had a significantly longer average fixation duration ($M = 701$ ms, $SD = 286$) compared to those without reading difficulties ($M = 462$, $SD = 132$), t (16) $= 3.106$, $p < .007$. Additionally, the individual Acadience Reading composite score correlated with the children's fixation duration ($r = .604, n = 53, p < .001$), (see Fig. 1).

The results also demonstrated a correlation between the total time required to perform the reading task and the composite score ($r = .619, n = 53, p < .001$), (see Fig. 2) and a

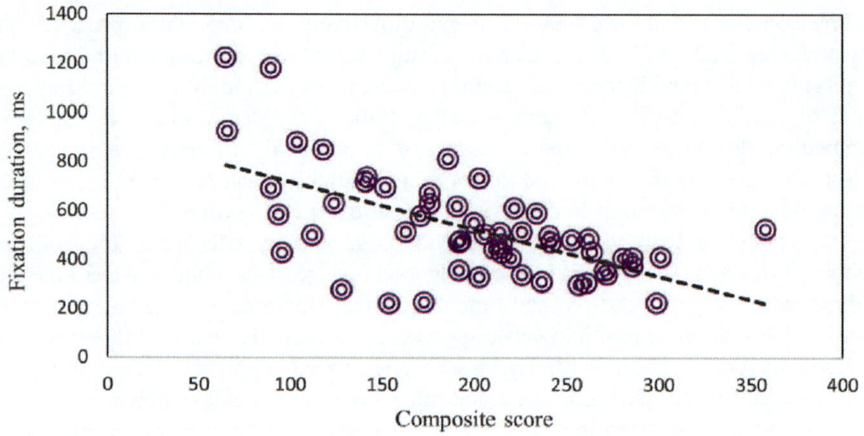

Fig. 1. A correlation between the fixation duration and the composite score.

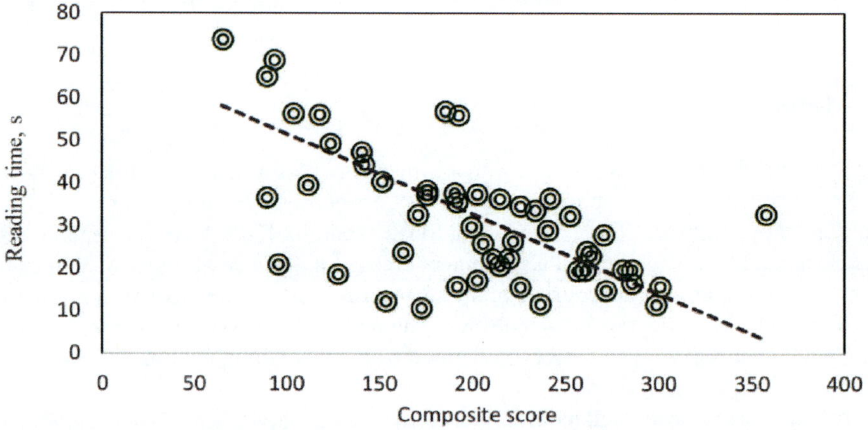

Fig. 2. A correlation between the reading time and the composite score.

moderate correlation was found between the number of eye fixations and the composite score ($r = .400$, $n = 53$, $p = .003$), (see Fig. 3).

The results of the correlation analysis demonstrated significant relationships between various eye movement parameters and the total reading time taken to perform the reading task. Specifically, a strong correlation was found between the fixation duration and total reading time ($r = .869$, $n = 53$, $p < .001$), indicating that children who had longer fixation durations took longer to complete the reading task. Additionally, a strong correlation was found between the number of eye fixations and total reading time ($r = .704$, $n = 53$, $p < .001$), indicating that children who had more eye fixations took longer to complete the reading task. Furthermore, a moderate positive correlation was found between the fixation duration and the total number of fixations ($r = .457$, $n = 53$, $p = .001$), indicating

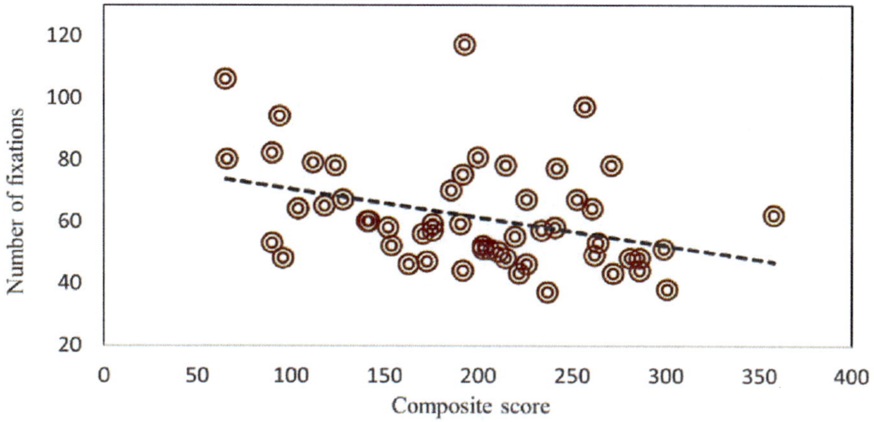

Fig. 3. A correlation between the number of eye fixations and the composite score.

that children who had longer fixation durations also tended to have a greater number of fixations.

4 Discussion

The method we have developed for assessing eye movements in children is based on objective eye movement evaluation using an eye tracker. During a specially designed reading task, we analysed fixational eye movement parameters, such as the number and duration of fixations. We found that the duration of fixations correlates more with the children's reading performance on the Acadience Reading test than the number of eye fixations. It is noteworthy that despite the short length of the reading task, which consisted of only 24 words, a positive correlation was observed between the total time taken to complete the task and reading performance. Even though all observed correlations are moderate, the findings are still notable.

We also analysed the relationship between fixation parameters. Our findings were consistent with previous research, as we observed that children with slower reading rate tended to perform more fixations and the duration of fixations was longer [12].

Previous studies have shown that a method based on eye movement analysis can be an effective in assessing text comprehension [13] and learners' second language proficiency [14]. Our results indicate that the eye-tracking method we have developed may be a reliable tool for both as a screening method and as a diagnostic method for assessing eye movements during reading. Using this method, the assessment of eye movement parameters during reading can be completed in a relative short period of time.

This study represents a pilot investigation, as we have tested our method only on second-grade school-aged children. Further research is necessary to determine the effectiveness of the method on other age groups.

Acknowledgements. We would like to express our gratitude to schools participating in the study (Marupe State Gymnasium, Marupe Elementary School, Riga Cultures Secondary School, and Kuldiga Center Secondary School), as well as speech therapists J.Hanzovska, L.Meiersone, M.Vorza, I.Petuhova, S.Depa, S.Jirgensone. The study is supported by the Latvian Council of Science (project No lzp-2021/1–0219, and SIA Mikrotikls and University of Latvia Foundation (project No 2260).

References

1. Katusic, S.K., Colligan, R.C., Barbaresi, W.J., Schaid, D.J., Jacobsen, S.J.: Incidence of reading disability in a population-based birth cohort, 1976–1982, Rochester. Minn. Mayo Clinic Proc. **76**(11), 1081–1092 (2001)
2. Dusek, W., Pierscionek, B.K., McClelland, J.F.: A survey of visual function in an Austrian population of school-age children with reading and writing difficulties. BMC Ophthalmol. **10**(16), 1–10 (2010)
3. Heinzle, J., Hepp, K., Martin, K.A.: A biologically realistic cortical model of eye movement control in reading. Psychol. Rev. **117**(3), 808–830 (2010)
4. Gran Ekstrand, A.C., Nilsson Benfatto, M., Öqvist Seimyr, G.: Screening for reading difficulties: comparing eye tracking outcomes to neuropsychological assessments. Front. Educ. **6**, 1–12 (2021)
5. Fraser, K.C., Fors, K.L., Kokkinakis, D., Nordlund, A.: An analysis of eye-movements during reading for the detection of mild cognitive impairment. In: Proceedings of the 2017 Conference on Empirical Methods in Natural Language Processing, pp. 1016–1026, Association for Computational Linguistics, Copenhagen, Denmark (2017)
6. Hindmarsh, G.P., Black, A.A., White, S.L., Hopkins, S., Wood, J.M.: Eye movement patterns and reading ability in children. Ophthalmic Physiol. Opt. **41**(5), 1134–1143 (2021)
7. Lefton, L.A., Nagle, R.J., Johnson, G., Fisher, D.F.: Eye movement dynamics of good and poor readers: then and now. J. Reading Behav. **11**, 319–328 (1979)
8. Liversedge, S.P., Paterson, K.B., Pickering, M.J.: Eye guidance in reading and scene perception. Elsevier Science Ltd, Amsterdam, The Netherlands (1998)
9. Raščevska, M., Umbraško, S., Vabale, A., Orlovska, M., Sokola–Nazarenko, M.: Lasītprasmes veicināšana: intervences nodarbības AcadienceTM (DIBELS Next) testa kontekstā. VISC, Riga, Latvia (2019)
10. Niehorster, D.C., Andersson, R., Nyström, M.: Titta: A toolbox for creating PsychToolbox and Psychopy experiments with Tobii eye trackers. Behav. Res. Methods **52**(5), 1970–1979 (2020). https://doi.org/10.3758/s13428-020-01358-8
11. Hessels, R.S., Niehorster, D.C., Kemner, C., Hooge, I.T.C.: Noise-robust fixation detection in eye movement data: Identification by two-means clustering (I2MC). Behav. Res. Methods **49**(5), 1802–1823 (2016). https://doi.org/10.3758/s13428-016-0822-1
12. Spichtig, A.N., Pascoe, J.P., Ferrara, J.D., Vorstius, C.A.: Comparison of eye movement measures across reading efficiency quartile groups in elementary, middle, and high school students in the U.S. J. Eye Movem. Res. **10**(4), pp. 1–17 (2017)
13. Lima Sanches, C., Augereau, O., Kise, K.: Estimation of reading subjective understanding based on eye gaze analysis. PLoS ONE **13**(10), e0206213 (2018)
14. Berzak Y., Katz B., Levy, R.: Assessing language proficiency from eye movements in reading. In: Proceedings of the 2018 Conference of the North American Chapter of the Association for Computational Linguistics: Human Language Technologies, vol.1, pp. 1986–1996, New Orleans, Louisiana (2018)

Method to Detect Trigger Pulling Errors in Shooting Sports: Smart Textile Application

Alexei Katashev[1](✉), Adelina Vevere[1], Alexander Oks[2], Galina Terlecka[2], Laima Saiva[3], Mareks Jansons[3,4], Natalya Dyachenko[2], and Patricija Seglina[3]

[1] Institute of Biomedical Engineering and Nanotechnologies, Riga Technical University, 6B Kipsalas St., Riga 1048, Latvia
Aleksejs.Katasevs@rtu.lv
[2] Institute of Design Technology, Riga Technical University, 6 Kipsalas St., Riga 1048, Latvia
[3] Latvian Academy of Sport Education, 333 Brīvības St., Riga 1006, Latvia
[4] National Guard Shooting Range, Riga, Latvia

Abstract. The correct trigger pull is one of the factors that significantly affect shooting results. DAid smart finger/glove prototype with integrated highly sensitive textile pressure sensors was proposed recently to offer a cost-effective solution for the recognition of trigger pull errors. The motivation of the present study is the further elaboration of data processing methods to enable the use of this device for the classification of trigger pull errors.

Aim: Develop a simple and undemanding method for recognition of the trigger pull error type in sport shooting using smart fingers/gloves as a sensing device.

Novelty: An original new method for recognition of the type of trigger pull errors is proposed. The method uses data obtained with the DAid smart finger/glove system.

Main results: Critical parameters of the trigger pull process were selected. The reference confidence intervals ($P = 0.99$) that characterize correct pull movements were determined for these parameters. Reference intervals were used for the recognition of the type of error. The method was verified by a professional shooter, simulating typical novices' errors during trigger pull. The sensitivity and specificity of the error detection in this limited study were about 82% and 100%, correspondingly.

ConclusionThe DAid smart finger/glove device, combined with the developed method for determining the trigger pull error, could provide reliable and objective data on the triggering process, that may be used as an effective tool to improve the effectiveness of training for shooters.

Keywords: Shooting Sports · Shooting Errors Recognition · Smart Textile · Textile Pressure Sensors

1 Introduction

1.1 Motivation

The performance of athletes in shooting sports is defined by such obvious aspects, as aiming, breathing, or position. Besides this, subtle and inconspicuous motions of palm and fingers that shooters do to pull a firearm's trigger may significantly affect the shot

Y. Dekhtyar and I. Saknite (Eds.): NBC 2023, IFMBE Proceedings 89, pp. 251–260, 2023.
https://doi.org/10.1007/978-3-031-37132-5_32

results, too [1]. Typical errors, such as excessively tightened forearm and palm muscles, leading to the tilt of the firearm just before the shot [2], could pass unnoticed both by the athlete and by the coach. The difficulty to detect incorrect trigger pull patterns is redoubled by the firearm's recoil. The recoil causes noticeable forearm movement and hereby conceals most of the trigger-pulling errors, so even experienced couches could not notice them. Tiny palm and finger movements, causing deviation of the pistol barrel by just 0.5 mm at the 25-m range lead to the deflection of a bullet by 8 cm, but at the 50-m range – already by 16 cm. [3]. The analysis of the bullet scattering pattern, in general, enables the detection of shooting errors [4], but this method is only effective for multiple shots and is not useful for single-shot analysis and for real-time "in situ" correction. Also, hit point position is the result of several factors, finger movement is just one of them.

Literature and market propose various systems, eligible for assessment of shooter's finger pressure and trigger pull training. Such systems range from the simplest finger spring-powered exercise devices [5] to the advanced finger force monitoring systems, designed either for real pistols, e.g., Beretta or Glock [6], or for the originally designed shooting training kits, such as SCATT [7]. In the lab conditions, researchers often use custom–made devices, based on commercial pressure sensors, mounted directly on the trigger [2].

These advanced systems have certain drawbacks. Mostly they are either designed for certain gun models or use mock weapons that differ from those athletes use in competition. Often, they are rigidly mounted on the gun and should be dismounted and re-attached to another gun when there is a need to switch to another person, that is especially important in sport shooting when guns are individually adjusted for each athlete. Finally, most of these systems are designed to measure the index finger pressure force only. Therefore, they could not evaluate the firearm handle grip. While pulling the trigger, the proper grip is as much important, as the correct index finger motion. It is the grip errors, such as excessive middle finger pressure, or unnecessary muscular tension, that causes the barrel of the weapon to deflect [4]. For the correct shot, the index finger should be the only finger that moves [8].

1.2 Approach

The alternative approach, discussed in this paper, proposes the wearable device, enabling the positioning of the finger and palm pressure monitoring sensors on the shooter's hand. "Smart" gloves or fingerstalls could be, for instance, equipped with elastomer-embedded silicon sensors [9] or graphite tensile-pressure sensors based on polydimethylsiloxane [10] to measure both index finger pressure on the trigger and grip pressure. However, commercially available sensors are often too rigid, inflexible, or just thick. Being tactilely perceptible, those sensors affect the sensitivity of the fingers and alter movement patterns, because shooters just feel them. A good alternative is entirely textile pressure or force sensors. They are flexible and do not disturb the shooter's fingers' movements. Smart gloves, equipped with such sensors, will not differ tactilely from any other type of gloves, that some shooters (biathletes, policemen, military) use regularly.

Previous research reported examples of smart gloves developed to measure fingers' pressure forces, such as grip monitoring gloves with embedded woven piezoresistive sensors [11], or basketball player gloves with embedded knitted piezoresistive pressure sensors [12]. Recently, authors introduced an originally developed prototype of the shooter's finger movement monitoring glove, which for simplicity was designed as just two fingerstalls for the index and middle fingers. In the previous paper, the authors demonstrated the prototype's ability to distinguish correct and wrong triggering patterns [13]. The present paper elaborates data processing technique, described in [13] and estimates the accuracy of the classification of correct and wrong trigger pull motions, bearing in mind future development of the tool for shooter's finger movement training.

2 Materials and Methods

2.1 Measurement System Design

The originally developed finger pressure monitoring system [13] included two fingerstalls and a data acquisition module (Fig. 1a). Resistive sensing elements were knitted from electrically conductive threads [14] and sewn to the finger cot with non-conductive yarn. The resistance of the sensing elements decreases with an increase in applied force or pressure [12]. Conductive paths were embroidered with electrically conductive silver-coated yarn [14] and ended with metal snaps for the connection to the data acquisition unit. The resistance of the unloaded sensors ranged from 25 to 170 kOhms. Such scattering could be caused by different initial deformation of the sensors' fabric during manual sewing.

The fingerstalls used in the present research had two sensors - on the distal (III) and middle (II) phalanges. The typical shooting technique [1] presumes that the trigger is pulled by the index finger's III phalanx. But some athletes, usually those who have fingers longer than other shooters, may use the II phalanx, too. Monitoring of the II phalanx pressure of the middle finger is important for the detection of excessive grip.

The original data acquisition unit (Fig. 1b) enables the connection of up to 8 resistive sensors with resistances up to 1024 kOhms. The measurement is based on the registration of the voltage drop over sensor resistance, fed by calibrated current. The unit uses 10-bit ADC, and the measurement range could be adjusted from 2 kOhm up to 1024 kOhm. The sampling rate is 160 samples/channel, which corresponds to 6.3 ms intervals between samples. This is generally compatible with firearm lock time (3–10 ms). To monitor the moment of the shot, the external microphone was connected (Fig. 1c). During the measurements, the microphone was attached to the gun barrel with adhesive tape. Although the sampling rate was too small to record shot sound, the acoustic channel still was able to provide an estimate of the shot moment with an accuracy $\pm$ 6.3 ms both for the dry shot and for the bullet shot. The trigger mechanism's click is detectable at the recording even for the dry fire shots (see Fig. 2). The data module communicates data via Bluetooth 2.1 wireless connection to the computer with custom-designed software that enables both real-time display and recording of the finger pressure waveforms. During measurements, the data acquisition unit is attached to the shooter's wrist. The rechargeable battery of the unit enables 8- hours of continuous operation.

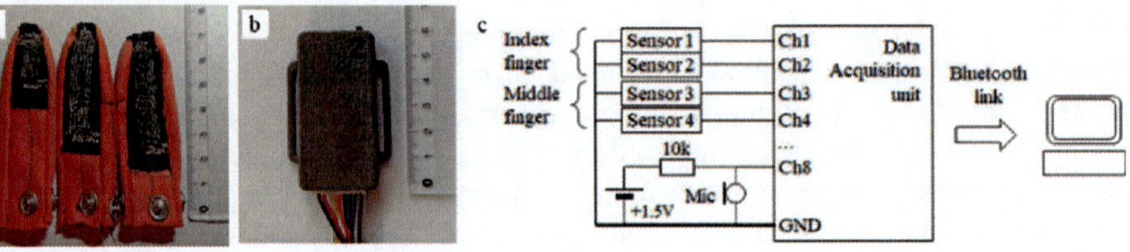

Fig. 1. Fingerstalls with one (left) and two (middle and right) textile pressure sensors (a), data acquisition module (b), and measurement scheme (c) [13].

2.2 Experimental Group

The present research was limited to the proof of concept of the designed fingerstalls and their application for the classification of the sport shooters' trigger pull movements. The experimental group consisted of just two volunteer shooters: a national champion-level shooter and a shooting coach. The shooters were selected because of their experience and ability to perform both correct shots and shots with a simulation of the typical novices' errors.

The first shooter used a pneumatic pistol Steyr LP 10 with a single-stage trigger and trigger pull effort of 4.9 N and a 9-mm Makarov pistol (PM) with a two-stage trigger and trigger effort of 19.6 N. The second shooter used Steyr LP 10 with a two-stage trigger and trigger effort of 4.90 N. and a similar 9-mm Makarov but with an effort of 24.52 N.

Each shooter performed a series of dry fire (with no bullet) and ordinary shots, trying to perform at their best. After the series of perfect shots, shooters simulated typical trigger pooling mistakes: (I) pulling the trigger with II phalanx of index finger; (II) tearing off the trigger at the beginning of the shot; (III) tearing off the trigger at the end of the free run; (IV) trigger tear off at the end of the pull; (V) early (<1 s) release of the trigger after shot; (VI) unstable jerky pressure of the middle finger. For each type of measurement, at least three shots in a series were performed.

Alongside, the second shooter performed a series of 13 dry shots with the 9-mm pistol, simulating different errors. This series was used to validate developed shot classification rules.

2.3 Data Processing

The pre-processing of the recorded waveforms included sliding median filtering with following moving average filtering to remove noise and spike-wise artifacts. The window for the filters varied from 20 to 80 points, and the result of the filtering was estimated visually to avoid loss of information due to the flattening of the stepwise fragments of the waveform. The delay, potentially introduced by the filters, was not compensated.

Figure 2 depicts typical correct single-shot waveforms for the III (distal) index finger phalanx sensor, II (middle) index finger phalanx sensor, and middle finger sensor. The spike in the arbitrary microphone signal indicates the instant of the shot. The typical shot waveform exhibits two distinguished stages: the pull stage, from the beginning of the trigger pull until shot, and the post-shot phase, from the instant of shot until the finger pressure to trigger is released. Although for the two-stage triggered pistols, the pull phase could be further divided into two sub-stages, one corresponds to idle trigger run, and another – to the increasing resistance of the trigger's spring, this division is unpractical, because the transition between two substages is not noticeable for the correct trigger pulling movement. The duration of the pull phase takes about 85% of the whole shot cycle for both single-stage and two-stage triggered pistols [16]. The post-shot phase is short and makes the remaining 15% (about 1 s in absolute units). This phase is characterized by the ability of the experienced shooter to maintain pressure over the trigger after the shot. A sharp increase at the end of the post-shot phase indicates the release of the trigger and the corresponding drop in finger pressure.

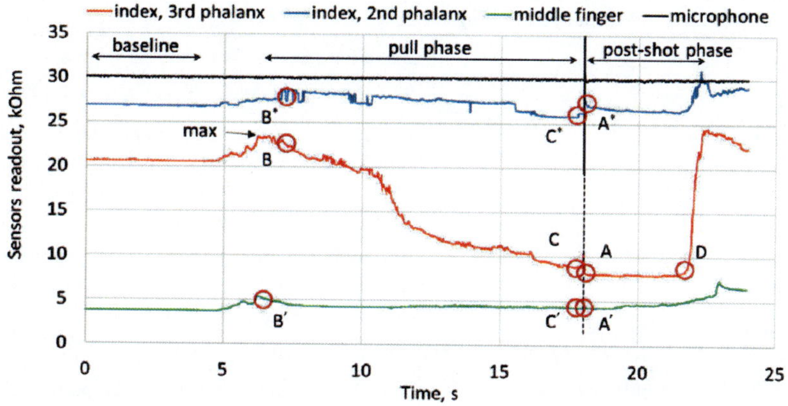

Fig. 2. Typical single-shot waveforms for the correct trigger pull. The characteristic points are A – the shot, B – start of the trigger pull, C – end of the trigger pull, and D – trigger release.

For any type of used firearm, the shot waveforms may be characterized by typical timing points (Fig. 2). Point A corresponds to the instant of the shot. Point B characterizes the beginning of the shot waveform, when applied trigger pressure starts to increase. Point B corresponded to the instant when the sensor readout decreased by 3% from the maximum on the left. The start of the shot was generally defined by the readouts from the III phalanx sensor, but if the shooter pulled the trigger with the II phalanx, either because of error or because of anatomical peculiarities (too long fingers), point B could be more easily detected on the II phalanx sensor waveform. Whatever waveform is used, point B was the same for index finger sensors. Point B' for the middle finger was defined independently, because, in most of the observed waveforms, it was delayed relative to point B. Point C corresponds to the end of the pull phase and activation of the trigger, and is defined as being by 0.3 s before the instant of the shot, here the typical time between triggering and shot itself was used. Point D corresponds to the release of the trigger and is characterized by a fast increase in the sensor's readout.

The waveforms for every single shot were limited to 20 s before point A and 4 s after point A. For the whole set of data, the longest pull phase was 19.55 s, but the longest post-shot phase was 3.93 s. If the end of the previous shot fell within these 20 s, the waveform was cut immediately after the end of the previous shot. Waveforms were normalized by the baseline value, calculated as an average of the waveform's first 100 datapoints (~ 0.63 s), which corresponds to the "no load" conditions before trigger pull.

The normalized waveforms were analyzed visually to find typical differences between correct and incorrect shots. For each single-shot waveform, a set of parameters was derived (Table 1) to be used to establish reference intervals for the correct shot.

Table 1. Parameters of the single shot waveform.

Parameter, unit	Description
Extracted from the waveform of index finger III phalanx sensor -	
Pull time, s	The time interval from point B to point C
Hold time, s	The time interval from point A to point D
Amplitude, arb. un	Relative change of signal amplitude over pull time, calculated using normalized maximal R_{max} and minimal R_{min} signal values: $A = (R_{max} - R_{min})/R_{max}$
Maximal slope at a shot, s^{-1}	The maximal absolute value (ignoring the sign) of the slope of the waveform, determined in the vicinity of point A $\pm$ 0.3 s
Extracted from the waveform of index finger II phalanx sensor -	
Average slope at the pull phase, s^{-1}	The slope of the waveform in the pull phase is calculated over the range from point B' to point C'
Standard deviation of the slope at the pull phase, s^{-1}	The standard deviation of the slope of the waveform in the pull phase, calculated over the range from point B to point C, $seconds^{-1}$
Maximal slope at a shot, s^{-1}	The maximal absolute value (ignoring the sign) of the slope of the waveform, determined in the vicinity of point A' $\pm$ 0.3 s
Extracted from the waveform of the middle finger sensor -	
Maximal slope at the pull phase, s^{-1}	The maximal absolute value (ignoring the sign) of the slope of the waveform, determined over the range from point B^* to point C^*

Comment: the slope was calculated for every point of the waveform, using a sliding backward window with a width of up to 100 points.

The visual comparison of the waveforms demonstrated that all the normalized waveforms, recorded for the correct shots, look similar for any shooter or pistol type. Therefore, parameters derived from different waveforms were considered as a sample. The reference intervals were calculated as sample-based either 95% or 99% confidence intervals. Further analysis demonstrated that a 99% interval provides a more accurate classification, therefore only this interval will be used from this point ahead.

Established reference intervals were used to formulate classification rules for the determination of shot correctness. Developed rules were validated, using 13 waveforms, recorded by the experienced coach with and without simulation of typical trigger-pulling mistakes.

3 Results and Discussion

3.1 Parameters of the Correct Shot

Table 2 summarizes threshold values for the parameters that correspond to the correct shot. The sensitivity of each parameter to the erroneous motion was estimated as a ratio of the number of incorrect shots, that had corresponding parameters out of the correct range, to the total number of incorrect shots. For each type of error, two parameters with maximal sensitivity were left to be used for shot classification. If the maximal sensitivity for the parameter did not exceed 0.5, this parameter was omitted. In Table 2, parameters that are eligible for classification are numbered.

Table 2. Waveform parameters sensitivity to the trigger pulling errors.

	Parameter	Correct shot threshold	Sensitivity to the error type					
			I	II	III	IV	V	VI
	Index finger III phalanx sensor							
1	Pull time, s	> 12.8	0.6	1.0	0.8			
2	Hold time, s	> 1.0					1.0	
3	Amplitude, a.u	> 0.43	1.0				0.7	
4	The maximal slope at the shot, s^{-1}	< 0.11		1.0	0.8	0.8		0.7
	Index finger II phalanx sensor							
	The average slope at the pull, s^{-1}	< 0.062						
	The standard deviation of the slope at the pull, s^{-1}	< 0.094						
5	The maximal slope at the shot, s^{-1}	< 0.22		1.0				
	Middle finger sensor							
6	The maximal slope at the pull, s^{-1}	< 0.25						0.8

Error types are: (I) pulling the trigger with II phalanx of index finger; (II) tearing off the trigger at the beginning of the shot; (III) tearing off the trigger at the end of the free run; (IV) trigger tear off at the end of the pull; (V) early (< 1 s) release of the trigger after shot; (VI) unstable jerky pressure of the middle finger.

3.2 Classification Rules

Classification rules were based on the established ranges for the shot parameters. The algorithm of the classification is summarized in Fig. 3. The names of parameters and types of errors correspond to those in Table 2.

```
IF (index III phalanx sensor amplitude) out of range (< 0.43)
THEN non-correct, error type I        END
ELSE
   IF (middle sensor maximal slope at pull) out of range (> 0.25)
   THEN non-correct, error type VI
   IF (index III phalanx sensor hold time) out of range (<1.0)
   THEN non-correct, error type V
   IF (index III phalanx sensor hold time) within the range (<0.11)
       THEN correct, no errors.
       ELSE
           IF (index III phalanx sensor hold time) within the range (>12.8)
           THEN non-correct, error type IV
           ELSE
               IF (index III phalanx sensor hold time) > 8.5 s
               THEN non-correct, error type III
               ELSE
                  IF (index II phalanx sensor maximal slope at shot)
                             out of range (> 0.22)
                  THEN non-correct, error type II
                  ELSE correct, no errors        END
```

Fig. 3. Trigger pull error classification algorithm.

3.3 Validation of the Classification Rules

Table 3 summarizes the results of the classification of the validation dataset. The classification was made following developed rules blindly: the classifier did not know what type of error was simulated. The results of validation demonstrated that the classification by criterion "correct/not correct" was accurate. All eleven incorrect shots were classified as incorrect, and two correct shots were classified as correct. As for the determination of the type of error, the correct type was recognized in 7 cases out of 11 (67%). In two cases, the simulated error was recognized, but the classifier decided that there is an additional error. In two cases the error was not recognized correctly.

Recognition of error II in addition to the correctly recognized error VI (excessive and unstable middle finger pressure) could be considered a success, too. Highly likely that the shooter, while simulating this type of error, may unconsciously make another error. According to the coach, excessive middle finger pressure could cause a tear-off of the trigger, i.e., error VI caused error IV, as it was recognized by the classifier using the proposed rules. If so, the rate of success could be estimated as 9 out of 11, or 82% of correctly recognized error types.

Table 3. Results of the classification of shots performed by coach with error simulation.

Test shot nr:	1	2	3	4	5	6	7	8	9	10	11	12	13
Shot:	nc	nc	nc	nc	nc	nc	c	nc	nc	nc	nc	nc	c
Simulated error	II	IV	V	I	III	VI	-	IV	I	III	VI	II	-
Classified as	nc	nc	nc	nc	nc	nc	c	nc	nc	nc	nc	nc	c
Detected error	II	IV	I	I	II	IV, VI	-	IV	I	III	IV, VI	II	-

c – correct shot, nc – non-correct shot.

3.4 Discussion

The results of the present study illustrate the potential of the textile-based shooter's trigger pull motion monitoring fingerstalls. The extraction of features of the recorded waveforms for both correct and incorrect shots enabled the development of classification criteria and rules for the recognition of the incorrect triggering techniques. Note, that the proposed features are close to the parameters that were described in the literature as important indicators of triggering: parameters, derived from normalized trigger force slope [2], and timing of triggering [15, 16].

The present research has obvious limitations. First, the incorrect motions were simulated by experienced shooters. Therefore, the first question that should be addressed in the future is to what extent simulated waveforms are like those recorded with inexperienced shooters, which errors are "natural". Another limitation is the small number of shooters and firearms. In the present research, the criteria for shot parameters were the same for any shooter/pistol combination, suggesting, that it is possible to elaborate some universal "good shot standard". Nevertheless, it may not be the case if more shooters or weapon types will be involved in the study.

Another reason to increase the number of shooters is the need to collect more data for the estimation of the accuracy of the classification algorithm. The present research gave a preliminary estimation of 100% sensitivity and 100% specificity for the "correct/incorrect" test, but the specificity was estimated using just two shots.

The resulting method is not neural network based; therefore, it does not require significant computing power and can provide real-time data flow with low memory controllers. Nevertheless, the use of a neural network for the classification of triggering errors could be considered as an alternative approach for shot waveform analysis.

4 Conclusion

The DAid smart finger/glove device, combined with the developed method for classification of the trigger pull error, could provide reliable and objective data on the triggering process, that may be used as a promising tool for improving the effectiveness of shooting training.

Acknowledgments. The research has been partially financed by the Riga Technical University research grant 01000–1.4-e/25 "Smart textile system for monitoring the pressure of the shooter'45s finger on the gun trigger".

References

1. Laaksonen, M.S., Finkenzeller, T., Holmberg, H.-C., Sattlecker, G.: The influence of physiobiomechanical parameters, technical aspects of shooting, and psychophysiological factors on biathlon performance: A review. J. Sport Health Sci. **7**(4), 394–404 (2018)
2. Sattlecker, G., Buchecker, M., Gressenbauer, C., Müller, E., Lindinger, S.: Factors discriminating high from low score performance in biathlon shooting. Int. J, Sports Physiol Perform. **12**(3), 377–384 (2017)
3. Todorović, Z.: ISSF Coach Course, Pistol Shooting. https://www.issf-sports.org, (Accessed 21 July 2021)
4. Cleckner, R.M.: Long Range Shooting Handbook: The Complete Beginner's Guide to Precision Rifle Shooting. North Shadow Press, Nashville TN (2016)
5. Trigger Trainer. http://www.thetriggertrainer.net/, (Accessed 15 April 2023)
6. Trigger Pull Trainer for Full Size Glocks. https://www.dvorakinstruments.com/, (Accessed 15 July 2021)
7. Zanevskyy, I., Korostylova, Y., Mykhaylov, V.: Aiming point trajectory as an assessment parameter of shooting performance. Human Movement **13**(3), 211–217 (2012)
8. Findley, B.: 7 trigger control errors and how to fix them. https://exclusive.multibriefs.com/content/7-trigger-control-errors-and-how-to-fix-them/law-enforcement-defense-security, (Accessed 15 April 2022)
9. Ying, M., et al.: Silicon nanomembranes for fingertip electronics. Nanotechnology **23**, 344004 (2012)
10. Wu Y., et al.: High resolution flexible strain sensors for biological signal measurements. In: Proceedings of 19th International Conference on Solid-State Sensors, Actuators and Microsystems (TRANSDUCERS), pp. 1144–1147 (2017)
11. Kim, G., Vu, C., Kim, J.: Single-layer pressure textile sensors with woven conductive yarn circuit. Appl. Sci. **10**, 2877 (2020)
12. Oks A.Jn., Oks A., Katashev A., Eizentals P.: Smart glove usage possibility for basketball training: proof of concept. In: Proceedings of International Scientific Conference Society Integration Education, vol. 4, p. 235, Rezekne (2019)
13. Vevere, A., et al.: Smart textile device for shooter's fingers movement monitoring. Technol. Health Care **30**(1), 217–229 (2022)
14. Oks A., Katashev A., Litvak J.: Knitted resistive fabric: properties and applications. In: Materials Science: Textile and Clothing Technology, RTU, Riga (2015)
15. Olsson, E., Laaksonen, M.S.: Key technical components for air pistol shooting performance. Int. J. Perform. Anal. Sport **21**(3), 348–360 (2021)
16. Ihalainen, S., Kuitunen, S., Mononen, K., Linnamo, V.: Determinants of elite-level air rifle shooting performance. Scand J. Med. Sci. Sports **26**, 266–274 (2016)

Influence of UV Radiation on Immobilization of Yeast Cells on the Surfaces of Mg- and Ti-Originated Alloys and Their CaP Coatings

Marks Gorohovs[1(✉)], Anna Bystrova[1], Yuri Dekhtyar[1], Marina Romanova[1], Gaļina Hrustaļova[2], Jürgen Schmidt[3], and Alina Vladescu[4,5]

[1] Riga Technical University, Institute of Biomedical Engineering and Nanotechnologies, Ķīpsalas Str 6B, Floor 3, Riga LV - 1048, Latvia
mark_gorohov@inbox.lv

[2] Institute of Microbiology and Biotechnology, University of Latvia, Jelgavas Str., 1-537, Riga 1004, Latvia

[3] INNOVENT E.V. Technology Development, Group-Leader Electrochemistry Ilmstraße 18, 07743 Jena, Germany

[4] National Institute of Research and Development for Optoelectronics, INOE 2000, 409 Atomistilor St., 077125 Magurele, Romania

[5] Research Center for Physical Materials Science and Composite Materials, National Research Tomsk Polytechnic University, Lenin Avenue 43, Tomsk 634050, Russia

Abstract. For the past few decades, many researches have been devoted to study and develop new biocompatible materials for medical implants. $MgCa_1$ and $TiAl_6V_4$ are examples of such materials. It is possible to adjust their physical and chemical parameters by coating the surface with calcium phosphates (CaP). In turn, exposing materials to UV radiation influencing the surface charge may contribute to cell immobilization and thus materials biocompatibility.

The purpose of this study was to investigate how UV radiation influences the immobilization of yeast cells on the surfaces of Mg- and Ti- originated alloys and their CaP coatings.

$MgCa_1$ and $TiAl_6V_4$ samples with different coatings (uncoated, CaP, Ca^+ and variations doped with Zn or Ga) were used. Surface roughness, FTIR spectras and surface electric potential (via electron work function measurements) were assessed before and after UV irradiation (for 30 and 60 min). *S. cerevisiae* were immobilized on the surfaces of samples and assessed its coverage using fluorescent microscopy.

Adherence of yeast cells was found to be more prominent for Ti samples, than to Mg samples. There was no conclusive link discovered between the samples' surface roughness and yeast cells' adherence to it. UV irradiation does not affect the chemical structure of the samples. Overly, UV irradiation has improved the cell immobilization on Mg- and Ti-originated alloys with different CaP coating, exceptions are only coatings with Zn or Ga dopings. Non-coated Mg- and Ti-originated alloys showed the best adherence of yeast cells.

Keywords: Coatings · UV radiation · Magnesium alloy · Titanium alloy · surface charge · yeast cells

Y. Dekhtyar and I. Saknite (Eds.): NBC 2023, IFMBE Proceedings 89, pp. 261–268, 2023.
https://doi.org/10.1007/978-3-031-37132-5_33

1 Introduction

New biocompatible materials for medical implants have been the focus of extensive research over the last few decades. Such materials include biodegradable MgCa and non-biodegradable $TiAl_6V_4$.

Both metals are highly biocompatible, have a high strength-to-weight ratio, and have an elastic modulus that is lower than that of other metals and more similar to that of human bone.

Calcium phosphate (CaP) coatings, potentially doped with Zn or Ga, can be applied to a surface to control the rate of corrosion, increase stiffness, manage cell adhesion, and promote bone regrowth [1].

It is important to assess the biocompatibility of Mg- and Ti-originated alloys with various protective coatings on living cells before using them as implants. Because they are eukaryotic, like human cells, yeast cells are excellent for these uses. In turn, exposing materials to UV radiation influencing the surface charge may contribute to cell immobilization [2].

The purpose of here presented study was to investigate how UV radiation influences the immobilization of yeast cells on the surfaces of Mg- and Ti- originated alloys and their CaP coatings.

2 Materials

2.1 Mg- and Ti-Originated Alloys Samples with and Without CaP Coatings

The samples were produced by INNOVENT e.V. in Jena, Germany, as part of "CoatDegraBac" research project. Coatings were formed using the micro-arc oxidation (MAO) method [3]. Mg samples were coated with CaP, CaP with Ga doping, CaP with Zn doping, Ca^+. Also, one set of Mg samples was not coated, but so-called "pickled", meaning that samples were chemically treated using acid to improve the corrosion resistance and induce biomineralization on the surface of material. Ti samples coatings were – CaP, Ca^+, and Ca + with Zn doping, also one set of samples was uncoated.

All samples look like pucks with 15 mm in diameter and various heights from ~ 3 to 5 mm.

2.2 Yeast Cells

Yeast cells for the research were obtained in form of dry baker's yeast from the grocery store. The method of its preparation is described further in Sect. 3.5.

3 Methods

3.1 Surface Roughness Measurements

Measurement to determine parameters of surface was done by using contact profilometer Mitutoyo Surfest SJ-500, which tracks the surface with stylus and record deviations in height.

3.2 Fourier Transform Infrared (FTIR) Spectroscopy

FTIR spectroscopy method uses infrared radiation, at some wavelengths it is absorbed by the sample more than at others, thus showing distinct bands on the output spectrum. These bands correspond to the different molecules' deformations present in the sample, showing the so-called "molecular fingerprint", thus, it is possible to define the components/molecules present in the material of the sample [2]. It may show whether any chemical bonds were broken due to exposure to UV radiation. It was performed before and after UV irradiation of samples.

For the research, a Bruker Tensor II spectrometer was used with a Platinum ATR module for examination of solid samples.

3.3 Photoemission Spectroscopy

The potential of the material surface can be calculated by knowing the work function that can be found based on the obtained data from the photoemission spectrometry. Electron work function is the minimal amount of energy needed to hit the electron out of the surface of the solid material. Therefore, an increase of EWF identifies that the surface has a negative charge.

To measure EWF, a custom designed photoelectron spectrometer was employed [2].

3.4 UV Irradiation

The electrical potential on a material's surface can be controlled using UV radiation. The outer membrane of yeast cells has a negative potential [4]. Increasing the potential difference between the sample surface and yeast cells by charging the substrate surface positively could make the yeast cells more electrostatically attracted [5].

Hamamatsu Lightning Cure LC8 UV light source with L10852 -01A type lamp (that has maximal irradiance of 4500 mW/cm^2 at 365 nm) was used in this research for irradiation of the samples. The distance from the source to the samples was maintained at 20 cm, all samples were placed in the same intensity region to evenly irradiate each sample without causing serious thermal damage to the surface.

In this research samples with no exposure with 30-min and 60-min exposures were compared.

3.5 Yeast Cell Preparation and Immobilization

To prepare a concentrated solution, about 6–7 mg of dried yeast powder was combined with 50 ml of distilled water. Mixing is performed using a vortex mixer until the solution has become homogeneous. Then it was centrifuged to separate yeast from the scraps, water changed and mixed again, then diluted with 500 ml of distilled water [6]. To verify the optical density and thus control the concentration of the prepared solution, a UV-VIS spectrometer was used at 600 nm. In this research, optical density was maintained at 0.175 ± 0.005 AV.

Samples on which cells must be immobilized are placed in a special holder, that is designed for 8 samples to be held at a time. The holder was placed in a petri dish and filled with 50 ml of prepared yeast substance and stirred for 60 min at 50 RPM. The solution was then removed using the syringe and samples were placed in the incubator at a temperature of 30 °C for drying.

3.6 Fluorescent Microscopy

Fluorescent microscopy is used to count the quantity of immobilized cells on the surfaces of the samples. Primulin was the fluorophore employed in the study, and its concentration was 1 mg/ml. Each sample received an equal amount of 15 l of primulin. About 15 min after applying the fluorophore, the examination must be performed.

Dead yeast cells' ruptured membranes allow the fluorophore to enter the interior of the cell, making it simpler to spot and identify immobilized cells.

The microscope used for the observation is Olympus BX51, equipped with Olympus DP71digital camera. To count the number of immobilized cells, images were processed using ImageJ software.

4 Results and Discussion

4.1 Surface Roughness Measurements

Ra values correspond to the height difference and *RSm* – to the horizontal distance between peaks (Fig. 1).

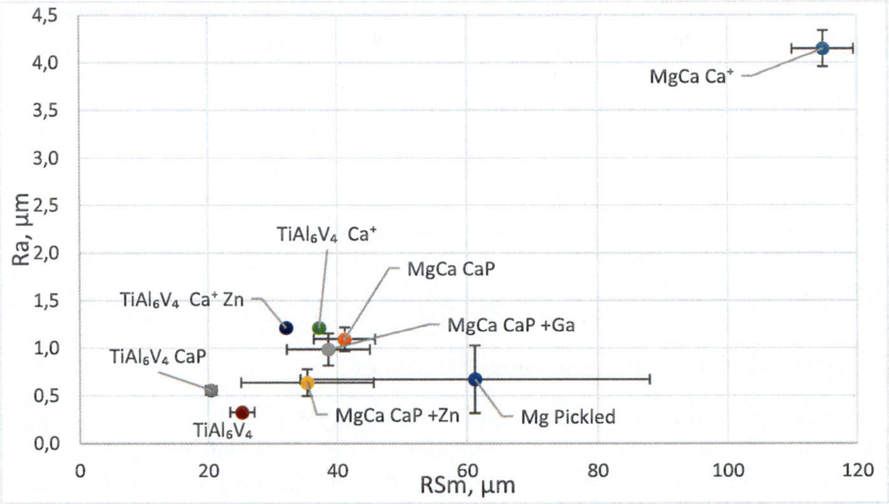

Fig. 1. Roughness values for all materials.

Compared with the uncoated substrate, CaP and Ca^{+} coatings lead to an increase of roughness. It is most likely due to the MAO deposition technique applied for the preparation of the coatings. High porosity is an expression of the strong gas formation, caused by the spark discharge on the surface during the deposition of the ions. However, addition of Zn or Ga into coating structures lead to a slight decrease of roughness. It may be because the formation of the dielectric was accompanied by a very fine and uniform spark discharge, whereas the other electrolytes had a coarser spark discharge and thus larger pore diameters [1].

Comparing yeast cells coverage areas for different materials (Sect. 4.4), there were not found any distinct correlation between surface roughness and cells adherence.

4.2 FTIR Measurements Before and After UV Irradiation

Acquired FTIR spectras were denoted by peaks of absorbance by its wave number. Example of MgCa CaP spectra with denotions are displayed on Fig. 2.

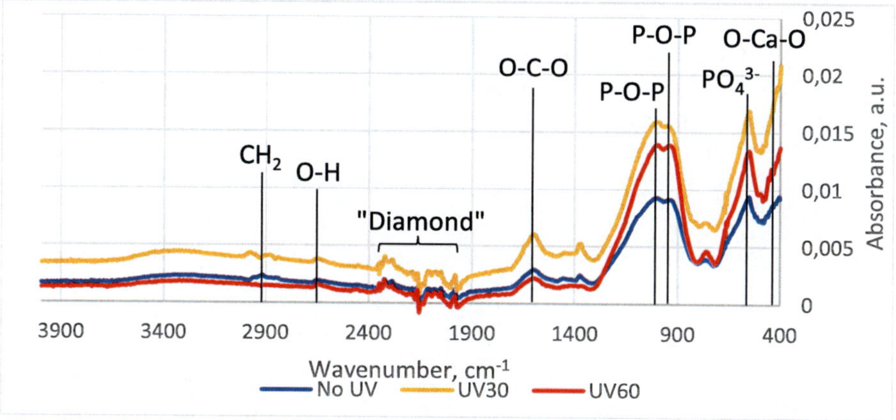

Fig. 2. FTIR, MgCa CaP.

When FTIR spectra were analyzed, no patterns that would suggest a certain alteration in the chemical structures of the substrate were discovered. The remaining peaks for all the materials exhibit a random change, but most of the wavenumber variances for each associated absorbance band fall within error bands.

4.3 Electrical Potential Measurements on the Surface Before and After UV Irradiation

The difference in work function values before and after UV represents a potential on the surface of the substrate. To show the change in the EWF values caused by UV exposure, pre-irradiation EWF values were subtracted from corresponding values after UV exposure. Results are displayed on Fig. 3.

Addition of coatings to Mg samples showed a negative EWF change, however, for Ti – a positive. Doping the coating with Zn showed the least change in the surface charge after UV irradiation.

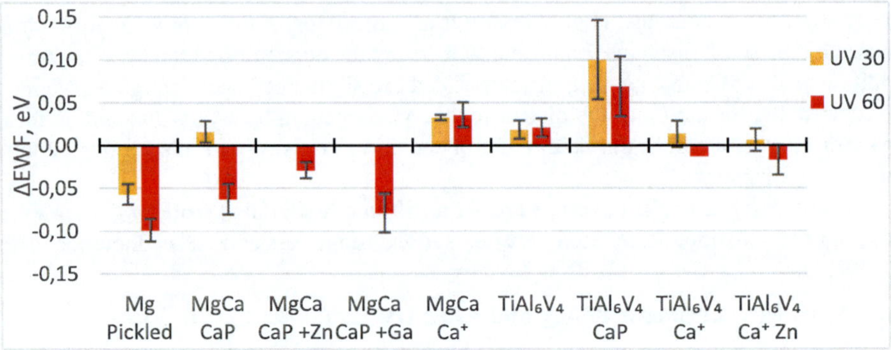

Fig. 3. EWF change for all samples.

4.4 Immobilized Yeast Cell Coverage Area

The photos of samples with immobilized yeast cells on their surfaces were taken using a fluorescence microscope with a built-in digital camera. Using the contrast and intensity of the luminescence of the cells against the image background, ImageJ software calculates the cell coverage of the substrate after processing the photos.

The data for all the samples at each exposure are shown in Figs. 4 and 5.

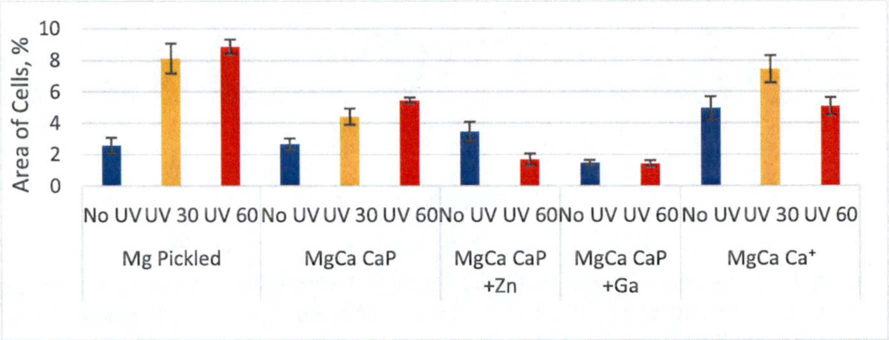

Fig. 4. Cells coverage areas at different exposure times, for Mg-based samples.

Adherence of yeast cells is more prominent for Ti samples, than to Mg samples. On Mg- and Ti-originated alloys with various CaP coatings, UV irradiation has generally improved cell immobilization; the only exceptions are coatings with dopings. For MgCa Ca^+, $TiAl_6V_4$, $TiAl_6V_4$ CaP, and $TiAl_6V_4$ Ca^+ with Zn doping, a 30-min exposure is ideal.

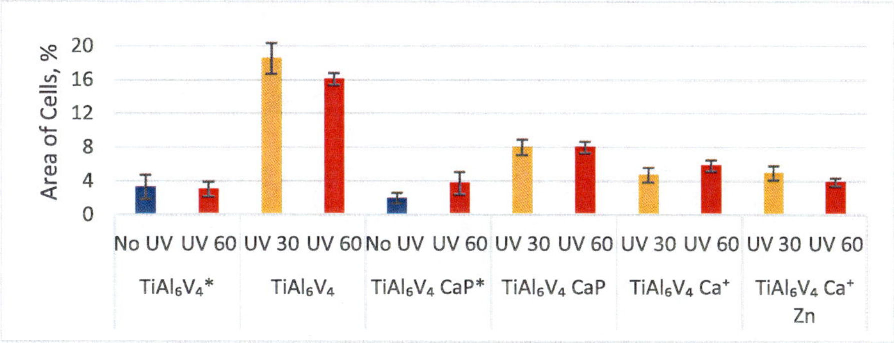

Fig. 5. Cells coverage areas at different exposure times, for Ti-based samples.

5 Conclusions

Adherence of yeast cells is more prominent for Ti samples, than to Mg samples.

Overly, UV irradiation improves the immobilization on the surfaces of Mg- and Ti-originated alloys and their CaP coatings. Although the surface is not exposed to UV, cell adhesion is improved if coatings are doped with Zn or Ga.

There was no conclusive link discovered between the samples' surface roughness and yeast cells' adherence to it.

UV irradiation does not change the molecular structures of Mg- and Ti-originated alloys.

Addition of coatings to Mg samples showed a negative EWF change, however, for Ti – a positive. Doping the coating with Zn showed the least change in the surface charge after UV irradiation.

The increase in photoelectron current for Mg samples enhances the adhesion of yeast cells to its surfaces. For the Ti sample, however, drops.

Yeast cells adhered to non-coated Mg- and Ti-originated alloys the best.

Acknowledgements. Presented work was part of ERA.Net RUS Plus – INNOVATION (Project RUS_INNO2017-031- CoatDegraBac) program.

References

1. Schmidt, J., et al.: Zn doped CaP coatings used for controlling the degradation rate of MgCa1 alloy: In vitro anticorrosive properties, sterilization, and bacteria/cell-material interactions. Colloid. Surf. B Biointerf. **222**, 113087 (2023) https://doi.org/10.1016/j.colsurfb.2022.113087
2. Gorohovs, M.: Influence of UV radiation on immobilization of yeast cells on the surfaces of Mg- and Ti- originated alloys and their CaP coatings. Bachelor's thesis. Riga Technical University (2021)
3. Schmidt, J.: Report about new coatings and their characterization of TiAl6V4 and MgCa1 – samples (2021)
4. Peña, A., Sánchez, N., Calahorra, M.: The plasma membrane electric potential in yeast: Probes, results, problems, and solutions: A new application of an old dye? (2017)

5. Baltacis, K., et al.: Physical fundamentals of biomaterials surface electrical functionalization. Materials **13**, 4575 (2020)
6. Gračova, V.: Rauga Šūnu Imobilizācija uz ar UV Apstarotas SiO2 Virsmas Atkarība no tās Struktūru Elementu Attālumiem. Riga Technical University (2019)

Experimental Comparison of Thin Film Surfaces to Detect Oral Salivary Biomarkers by Surface-Enhanced Raman Spectroscopy

Miia Hurskainen(✉), Hannu Korhonen, Sami Myllymaa, and Reijo Lappalainen

Department of Technical Physics, University of Eastern Finland, 70211 Kuopio, Finland
miia.hurskainen@uef.fi

Abstract. Six potential thin film candidates, Bi, Ti, Al_2O_3, Ag, Au, and C, were prepared on Si substrate to determine the most suitable ones to detect three known salivary biomarkers, L-Fucose, L-Proline, and N-acetylneuraminic acid from surface-enhanced Raman spectrum. In the two-phase study, 5 mg of each biomarker substance was first diluted into 1 ml of Millipore water and then into 1 ml of artificial saliva. Artificial saliva was used to ensure the molecular homogeneity of the samples. The normalized and smoothed spectra were area normalized, and cubic k-nearest neighbors' confusion matrixes and F1-scores were computed, in both phases. The results show that laser-deposited Bi and Al_2O_3 thin films along with the typical Au and Ag are useful in detecting biomarkers from saliva. Compositional and microtextural modifications could further improve the performance of thin films.

Keywords: Surface-enhanced Raman spectroscopy · Thin films · Salivary Biomarkers

1 Introduction

Surface-enhanced Raman spectroscopy (SERS) detects weak Raman signals from substances with low concentrations or low Raman scattering cross-sections as SERS amplifies Raman signals with metallic surfaces that are roughened on the nanoscale. Typical metallic surfaces used for SERS are silver and gold [1]. SERS is used to detect substances based on their unique "fingerprint" region in the spectrum, and it is based on the inelastic scattering of light. Most of the scattered light has the same frequency as the incident light but some of the scattered light has different frequencies. Only one in 10^8 photons undergo Raman scattering [1, 2].

One of the challenges detecting salivary biomarkers is that the chosen method must be able to recognize the disease specific biomarkers from the plethora of interfering substances to be effective [3]. Biomedical and forensic studies use SERS [2, 4–9] to identify the different pathophysiological biomarkers or illegal substances like cocaine [8, 9] from human saliva. Feng et al. [5] used SERS to study salivary proteins to differentiate malignant and benign breast cancer tumors. Hernández-Arteaga et al. [6] used SERS

Y. Dekhtyar and I. Saknite (Eds.): NBC 2023, IFMBE Proceedings 89, pp. 269–277, 2023.
https://doi.org/10.1007/978-3-031-37132-5_34

to diagnose breast cancer based on elevated N-acetylneuraminic acid levels in saliva. Hernández-Cedillo et al. [7] used SERS to study sialic acid levels in periodontitis and gingivitis. An identification method for N-acetylneuraminic acid SERS scattering on silver nanoparticle surface was developed by Vinogradova et al. [1]. They used solutions of different concentrations of N-acetylneuraminic in DI water meaning that studies with biological fluids like saliva, serum or blood are needed to confirm the results.

Along with systemic diseases, oral diseases benefit from saliva-based diagnosis. The common methods for diagnosing oral diseases are visual inspection, histopathological analysis to tissue biopsy, chemiluminescence and tissue autofluorescence [2, 4, 10]. Visual inspection, chemiluminescence and tissue fluoresce can detect lesions but cannot differentiate between pre-malignant or malignant lesions [10]. Tissue biopsy is highly invasive method [10]. Oral squamous cell carcinoma (OSCC) is among the six most common cancers worldwide and 90% of the diagnosed oral cancers are OSCCs [3, 11]. L-Fucose, L-Proline, and N-acetylneuraminic acid levels change with progressing oral diseases such as OSCC [12, 13], gingivitis or periodontitis [7]. Changing N-acetylneuraminic acid levels in saliva are also associated with Sjögren's syndrome.

Saliva acts as an analyte for Raman and other vibrational spectroscopies for diagnosing oral diseases [14] as saliva is in proximity with oral cavity and it is mainly produced by secretory glands in oral cavity [15]. That, and the non-invasive collection makes saliva potentially better diagnostic fluid for oral cancers than blood.

Ultra short-pulsed laser deposition (USPLD) technique is a well-controlled physical vapor deposition method to deposit a wide variety of thin films on solid substrate materials with good adhesion. In this study, we aimed to clarify the potential of four USPLD thin films i.e., carbon (C), bismuth (Bi), titanium (Ti), and alumina (Al_2O_3) in detecting biomarkers from saliva. In addition to conventionally used thin film materials Au, Ag, and TiO_2 [1, 14] we wanted to compare the potential of four other biomaterial surfaces, i.e., carbon (C), bismuth (Bi), titanium (Ti), and alumina (Al_2O_3).

2 Materials and Methods

2.1 Preparation of Thin Films

Deposition of different thin film materials on high purity silicon substrates were carried out using ultra short-pulsed laser deposition (USPLD) with a high-energy ultrafast Tangerine fs fiber laser (Amplitude Systemes, Pessac, France) as described in detail in [16]. For Au deposition we utilized direct-current magnetron sputtering method (Stiletto Serie ST20, AJA International Inc., North Scituate, MA, US) with a high-purity (99.6% up) Au target. Physical plasma cleaning treatment of the samples was conducted just before deposition using an argon ion sputtering unit (SAM-7kV, Minsk, Belarus). The purity of noble argon was 99.9999% (Scientific grade Argon 6.0, Oy AGA Ab, Espoo, Finland). Six different thin film materials were deposited in a similar manner. The deposition parameters in USPLD process were adjusted to achieve good adhesion and smooth surface quality. To reach film thicknesses of about 200 − 400 nm we used a 0.3 ps pulse length and 44 μJ pulse energy with an average power of 22 W. This thickness offers a uniform durable thin layer on smooth surfaces like Si. Repetition rate of 0.5 MHz was to reach a stable deposition at a high-vacuum pressure of about 0.5 mPa.

2.2 Preparation of Sample Solutions

Five independent solutions of each L-Fucose (L-(-)-Fucose ≥ 99%, Sigma Aldrich, Saint Louis, MO, USA), L-Proline (L-Proline-Reagent Plus* ≥ 99%, Sigma Aldrich, Saint Louis, MO, USA), and N-Acetylneuraminic acid (N-Acetylneuraminic acid, synthetic, ≥ 95%, Sigma Aldrich, Saint Louis, MO, US) were prepared by mixing 5 mg of the substance into 1 ml of Millipore (18.2 Ω) water in the first phase and into 1 ml of artificial saliva (Artificial Saliva for Medical and Dental Research, Pickering Laboratories, CA, USA) in the second phase. Artificial saliva was chosen instead of collecting human saliva to ensure the molecular consistency in each sample. N-acetylneuraminic acid is a predominant form of sialic acid (SA) family found in humans [17], so it will be referred as SA from this point forward. A total of 15 solutions were prepared in both phases. Five 5 μl drops were pipetted on top of the thin film surface and left to dry out at a room temperature (21.6 °C) for 60 min. This procedure was repeated for each of the three solutions L-Fucose, L-Proline, and SA.

2.3 Surface-Enhanced Raman Spectroscopy

Raman spectrum (50.19 cm^{-1} – 3399.94 cm^{-1}) was collected using Thermo Nicolet Raman Imaging Microscope DXR2xi (Thermo Fisher Scientific, WI, USA) with 532 nm laser (DXR 532 nm filter, full range grating, confocal pinhole 25 μm, EM gain on, depolarized) and 5.0 mW laser power and exposure time of 0.2 s. Three 20 nm × 20 nm areas were measured from each drop, and the mean spectra from each of the three areas was calculated with Thermo Scientific™ OMNIC™xi Software (Thermo Fisher Scientific, WI, USA). This means that a total of 15 spectra from each substance per each thin film were acquired.

2.4 Spectral and Multivariate Analysis

An average spectrum was calculated out of the raw spectra. A standard normal variate (SNV) was used to normalize average spectrum. A Savitzky-Golay filter (degree = 3, window = 31) was used to smooth the SNV spectrum. Area normalization was done to the smoothed spectrum. To distinguish the different spectra of the three biomarkers from water, artificial saliva and from each other, we used a cubic k-nearest neighbors (kNN) classification algorithm. The data was cross-validated (k-fold = 5) a hundred times. To validate the performance of the cubic kNN algorithm the F1-score was calculated with confusionStats(group, grouphat) function by Cheong [18]. The calculations were done using MATLAB 2019a (The MathWorks, Natick, MA, USA, 2019).

3 Results

The smoothed and area normalized average spectra of all aqueous solutions and spiked salivas are presented in Figs. 1 and 2. A strong peak around 500 cm^{-1} belonging to Si [19] appears on all spectra except on the Ag thin film surface. Another peak belonging to Si appears between 1062 – 1014 cm^{-1} [20]. For SA spectrum the silicon peak appears strong only on Bi and Al_2O_3 thin films. Peak positions and their corresponding molecular groups are presented on Table 1. Major peaks around 3040 – 2800 cm-1 are assigned to C-H stretch [21] for all three substances. On C thin films, all three substances have a wide peak around 1700–1000 cm^{-1} that overrides other peaks. All three substances have a peak around 1140 – 1090 cm^{-1} that is assigned to out-of-phase C-C stretch and to C-O stretch [1, 20, 22].

The four most suitable thin films are Bi, Au, Ag, and Al_2O_3 based on both the visual inspection of the spectra in Fig. 1 and the confusion matrix of cross-validation of the aqueous solutions in Fig. 3a. The overriding C peak (Fig. 1e) makes C thin film unsuitable for SERS detection of biomarkers in our study. Thus, these four surfaces were selected for further measurements with spiked artificial saliva, even though the F1-score for SA measured on C was higher than on Bi.

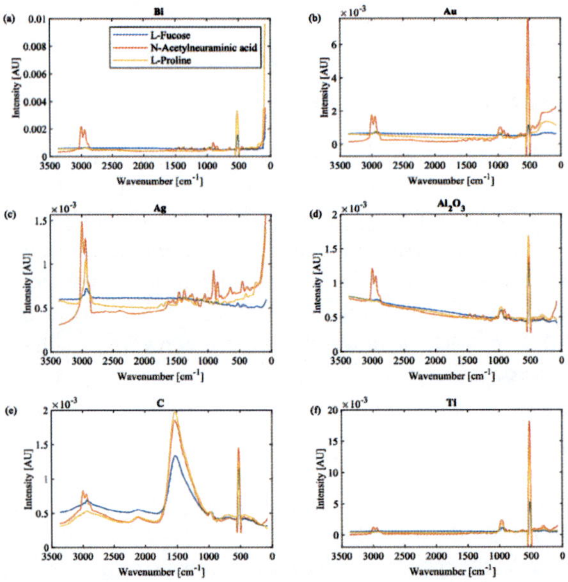

Fig. 1. The smoothed, area normalized average spectra of aqueous solutions.

3.1 Multivariate Analysis

Confusion matrixes for both aqueous solutions and spiked artificial saliva and their corresponding F1-scores are presented in Fig. 3. Validation accuracies for aqueous solutions and artificial salivas were 99.53% and 99.87%, respectively. Confusion matrix defines the performance of the classification algorithm. It shows the number of observations in each cell's true and predicted classes.

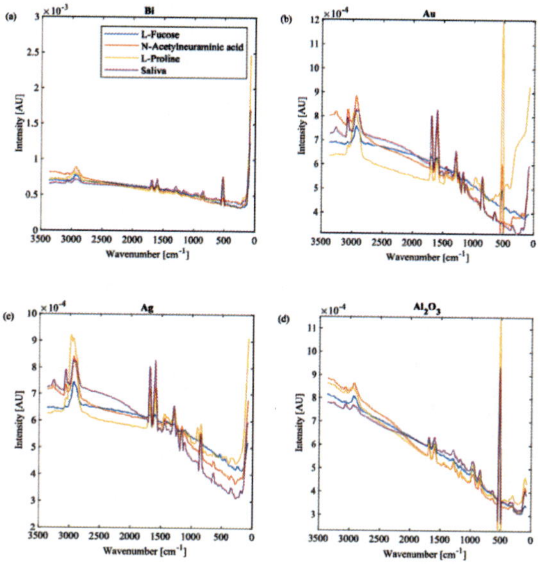

Fig. 2. The smoothed, area normalized average spectra of spiked salivas.

Table 1. Peak assignments for each substance both in aqueous solutions and spiked salivas.

Material	Raman shift [cm^{-1}]		Peak assignment	Reference
	Aqueous solutions	Spiked salivas		
SA	3040 – 2800	3000 – 2831	ν(C-H_2)	[21]
L-Proline	–	1740 – 1658	ν(C-H)	[1]
L-Fucose	–	1718 – 1660	ν(C-H)	[1]
SA	1656 – 1594	1658 – 1641	ν(C = O), Amide I	[23]
SA	1569 – 1508	1556 – 1548	ν(C-N)	[1]
L-Fucose	1475 – 1429	1477 – 1411	δ(C-H_2)	[20, 24]
L-Proline	1484–1348	1490 – 1411	δ(C-H)	[20, 22]
SA	1403 – 1344	1411 – 1346	γ(C-H_2), δ(C-O-H)	[1, 25]
L-Proline	1330 – 1270	1344 – 1207	medium C-H, Amide III	[22]
L-Proline	1270 – 1153	1270 – 1209	β(N-H)	[22]
SA	1263 – 1205	1257 – 1207	ν(C-N), Amide III	[1]
L-Fucose, L-Proline	1153 – 1064	1139 – 1068	out-of-phase ν(C-C), ν(C-N) and ν(C-O)	[20, 22]

(continued)

Table 1. *(continued)*

Material	Raman shift [cm^{-1}]		Peak assignment	Reference
	Aqueous solutions	Spiked salivas		
SA	1106 – 1058	–	δ(C-H), δ(C-O-H), ring asym ν(C-O-C)	[26]
SA	1004 – 925	1014 – 910	ring breathing	[1]
SA	971 – 944	973 – 881	γ(N-H), ν(C-O)	[1]
L-Proline	964 – 819	964 – 819	ν(N-H)	[22]
SA	946 – 869	946 – 869	ν(C-O-C)	[1]
L-Fucose	883 – 830	883 – 830	C-O-C ring	[20]
SA	813 – 761	–	δ(C–C–O), δ(C–C–H), δ(O–C–H) + γ(H)	[27]
SA	879 – 802	879 – 802	C-H vibration	[1]
SA	669 – 590	667 – 611	ring δ, ρ(C-H_2)	[28]
L-Proline	611 – 592	611 – 592	τ(C-C)	[20, 22]
L-Fucose	478 – 162	478 – 162	aliphatic C-C	[24]

*β = in-plane-bend, γ = wagging, δ = deformation, ν = stretch, ρ = rocking, τ = twisting

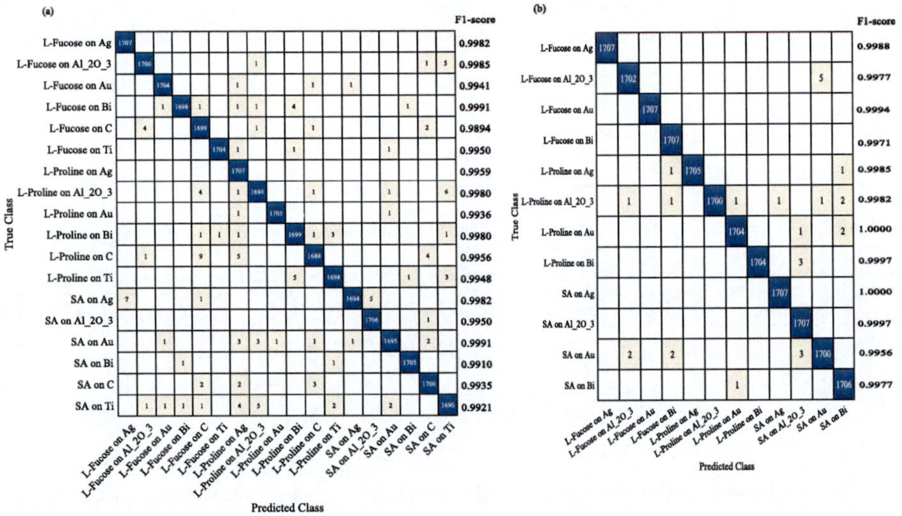

Fig. 3. Cubic k-nearest neighbor's confusion matrixes and F1-scores of (a) aqueous solutions, and (b) spiked salivas. Confusion matrix defines the performance of the classification algorithm by showing the number of observations in each cell's true and predicted classes.

4 Discussion

In our experimental study, we found out that four out of the six thin films are well suitable in detecting salivary biomarkers: L-Fucose, L-Proline and SA from aqueous solutions and spiked artificial saliva. In case of both aqueous L-Fucose and L-Proline solutions, the best thin films were Bi, Au, Ag and Al_2O_3. In case of the aqueous SA solutions, the average spectrum on C thin film had better F1-score (0.9935) than Bi (0.9910). The spectra of all three biomarker solutions are dominated by a strong characteristic amorphous carbon peak around $1700 - 1000\ cm^{-1}$ (Fig. 1e) on the C thin film making it unusable to detect these biomarkers with SERS. A strong Si peak around 500 cm^{-1} appeared on all thin films except for spectra measured on Ag thin film. The intensity of the Si peak is dependent on the thickness and the transparency of the thin film.

The confusion matrix in Fig. 3a shows that the F1-scores for average spectrum measured on Ti is higher than those measured on Au for L-Fucose and L-Proline aqueous solutions. However, spectra measured on Au have an equal or a higher number of correct classifications than Ti.

For the second phase of the study, we decided to use the four thin films: Bi, Au, Ag and Al_2O_3. The decision was based on the visual inspection of the spectra in Fig. 1a-f that shows a strong C peak overpowering the spectra. To confirm the decision a classification algorithm, a cubic kNN was used. The confusion matrix and F1-scores were computed to evaluate the performance of the cubic kNN classification in phase one. Artificial saliva was spiked with L-Fucose, L-Proline and SA. Saliva consists of $94 - 99\%$ water [15] but as an ultrafiltration of blood, it contains amino acids, proteins, antioxidants, nucleic acids, enzymes, and primary metabolites [2, 15]. Artificial saliva was chosen over human saliva samples to ensure the molecular consistency of the saliva and the spiked samples.

5 Conclusions

For our study, we chose six thin films that can be used in SERS studies to detect known salivary biomarkers. We focused on known oral disease biomarkers, L-Fucose, L-Proline and SA. Further studies are needed to collect saliva from healthy volunteers and spike them with different amounts of biomarkers to quantify the amounts of substance based on the peak intensities. Microtextural and compositional modifications to the thin films could further improve their performance.

Acknowledgements. We thank PhD Emilia Uurasjärvi for her help with Raman spectroscope. The first author received personal grants from K. Albin Johansson Foundation and Instrumentarium Science Foundation.

References

1. Vinogradova, E., Tlahuice-Flores, A., Velazquez-Salazar, J.J., Larios-Rodriguez, E., Jose-Yacaman, M.: Surface-enhanced Raman scattering of N-acetylneuraminic acid on silver nanoparticle surface. J. Raman Spectrosc. **45**, 730–735 (2014)

2. Rekha, P., Aruna, P., Brindha, E., Koteeswaran, D., Baludavid, M., Ganesan, S.: Near-infrared Raman spectroscopic characterization of salivary metabolites in the discrimination of normal from oral premalignant and malignant conditions. J. Raman Spectrosc. **47**, 763–772 (2016)
3. Wu, J.Y., et al.: Potential biomarkers in saliva for oral squamous cell carcinoma. Oral Oncol. **46**, 226–231 (2010)
4. Calado, G., Behl, I., Daniel, A., Byrne, H.J., Lyng, F.M.: Raman spectroscopic analysis of saliva for the diagnosis of oral cancer: A systematic review. Transl Biophotonics. **1**, 1–10 (2019)
5. Feng, S., Huang, S., Lin, D., Chen, G., Xu, Y., Li, Y., Huang, Z., Pan, J., Chen, R., Zeng, H.: Surface-enhanced Raman spectroscopy of saliva proteins for the noninvasive differentiation of benign and malignant breast tumors. Int. J. Nanomed. **10**, 537 (2015)
6. Hernández-Arteaga, A., de Jesús Zermeño Nava, J., Kolosovas-Machuca, E.S., Velázquez-Salazar, J.J., Vinogradova, E., José-Yacamán, M., Navarro-Contreras, H.R.: Diagnosis of breast cancer by analysis of sialic acid concentrations in human saliva by surface-enhanced Raman spectroscopy of silver nanoparticles. Nano Res. **10**, 3662–3670 (2017)
7. Hernández-Cedillo, A., García-Valdivieso, M.G., Hernández-Arteaga, A.C., Patiño-Marín, N., Vértiz-Hernández, Á.A., José-Yacamán, M., Navarro-Contreras, H.R.: Determination of sialic acid levels by using surface-enhanced Raman spectroscopy in periodontitis and gingivitis. Oral Dis. **25**, 1627–1633 (2019)
8. Dana, K., Shende, C., Huang, H., Farquharson, S.: Rapid analysis of cocaine in saliva by surface-enhanced Raman spectroscopy. J. Anal. Bioanal. Tech. **6**, 1–5 (2015)
9. Dies, H., Raveendran, J., Escobedo, C., Docoslis, A.: Rapid identification and quantification of illicit drugs on nanodendritic surface-enhanced Raman scattering substrates. Sens. Actuat. B Chem. **257**, 382–388 (2018)
10. Carreras-Torras, C., Gay-Escoda, C.: Techniques for early diagnosis of oral squamous cell carcinoma: Systematic review. Med. Oral Patol. Oral Cir. Bucal. **20**, e305–e315 (2015)
11. Radhika, T., Jeddy, N., Nithya, S., Muthumeenakshi, R.M.: Salivary biomarkers in oral squamous cell carcinoma – An insight (2016)
12. Mikkonen, J.J.W., Singh, S.P., Herrala, M., Lappalainen, R., Myllymaa, S., Kullaa, A.M.: Salivary metabolomics in the diagnosis of oral cancer and periodontal diseases (2016)
13. Mikkonen, J.J.W., et al.: Potential role of nuclear magnetic resonance spectroscopy to identify salivary metabolite alterations in patients with head and neck cancer. Oncol. Lett. **16**, 6795–6800 (2018)
14. Hackshaw, K.V., Miller, J.S., Aykas, D.P., Rodriguez-Saona, L.: Vibrational spectroscopy for identification of metabolites in biologic samples. Molecules **25**, 4725 (2020)
15. Sivadasan, P., et al.: Human salivary proteome - A resource of potential biomarkers for oral cancer. J. Proteomics **127** (2015)
16. Korhonen, H.: Novel coatings for tools, instruments and probe. **116** (2019)
17. Schauer, R., Kelm, S., Reuter, G., Roggentin, P., Shaw, L.: Biochemistry and role of sialic acids. In: Rosenberg, A. (ed.) Biology of the Sialic Acids, pp. 7–67. Springer, Boston (1995)
18. https://www.mathworks.com/matlabcentral/fileexchange/46035-confusionmatstats-group-grouphat. Accessed 25 Apr 2023
19. Senez, V., Armigliato, A., De Wolf, I., Carnevale, G., Balboni, R., Frabboni, S., Benedett, A.: Strain determination in silicon microstructures by combined convergent beam electron diffraction, process simulation, and micro-Raman spectroscopy. J. Appl. Phys. **94**, 5574–5583 (2003)
20. Notingher, I.: Raman spectroscopy cell-based biosensors. Sensors **7**, 1343–1358 (2007)
21. Larkin, P.: Basic principles. Infrared Raman Spectrosc. 7–25 (2011)
22. De Gelder, J., De Gussem, K., Vandenabeele, P., Moens, L.: Reference database of Raman spectra of biological molecules. J. Raman Spectrosc. **38**, 1133–1147 (2007)

23. Stewart, S., Fredericks, P.M.: Surface-enhanced Raman spectroscopy of amino acids adsorbed on an electrochemically prepared silver surface. Spectrochim. Acta A Mol. Biomol. Spectrosc. **55**, 1641–1660 (1999)
24. Nguyen, E.P., Chrimes, A.F., Brkljaca, R., Ou, J.Z., Berean, K., Zhuiykov, S., Kalantarzadeh, K.: Assessment of a Raman micro-spectroscopy/microfluidics unit using a model E. coli/glucose bio-system. In: IEEE International Conference on Nano/Molecular Medicine and Engineering, NANOMED, pp. 157–162 (2013)
25. Wu, H., Volponi, J.V., Oliver, A.E., Parikh, A.N., Simmons, B.A., Singh, S.: In vivo lipidomics using single-cell Raman spectroscopy. Proc. Natl. Acad. Sci. U. S. A. **108**, 3809–3814 (2011)
26. Oliveira, A.P., Bitar, R.A., Silveira, L., Zângaro, R.A., Martin, A.A.: Near-infrared Raman spectroscopy for oral carcinoma diagnosis. Photomed. Laser Surg. **24**, 348–353 (2006)
27. Feofanov, A.V., et al.: Study of sialylated neoglycoconjugates by surface-enhanced Raman scattering spectroscopy. Bioorg. Khim. **23**, 917–918 (1997)
28. Vandenabeele, P., Wehling, B., Moens, L., Edwards, H., De Reu, M., Van Hooydonk, G.: Analysis with micro-Raman spectroscopy of natural organic binding media and varnishes used in art. Anal. Chim. Acta **407**, 261–274 (2000)

Differentiation of Rabbit Stem Cells on Gelatin for Tissue Engineering Applications

Povilas Barasa[1(✉)], Andrius Buivydas[1], Emilija Baltrukonytė[1], Natalija Krestnikova[1], Aivaras Grybas[2], Ramunė Žilinskaitė-Tamašauskė[3], and Virginija Bukelskienė[1]

[1] Department of Biological Models, Institute of Biochemistry, Life Sciences Center, Vilnius University, Vilnius, Lithuania
povilas.barasa@gmc.vu.lt
[2] Urology Center, Vilnius University Hospital Santaros Klinikos, Vilnius, Lithuania
[3] Children's Surgery, Orthopaedic and Traumatology Centre, Vilnius University Hospital Santaros Klinikos, Vilnius, Lithuania

Abstract. 3D bioprinting is a novel technology allowing for the precise and controlled deposition of cells and extracellular matrix in order to manufacture artificial tissues. Gelatin is a material commonly used for various bioprinting procedures, owing to its suitability for cell attachment, biocompatibility and optimal physical properties for the printing procedure. In this research, we aimed to compare the smooth muscle and endothelial differentiation potential of rabbit adipose (ASC) and buccal mucosa (BMSC) stem cells cultivated on gelatin. These types of cells could be used in various tissue engineering applications, such as the bioprinting of larger blood vessels or urethral tissue. In the case of smooth muscle differentiation, both types of cells (ASC and BMSC) show reduced proliferation rates indicating commitment to differentiation. Both cell types showed similar patterns of smooth muscle differentiation-related gene expression: ASC and BMSC displayed increased intensity of *Acta2* and *Tagln* gene expression 10 days after differentiation induction; ASC also had increased level of *Cald1* expression 5 days after differentiation induction. The appearance of alpha smooth muscle actin stress fibers in ASC cells days after differentiation induction was further confirmed by immunocytochemistry. In the case of epithelial differentiation, both cell types showed similar changes in proliferation intensity 10 days after differentiation induction, as well as no clear differences in gene expression patterns. Thus, we conclude that both stem cell types, when cultivated on gelatin, are similarly suitable for epithelial differentiation, however, ASC would be more suitable for smooth muscle differentiation.

Keywords: stem cells · tissue engineering · gelatin

1 Introduction

Although the body has capacities to regenerate damaged tissue, if the injury is substantial, full recovery becomes impossible. In those cases, allograft or xenograft replacement of the damaged tissue may be carried out. However, in some cases, such as when source

Y. Dekhtyar and I. Saknite (Eds.): NBC 2023, IFMBE Proceedings 89, pp. 278–286, 2023.
https://doi.org/10.1007/978-3-031-37132-5_35

tissue for the graft is scarce or complicated to obtain, engineered artificial tissue could be used [1]. Artificial tissues manufactured from the patient's differentiated stem cells and biocompatible scaffolding material are attractive for several reasons: first, the technology allows the manufacture of grafts with sizes suitable for individual needs; second, to establish stem cell cultures for tissue engineering, minimal inconvenience and minor intervention is needed on behalf of the patient; third, the engineered tissue can eventually completely integrate into the patient's organism, allowing for full recovery of the injured tissue. For these reasons, tissue engineering is seen as an innovative and highly effective method of treatment that may eventually significantly improve clinical outcomes.

In the field of tissue engineering, new and improved biomaterials and sources for adult stem cells are constantly sought after [2]. The biomaterial used in this application should be biocompatible, easily synthesizable into an appropriate-sized scaffold, and provide the required chemical and mechanical cues for the stem cells grown inside it to differentiate into cells of the engineered tissue. Gelatin and methacrylated gelatin are a very attractive, and, thus, widely used biomaterials for tissue engineering, especially when 3D bioprinting is employed [3]. The components of gelatin – collagens – mimic the natural extracellular matrix, provide the molecular cues necessary for cell attachment and survival and are easily enzymatically or photo-crosslinked to form tissue-like materials with encapsulated cells. The stem cells used for tissue engineering applications should be easily accessible in the organism and have good proliferation rates and differentiation potential. Adipose stem cells have shown a spectrum of applications in the fields of tissue engineering and regenerative medicine, owing to their relatively simple isolation from excised adipose tissue or lipoaspirates, multipotency of differentiation into various mesenchymal tissues, such as bone or muscle, and beneficial immune modulation and anti-inflammatory properties. Buccal mucosa stem cells, on the other hand, are a novel source of stem cells and their potential in the field of tissue engineering is yet to be fully investigated.

The rabbit (*Oryctolagus cuniculus*) is a laboratory animal model occasionally used in tissue engineering research, for example, to evaluate *in vivo* the application of efficacy of artificial tissues for urethral repair [4]. Constructing artificial urethral tissue, comprised of a smooth muscle and urothelial layers, would require the use of stem cells that optimally differentiate into smooth muscle and epithelium. Therefore, in this research, we compared the smooth muscle and epithelial differentiation potentials of rabbit adipose stem cells (ASC) and rabbit buccal mucosa stem (BMSC) cells.

2 Materials and Methods

2.1 Cell Isolation and Cultivation

Stem cells of adipose tissue (ASC) were isolated as follows: a ~10 mm^3 piece of subcutaneous fat was excised from the subcutaneous part of a New Zealand rabbit's genital are. The tissue piece was transported to the cell culture facility in DMEM (Gibco) with 300 U/ml penicillin 300 μg/ml streptomycin (Gibco). The tissue was washed for 10 min in HBSS with calcium and magnesium (Gibco) and subsequently digested for 2 h in Collagenase I (Gibco) prepared in HBSS according to manufacturer's instructions. Subsequently, the digested tissue with the cell suspension was centrifuged at 600 g for

10 min. The supernatant was removed, then the cells were resuspended in regular culture medium (DMEM with 10% FBS (Gibco) and 1% penicillin-streptomycin) and seeded into 24-well plates (TPP), then transferred into the cell incubator. The medium was removed after 2 days and fresh media was added.

Stem cells of buccal mucosa tissue (BMSC) were isolated from a ~10 mm^3 piece of rabbit buccal mucosa. The procedure was identical to adipose stem cell isolation, as described above. The culture medium used for buccal mucosa stem cell cultivation contained the following: DME/F12 base medium (Gibco), 5% FBS, 1% insulin-transferrin-selenium (Gibco), 1% penicillin-streptomycin, 10 ng/ml βNGF (Peprotech), 5 ng/ml EGF (Peprotech) and 200 ng/ml hydrocortisone (Sigma-Aldrich).

The containers with cells were kept in a cell culture incubator with humidified 5% CO_2 atmosphere and 37 °C temperature (Thermo Fisher). The monolayer cell cultures were split every 2–3 days as follows: once the cultures were 90% confluent, the spent media was removed and the containers washed twice with PBS (Roth). 0.1% trypsin (Gibco) in EDTA (Roth) solution was added and the plates or flasks transferred into 37 °C cell culture incubator for 3–5 min. Once the monolayer was dispersed, fresh media were added and the suspension divided into 2 to 3 new plates or flasks, depending on the amount of cells.

2.2 Gelatin Coating of Tissue Culture Surfaces

Food-grade gelatin was used to prepare a 2% gelatin solution in PBS. The solution was autoclaved after preparation. 1 ml of the solution was added into the wells of 12-well plates (TPP) used for the experiments, and the plates were incubated in 37 °C incubator overnight to allow the adsorbtion of gelatin. The solution was removed after incubation, and the gelatin coated plates would then be used for cell culture experiments.

2.3 Differentiation of Rabbit Stem Cells on Gelatin

Rabbit stem cells were differentiated towards smooth muscle and epithelium using biochemical inductors in the culture media. The contents for the smooth muscle differentiation medium are thus: low glucose DMEM (Gibco), 1% Ultroser (Sartorius), 10 ng/ml TGFβ1 (Peprotech), 1% penicillin-streptomycin. The contents for the epithelial differentiation medium are as follows: low glucose DMEM (Gibco), 0.5% Ultroser (Sartorius), 2.5 μg/ml retinoic acid (Peprotech), 10 ng/ml KGF (Peprotech), 10 ng/ml HGF (Peprotech), 20 ng/ml EGF, 500 ng/ml hydrocortisone and 1% penicillin-streptomycin. As a negative control, cells that hadn't induced differentiation were cultivated on gelatin in parallel, using the regular culture media with the FBS exchanged for Ultroser at 2% for ASC and 1% for BMSC. The media were removed and fresh media added every 2–3 days.

2.4 Cell Proliferation Measurement

Cell proliferation rate was assessed via MTT assay. MTT powder (Sigma-Aldrich) was dissolved in PBS at a concentration of 0.5 mg/ml and sterile filtered through

0.22 μm syringe filter (Millipore). Cells on gelatin-coated surfaces differentiated for 5 and 10 days, as well as control cells that had not induced differentiation for 1, 5 and 10 days, were washed twice with PBS after media removal. 1 ml of MTT solution was added to the wells and plates were incubated for 1 h in 37 °C incubator. MTT solution was removed, and the formazan crystals that were produced dissolved in 440 μl DMSO (Sigma-Aldrich) for 5 min at room temperature. 200 μl samples were transferred to a 96-well plate (TPP) in duplicate. Absorption of 540 nm (formazan) and 650 nm (background) length waves was measured via Varioskan plate reader (Thermo Fisher). The measured formazan absorption was representative of the proliferation rate of the cells.

2.5 Quantification of Gene Expression

Levels of gene expression (*Ocln, Cdh1, Ck14* for epithelial differentiation and *Tagln, Acta2* and *Cald1* for smooth muscle differentiation) was measured by quantitative PCR. The RNA was isolated from the cells using TriZOL reagent (Thermo Fisher) according to manufacturer's instructions. RNA quality was assessed via RNA electrophoresis in 2% agarose gel. cDNA was synthesized using the Maxima H Minus cDNA synthesis kit with dsDNAse (Thermo Fisher) according to manufacturer's instructions. qPCR was performed using the Luminaris Color HiGreen qPCR Master Mix (Thermo Fisher) according to manufacturer's instructions. Expression level of *Gapdh* was measured to calculate relative expression of the genes of interest.

2.6 Immunocytochemistry

Samples for immunocytochemistry were first washed twice with PBS and then fixed with 4% paraformaldehyde for 15 min in room temperature. Permeabilization was performed using a 0.1% Triton X (Sigma-Aldrich) solution in PBS (PBST) for 30 min, followed by blocking with 10% goat serum (Sigma-Aldrich), 0.3 M glycin (Sigma-Aldrich) and 1% BSA (Sigma-Aldrich) solution in PBST for 4 h. After removal of blocking solution, samples were incubated with primary antibodies diluted in 1% BSA solution in PBST overnight in + 4 °C. Following incubation, the primary antibody solution was removed, the samples were washed three times with PBS and incubated with diluted secondary antibodies for 4 h in room temperature. Secondary antibody solution was removed, samples were washed two times with PBS and incubated with 1μg/ml DAPI solution in PBS for 15 min. DAPI solution was removed, the samples were washed two times with PBS and then imaged using Leica SP8 STED confocal microscope.

3 Results and Discussion

3.1 Cell Proliferation Measurement

Rabbit adipose stem cells (ASC) and buccal mucosa stem cells (BMSC) were differentiated with chemical induction on gelatin for a period of 10 days, with negative control (undifferentiated) cells cultivated in parallel. The proliferation rate of the cells was measured via MTT assay at three time points: after seeding the cells, prior to differentiation induction; 5 days after differentiation induction and 10 days thereafter (Fig. 1).

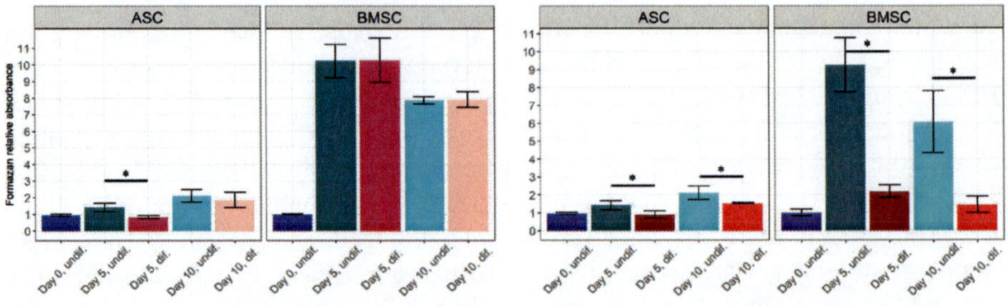

Fig. 1. Cell proliferation rates during epitheliogenic (A) and myogenic (B) differentiation. The intensity of proliferation was determined via MTT assay, with the MTT oxidation product (formazan) absorbance representing the proliferation rate of the cells. Calculations were made for the ratio between each group and the day 0 measurements. Averages (columns) and SD (crossbars) displayed. $N = 6$. Star (*) denotes statistically significant ($p < 0.05$) differences between the cell groups, evaluated using the t test.

Among the cells that were differentiated towards the epithelium, only ASC displayed a reduced proliferative rate 5 days after induction of differentiation when compared to undifferentiated cells (Fig. 1, A). The maintenance proliferation rate of cells differentiated towards the epithelial lineage has been observed by other researchers [5], and our findings are similar in the case of BMSC, but not ASC. In the case of myogenic differentiation, significant reduction in proliferation rates was observed in all cases: both ASC and BMSC, after 5 and 10 days of differentiation induction (Fig. 1, B). A significant reduction of the proliferation rate is a sign of commitment towards smooth muscle differentiation [6], therefore, both cell types exhibited similar signs of induced myogenic differentiation.

3.2 Gene Expression Evaluation

Rabbit adipose stem cells (ASC) and buccal mucosa stem cells (BMSC) were differentiated with chemical induction on gelatin for a period of 10 days, with negative control (undifferentiated) cells cultivated in parallel. The expression of genes related to myogenic and epitheliogenic differentiation was measured on the initial day of the experiment, as well as after 5 and 10 days (Fig. 2). Gene expression in undifferentiated cells was evaluated in parallel.

The *Acta2* gene codes for alpha smooth muscle actin, one of the crucial components of the cytoskeletal contraction mechanism of smooth muscle cells [7]; *Tagln* codes for the transgelin transcription factor crucial for smooth muscle differentiation [7]; meanwhile, *Cald1* codes for a calmodulin-binding protein and is expressed during the middle stage of smooth muscle differentiation [8]. Thus, the increase of the expression of these genes would indicate stem cell commitment towards the smooth muscle lineage.

In the case of epithelial differentiation, both *Ocln* and *Cdh1* code for proteins found in tight junctions – the physical connections between epithelial cells – and are indicators of late-stage epithelial differentiation [9]. We have not found any reliable indicators of middle-stage epithelial differentiation, however, *Ck14*, coding for cytokeratin 14, has been suggested as an indicator for epithelial differentiation [10]. An initial upregulation of the expression of this gene, followed by attenuation, could be indicate the dynamics of differentiation of stem cells towards the epithelial lineage.

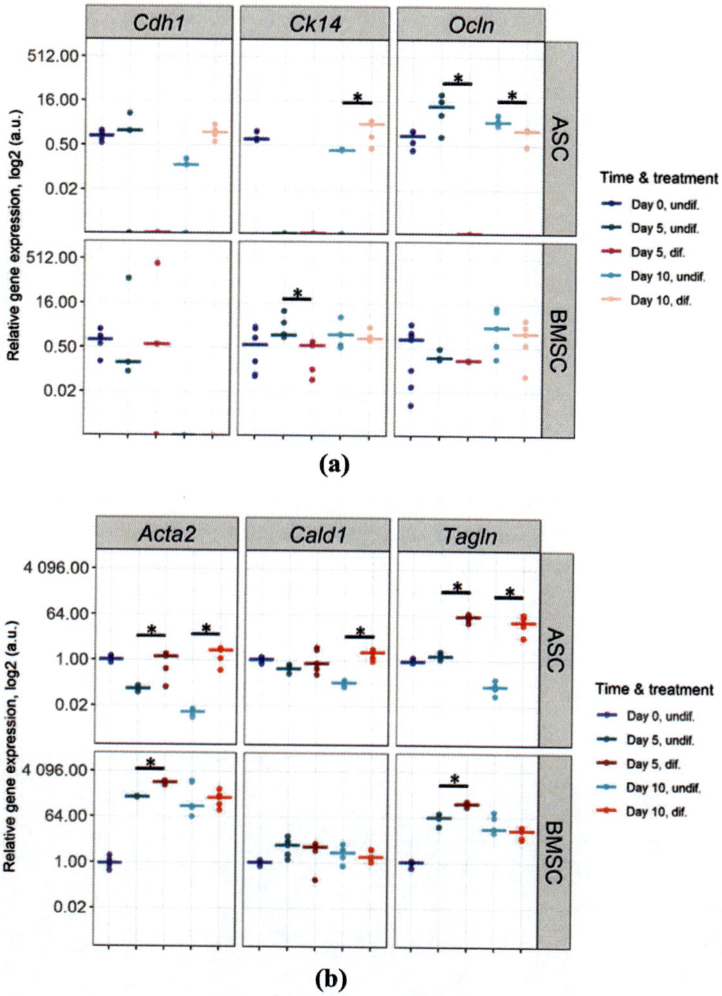

Fig. 2. Gene expression measurements during stem cell cell epitheliogenic (A) and myogenic (B) differentiation on gelatin. Medians (bars) and all measurement results (dots) displayed. Calculations were made for the ratio between each group and the day 0 measurements. N = 4–6. Star (*) denotes statistically significant ($p < 0.05$) differences between the cell groups, evaluated using the Wilcoxon test.

ASC displayed significant differences between differentiated and undifferentiated cells when the gene expression intenisities of *Ck14* (at day 10) and *Ocln* (at days 5 & 10) were measured. Meanwhile, BMSC only showed a significant reduction of *Ck14* 5 days after differentiation induction, which was not consistent at the last time point of measurement (10 days). In general, neither of the cell types showed substantial differences in differentiation intensity towards the epithelial lineage when gene expression was evaluated.

The evaluation of gene expression during differentiation of stem cells towards smooth muscle showed a clearer picture. ASC cells displayed significantly higher expression of all three smooth muscle differentiation-related genes 10 days after differentiation induction. Both cell types – ASC and BMSC – also showed increased expression of *Acta2,* coding for alpha smooth muscle actin already 5 days after differentiation induction, and increased *Tagln* expression at that same time point. These results are similar to findings by other researchers, such as when bone marrow stem cells were differentiatied [8]. In summary, although both cell types displayed gene expression changes indicative of commitment towards the smooth muscle lineage, ASC cells showed more significant changes in gene expression levels and could be more suited for smooth muscle differentiation than BMSC cells.

3.3 Immunocytochemistry

To confirm successful differentiation of the cells towards the smooth muscle lineage, immunocytochemistry imaging to detect alpha smooth muscle actin (αSMA, marker of smooth muscle) was performed (Fig. 3). Although diffuse distribution of αSMA is present in various cell types, formation of clear stress fibers comprised of αSMA is essential for the contractile function of smooth muscle [11]. We chose to evaluate only ASC cells differentiated towards the smooth muscle lineage as they displayed more significant changes in gene expression during smooth muscle differentiation.

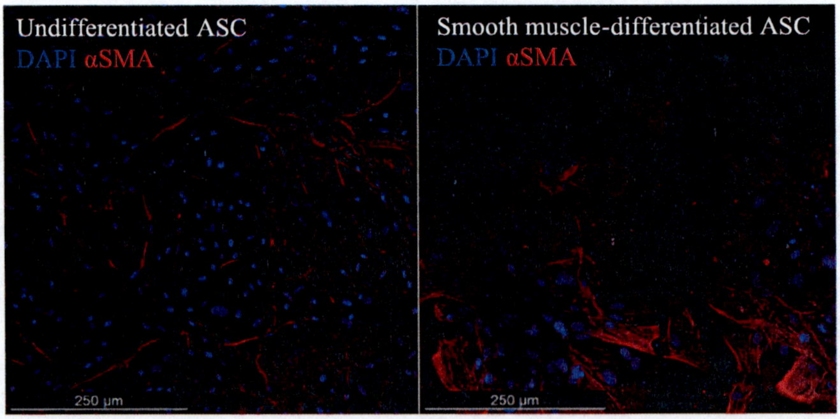

Fig. 3. Immunocytochemistry imaging of undifferentiated rabbit ASC (left) and after 5 days of myogenic differentiation (right) on gelatin. Formation of clear, straight αSMA stress fibers (right, bottom) indicates differentiation towards smooth muscle lineage.

The immunocytochemistry imaging results showed formation of αSMA stress fibers in ASC cells differentiated towards the smooth muscle lineage [11]. This result further confirms the suitability of ASC for use in smooth muscle differentiation on gelatin.

4 Conclusions

Both rabbit ASC and BMSC cells cultivated on gelatin have similar changes in proliferation rates when differentiated towards either myogenic or epitheliogenic direction with chemical induction of differentiation. When changes in differentiation-related gene expression were evaluated, in the case of epithelial differentiation, no clear pattern of changes was observed for either *Ocln*, *Cdh1* or *Ck14* gene expression. On the other hand, both cell types displayed an increased intensity of *Acta2* and *Tagln* gene expression, with ASC cells also showing increased *Cald1* gene expression 10 days after differentiation induction. ASC cell commitment towards smooth muscle lineage was further confirmed by immunocytochemistry, with clear formation of αSMA stress fibers in the microscopic images. From our findings, both cell types, when cultured on a gelatin substrate, are similarly suitable towards epithelial differentiation, but ASC cells are more suitable for smooth muscle differentiation.

Acknowledgements. This project has received funding from European Regional Development Fund (01.2.2-LMT-K-718-03-0087) under grant agreement with the Research Council of Lithuania (LMTLT).

References

1. Khademhosseini, A., Langer, R.: A decade of progress in tissue engineering. Nat. Protoc. **11**(10), 1775–1781 (2016). https://doi.org/10.1038/nprot.2016.123
2. Zhu, W., Ma, X., Gou, M., Mei, D., Zhang, K., Chen, S.: 3D printing of functional biomaterials for tissue engineering. Current Opin. Biotechnol. **40**, 103–112 (2016). https://doi.org/10.1016/j.copbio.2016.03.014
3. Yin, J., Yan, M., Wang, Y., Fu, J., Suo, H.: 3D Bioprinting of low-concentration cell-laden gelatin methacrylate (GelMA) Bioinks with a two-step cross-linking strategy. ACS Appl. Mater. Interfaces **10**(8), 6849–6857 (2018). https://doi.org/10.1021/acsami.7b16059
4. Li, C., Xu, Y.M., Liu, Z.S., Li, H.B.: Urethral reconstruction with tissue engineering and RNA interference techniques in rabbits. Urology (2013). https://doi.org/10.1016/j.urology.2013.01.041
5. Li, H., et al.: Epithelial-differentiated adipose-derived stem cells seeded bladder acellular matrix grafts for urethral reconstruction: an animal model. Tissue Eng. Part A **20**(3–4), 774–784 (2014). https://doi.org/10.1089/ten.tea.2013.0122
6. Shi, N., Chen, S.Y.: Smooth muscle cell differentiation: model systems, regulatory mechanisms, and vascular diseases. J. Cellular Physiol. **231**(4), 777–787 (2016). https://doi.org/10.1002/jcp.25208
7. Chakraborty, R., Saddouk, F.Z. Carrao, A.C., Krause, D.S., Greif, D.M., Martin, K.A.: Promoters to study vascular smooth muscle. Arterioscler. Thromb. Vasc. Biol. **39**(4), 603–612 (2019). https://doi.org/10.1161/ATVBAHA.119.312449

8. Swaminathan, G., Gadepalli, V.S., Stoilov, I., Mecham, R.P., Rao, R.R., Ramamurthi, A.: Pro-elastogenic effects of bone marrow mesenchymal stem cell-derived smooth muscle cells on cultured aneurysmal smooth muscle cells. J. Tissue Eng. Regen. Med. **11**(3), 679–693 (2017). https://doi.org/10.1002/term.1964
9. Balda, M.S., Matter, K.: Tight junctions as regulators of tissue remodeling. Current Opin. Cell Biol. **42**, 94–101 (2016). https://doi.org/10.1016/j.ceb.2016.05.006
10. Alam, H., Sehgal, L., Kundu, S.T., Dalal, S.N., Vaidya, M.M.: Novel function of keratins 5 and 14 in proliferation and differentiation of stratified epithelial cells. Mol. Biol. Cell **22**(21), 4068–4078 (2011). https://doi.org/10.1091/mbc.E10-08-0703
11. Gu, W., et al.: Smooth muscle cells differentiated from mesenchymal stem cells are regulated by microRNAs and suitable for vascular tissue grafts. J. Biol. Chem. **293**(21), 8089–8102 (2018). https://doi.org/10.1074/jbc.RA118.001739

Study of the Effect of Surfactant Decamethoxin on the Interaction of Cu (II) and Zn (II) with Lysozyme in Solution Using MALDI-ToF Mass Spectrometry

Igor I. Gerashchenko and Taras Yu. Gromovoy(✉)

Chuiko Institute of Surface Chemistry, NAS of Ukraine, 17 General Naumov Str., Kyiv 03164, Ukraine
grota@ukr.net

Abstract. The patterns of complexation of Cu (II) and Zn (II) with lysozyme without and in the presence of the surfactant decamethoxin were studied. It has been established that with an increase in the concentration of metal ions, complexes are formed with a stoichiometric ratio of 1:1 to 1:4 for the lysozyme-zinc system and from 1:1 to 1:5 for the lysozyme-copper system, and lysozyme remains, which does not entered into interaction. It was shown that the binding of the second atom of zinc and copper with the corresponding 1:1 complex occurs more actively than with the native lysozyme molecule (positive cooperative effect). The introduction of decamethoxin neutralizes this effect and leads to an increase in the average number of metal atoms interacting with lysozyme. We explain the increase in the involvement of lysozyme in the process of complex formation by the influence of decamethoxin on the folding process, which manifests itself in an increase in the resistance of lysozyme to changes in conformation as a result of interaction with metals. The possibility of using the MALDI-ToF MS method for assessing the interaction of metal ions with proteins was shown.

Keywords: decamethoxin · Cu (II) and Zn (II) ions · lysozyme · complexation · MALDI-ToF MS · cooperative effect

1 Introduction

It is known that surfactants are able to regulate conformation [1–3] and influence the process of protein folding [4, 5]. Cationic surfactant decamethoxin (DCM), which belongs to bis-quaternary ammonium compounds, is used as an antiseptic for the treatment of purulent wounds [6], in acute conjunctivitis and after ophthalmic operations in the form of a 0.02% aqueous solution (Decasan®), as well as in the composition of eye drops, which determines its possible interaction with the lysozyme of the lacrimal fluid. It is obvious that the consequence of such an interaction will be a change in conformation and, consequently, properties of lysozyme (enzymatic activity) as shown in [7]. Another factor affecting the properties and conformation of proteins is complex formation with

Y. Dekhtyar and I. Saknite (Eds.): NBC 2023, IFMBE Proceedings 89, pp. 287–292, 2023.
https://doi.org/10.1007/978-3-031-37132-5_36

metal ions. Some metals are included in enzymes as a cofactor and take part in various biological processes [8, 9]. Of interest is the interaction in the ternary system, where, together with lysozyme, the solution contains DCM and Cu (II) or Zn (II) atoms, which have a pronounced complexing ability. Copper and zinc atoms are present in biological fluids, and their interaction with lysozyme has previously been well studied by various methods by other authors. These systems were used by us as models for studying the effect of surfactants on the processes of complex formation. In this work, we demonstrate the possibilities of the mass spectrometry method in the analysis of relative quantitative changes in the contributions of stoichiometric complexes during the interaction of metal atoms with lysozyme in various media.

2 Materials and Methods

The studies were carried out using an Autoflex II LRF 20 mass spectrometer (Bruker Daltonics) equipped with the MALDI-ToF method in the linear mode of the instrument. Matrix for mass spectrometric studies was prepared according to the standard method: 12 mg of synapic acid (Fluka) was dissolved in 1 ml of a mixture of water-isopropanol, 1:1. Chemically pure $CuSO_4$ and $ZnCl_2$ salts served as the source of the corresponding ions. The examined sample was applied to a standard steel target, which was dried under ambient conditions. Each total mass spectrum obtained is the sum of 700 individual spectra.

In the first series of experiments, we studied the dependence of the complexation of lysozyme (0.00007 M) on the concentration of Cu (II) or Zn (II) ions, exceeding the concentration of lysozyme in solution by 1; 3; 6; 12; 24; 50; 100; 250 and 500 times.

In the second series of studies, DCM was added to solutions at a constant molar ratio of lysozyme:DCM = 1:6, close to that in human tear fluid when using Decasan®. The fact of protein binding to the metal was determined by the appearance in the mass spectrum of new peaks corresponding to the calculated stoichiometric complexes. Based on the obtained and normalized mass spectra according to [10], the relative contributions of each formed stoichiometric complex Lys-Me; Lys-Me_2; Lys-Me_3.... Lys-Me_n were defined. In order to study the interaction of zinc and copper atoms with lysozyme and its complexes, we proposed an approach based on the statistical change in the relative contributions of the complexes, which consists in constructing individual curves reflecting the interaction of atoms with the corresponding complexes. These curves reflect the relative percentage of the corresponding stoichiometric complexes that have interacted with the metal. They demonstrate the relative reactivity of the respective stoichiometric complex under given conditions. The higher this curve is, the higher the reactivity of this complex. Based on the relative contributions of the complexes, curves were constructed reflecting the relative number of metal atoms (x 0.01) associated with the lysozyme molecule (general interaction curve).

We proceeded from the fact that the attachment of metal atoms to lysozyme occurs through energetically nonequivalent centers. With a more energetically favorable addition of the next atom (positive cooperative effect), the curve reflecting this process will be located above the curve reflecting the formation of the previous, less energetically favorable complex, and if the addition of the next metal atom becomes less energetically favorable, then below it (negative cooperative effect) [11, 12]. As is known, one of the causes of the cooperative effect is such a change in protein conformation that creates conditions for favorable or unfavorable binding with the subsequent ligand [13, 14]. The phenomenon of cooperativity and the formation of the first complex 1:1 are decisive for the subsequent quantitative interaction of the protein with metal atoms.

3 Results and Discussion

In a blank experiment in the absence of metal compounds in solution, we did not record the formation of complexes of lysozyme with DCM. In the main series of experiments, mass spectra revealed peaks related to stoichiometric complexes of lysozyme with copper with stoichiometry from 1:1 to 1:5 and with zinc from 1:1 to 1:4, from which we can conclude that lysozyme interacts more actively with copper (II) than with zinc, which is consistent with the data [15, 16]. The presence of DCM in the solution leads to a more active interaction of both metals with lysozyme, which manifests itself in the formation of complexes of the same stoichiometry at a lower concentration of copper and zinc compounds in the solution. In the case of zinc, the activating effect of DCM is more pronounced (Fig. 1). It is also seen that the addition of the second metal atom proceeds with a positive cooperative effect. The addition of DCM narrows the range of concentrations at which a positive cooperative effect is manifested, which is most pronounced in the case of zinc (II) (Fig. 2).

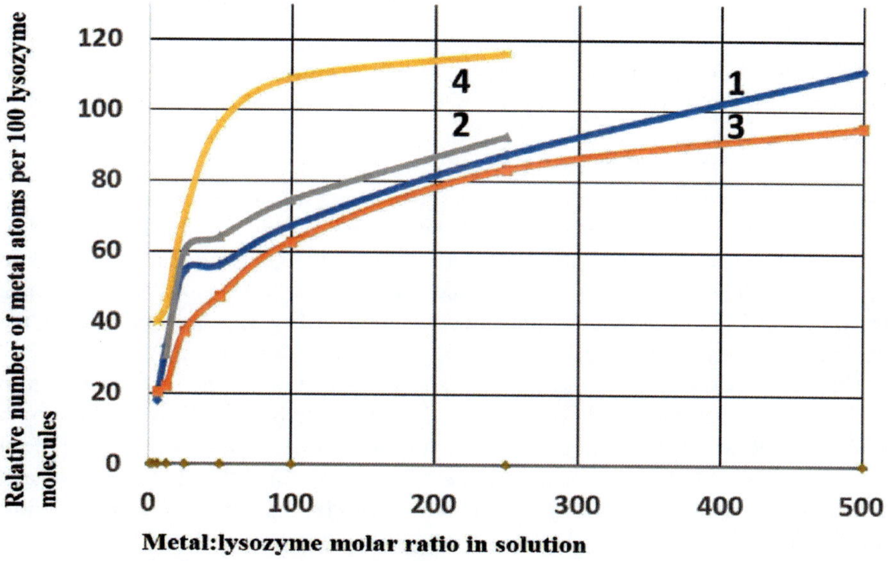

Fig. 1. Curves showing the average number of metal ions bound to lysozyme depending on the molar ratio of metal:lysozyme: Cu (II) ions without (1) and in the presence of DCM (2); Zn (II) ions without (3) and in the presence of DCM (4).

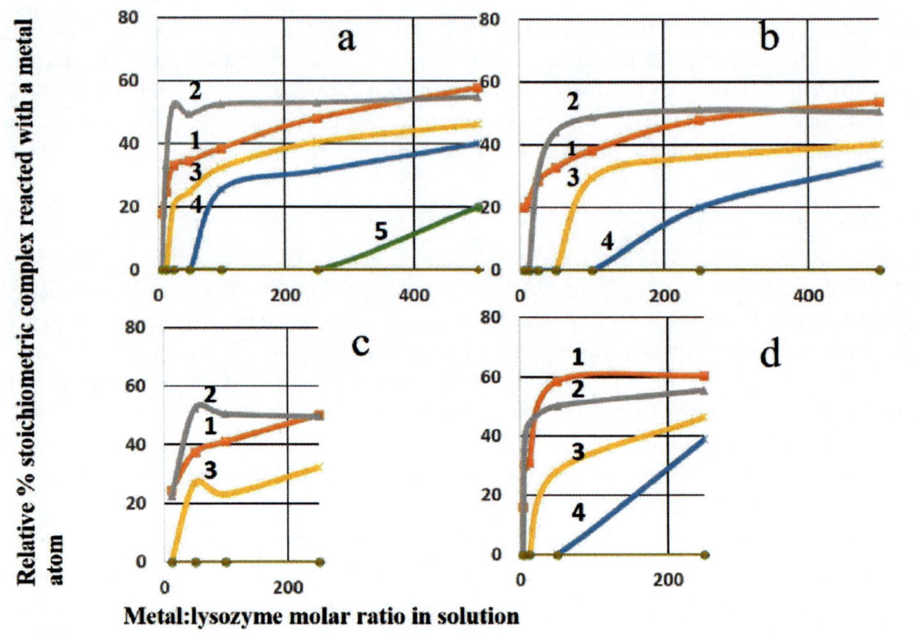

Fig. 2. Curves reflecting the relative percentage of the stoichiometric complex lysozyme:metal atom as 1:1 (1), 1:2 (2), 1:3 (3), 1:4 (4) 1:5 (5) forming with the Cu(II) ions without (a) and in the presence of DCM (b); forming with the Zn(II) ions without (c) and in the presence of DCM (d).

The addition of DCM in the case of both metals leads to a significant increase in the proportion of reacted lysozyme. This is shown in Fig. 2 general curves (1). Curve (1) corresponding to the formation of the 1:1 complex reflects the relative percentage of intact lysozyme that interacts with the metal atom. However, even at a ratio of lysozyme:metal = 1:250 and 1:500 some amount of protein remains in a free (unreacted) state.

We note the presence of two phases in the process of reducing the amount of free lysozyme (this is an inverse relationship with the amount of lysozyme that has entered into interaction). The first phase is associated with its sharp decrease, is in the range of ratios up to 1:50. In the second phase, after the ratio of lysozyme:metal = 1:50, the decrease in the percentage of unreacted lysozyme slows down (Fig. 2).

This can be explained by the fact that part of the free lysozyme is in solution in native, and some in denatured form, which does not take part in the interaction with metals. The introduction of surfactants activates the interaction of lysozyme with metal atoms due to the partial transition of the protein from the denatured to the native state and the simultaneous slowing down of its denaturation under the influence of metal atoms (stabilization effect) [17, 18].

We have analyzed the change in the relative number of metal atoms in the composition of stoichiometric complexes. In this case, we did not statistically take into account unreacted lysozyme, as was taken into account in the curves in Fig. 1. It is shown that at the maximum ratio of lysozyme:metal = 1:250, the relative number of copper atoms bound into complexes reaches 1.81 versus 1.73 for zinc. However, when DCM is added, in the case of copper, this average number drops to a value of 1.65, and in the case of zinc, on the contrary, it increases to 1.91.

As shown earlier, due to the interaction of lysozyme with copper [19], conformational changes in the protein occur in the range of lysozyme:metal ratios between 1:50 and 1:250 and are accompanied by a knee on the interaction curve. In our case, approximately in the same concentration range, a number of processes occur associated with a decrease in the influence of the positive cooperative effect, a decrease in the drop in the relative concentration of unreacted lysozyme, and the overall interaction of lysozyme with metal atoms. The addition of DCM leads to a change in the range of ratios at which a positive cooperative effect is observed without a significant change in the range of ratios at which there is a sharp drop in the concentration of unreacted lysozyme. This indicates the ability of DCM not only to promote lysozyme folding, but also to prevent a critical change in protein conformation as a result of the denaturing effect of copper and zinc. This is indirectly confirmed by our early results, which show that the presence of ethonium or DCM surfactants in the medium contributes to the preservation of the enzymatic activity of trypsin and chymotrypsin after their thermal denaturation [20].

4 Conclusion

Thus, the addition of DCM at concentrations corresponding to the Decasan® preparation leads to a change in the physicochemical properties of lysozyme, namely, to an increase in the number of zinc and copper atoms interacting with it. In the case of copper, this increase is achieved due to the advanced formation of complex 1:1 against the background of a decrease in the maximum stoichiometric ratio and the average number of copper atoms in the formed complexes. In the case of zinc, an increase in the formation of complex 1:1 is also observed, but with an increase in the average number of zinc atoms in stoichiometric complexes. We associate the increase in the involvement of lysozyme in the process of complexation with the effect of DCM on the folding process and its protective properties, which manifests itself in an increase in the resistance of lysozyme to changes in conformation as a result of interaction with metals. It is likely that different functional groups choose Cu (II) and Zn (II) as sites of interaction, so the nature of the change in conformation and the overall response of the system is different.

References

1. Ishrat, F.: Current trends in protein-surfactant interactions: A review. J. Mol. Liq. **341**, 117344 (2021)
2. Guerrero-Hernández, L.: Gemini and bicephalous surfactants: a review on their synthesis, micelle formation, and uses. Int. J. Mol. Sci. **23**(3), 1798 (2022)
3. Zakharova, L.: Cationic surfactants: self-assembly, structure-activity correlation and their biological applications. Int. J. Mol. Sci. **20**(22), 5534 (2019)
4. Otzen, D.: Protein–surfactant interactions: a tale of many states. Biochimica et Biophysica Acta (BBA)-Proteins and Proteomics **1814**(5), 562–591 (2011)
5. Leonardo, M.: Interaction between surfactants and proteins. PUBVET **16**(05), 1–8 (2022)

6. Hlaholieva, A.: Comparison of the effectiveness of preventive systemic administration of antibiotics with topical use of decamethoxin during clean surgical interventions. Perioperacina Med. **2**(2), 8–13 (2019)
7. Godovanets, O.: Clinical and immunologic assessment of a complex of therapeutic-preventive measures concerning chronic catarrhal gingivitis in children with comorbid diabetes mellitus. Wiad Lek **73**(2), 298–301 (2020)
8. Wittung-Stafshede, P.: Role of cofactors in protein folding. Acc. Chem. Res. **35**(4), 201–208 (2002)
9. Ward, W.: Nucleic acid catalysis: metals, nucleobases, and other cofactors. Chem. Rev. **114**(8), 4318–4342 (2014)
10. Croguennec, T.: Iron and citrate interactions with hen egg white lysozyme. Food Chem. **68**(1), 29–35 (2000)
11. Rong, C.: Homogeneous molecular systems are positively cooperative, but charged molecular systems are negatively cooperative. J. Phys. Chem. Lett. **10**(8), 1716–1721 (2019)
12. Hunter, C.: What is cooperativity? Angew. Chem. **48**(41), 7488–7499 (2009)
13. Koshland, D.E., Jr.: Comparison of experimental binding data and theoretical models in proteins containing subunits. Biochemistry **5**(1), 365–385 (1966)
14. Koshland, D.E., Jr.: The structural basis of negative cooperativity: receptors and enzymes. Curr. Opin. Struct. Biol. **6**(6), 757–761 (1996)
15. Moreau, S.: Hen egg white lysozyme-metal ion interactions: investigation by electrospray Ionization Mass Spectrometry. J. Agric. Food Chem. **43**(4), 883–889 (1995)
16. Schmidt, A.: Influence of one-and two-dimensional gel electrophoresis procedure on metal–protein bindings examined by electrospray ionization mass spectrometry, inductively coupled plasma mass spectrometry, and ultrafiltration. Talanta **85**(2), 1118–1128 (2011)
17. Betz, S., F.: Disulfide bonds and the stability of globular proteins. Protein Sci. **2**(10), 1551–1558 (1993)
18. Imoto, T.: A study of the native-denatured ($N \rightleftharpoons D$) transition in lysozyme I. Detection of the transition by product analyses of protease digests. Biochimica et Biophysica Acta (BBA)-Protein Structure **336**(2), 264–269 (1974)
19. Abad, C.: Dilatometric, refractomeyric and viscometric study of lysozyme-cation interaction. Biophys. Chem. **14**(3), 293–300 (1981)
20. Gerashchenko, I.: Stabilization of proteolytic enzymes in solutions of some bis-quaternary ammonium compounds. Pharm. J. **5–6**, 112–115 (1996)

Weak Electron Emission of Nanodiamond Irradiated with High Energy Electrons

Galina Boka[1], Yuri Dekhtyar[2], Mirko Rocca[2](✉), Artur Sokolov[2], and Hermanis Sorokins[2]

[1] Oncological Center of Latvia, Hipokrat str. 4, 1038 Rīga, Latvia
[2] Riga Technical University, Kipsala str. 6A, 1048 Rīga, Latvia
Mirko.Rocca@edu.rtu.lv

Abstract. Diamond is among the best solid-state dosimeters for measuring ionising radiation. Diamond nanoparticles (DNPs) could be used to estimate doses at nanovolumes, which is desirable for determining ionising radiation absorption by a single structural nanounit of DNA. However, there is no research aimed at verifying the possibility of exploiting DNP for nanodosimetry. To measure a DNP response to radiation, weak electron emission spectroscopy could be applied. This article is based on understanding diamond nanoparticles' electron emission behaviour. First, it's necessary to understand their electron emission without any type of radiation, then compare the results to the nanoparticles with the radiation treatment. For the photoemission (PE) technique, diamond nanoparticles are irradiated by UV photons to promote photoelectron emission, which has an energy of 4.0–6.0 eV, showing a proportional response of the emitted electrons to the increasing amount of radiation. It was found out that the PE signals reach an energy maximum between 5.2 eV and 5.9 eV; furthermore, the photothermal-stimulated emission (PTSE) highlights that the diamond nanoparticles reach their energy maximum value around a temperature of 530 to 570 °C. Concluding with a comparison between the PE and PTSE data and the electron-radiated diamond nanoparticles with a dose of 20 Gy, 40 Gy, 60 Gy, 80 Gy and 100 Gy, used for clinical purposes, it was found out that the diamond nanoparticles have similar signals to the PE and PTSE, which suggests that the diamond nanoparticles could have the potential to be used as a nano-dosimeter.

Keywords: diamond nanoparticles · photoemission (PE) · photothermalstimulated emission (PTSE) · electron radiation

1 Introduction

Individual monitoring of exposure to ionising radiation is a vital subject for anyone working with radioactive materials and has been an active field of research and development for well over a hundred years [1]. The first generation of such individual monitoring devices (dosimeters) were macroscale devices, which allowed for the detection and quantification of ionising radiation that came within the detection volume of the dosimeter [2]. However, radiation absorption is medium-dependent, i.e., the dose delivered to some

Y. Dekhtyar and I. Saknite (Eds.): NBC 2023, IFMBE Proceedings 89, pp. 293–303, 2023.
https://doi.org/10.1007/978-3-031-37132-5_37

object after its exposure to ionising radiation depends on the composition of that object. If the objects' composition is non-uniform or spatially anisotropic, with each component having its own mass attenuation coefficient, then the use of large-volume external dosimeters will not reflect the actual dose absorbed by the object. This is especially important for biological organisms, which are inherently non-uniform and anisotropic in their composition on multiple scales. Furthermore, when high-energy particles enter an object, they begin to scatter, losing energy in the process and generating new lower-energy particles that behave much in the same way until there is no more kinetic energy to lose [3]. This creates not a single site of exposure but rather an affected volume with each of its unit volumes receiving a different dose. Thus, to account for this as well as to be able to see the spatial distribution of incident radiation, a different, more refined approach to dosimetry was required. As a response to this need, microdosimetry emerged as a subfield of dosimetry. It studies the distribution of energy deposition in microscopic volumes [4, 5], starting with the work of Rossi et al. in the 1950s [6], where the possible difference between the macroscopic absorbed dose and the energy deposition in microscopic structures was first considered. Since then, microdosimetry has matured and now has a wide range of approaches for measuring the dose of absorbed ionising radiation at the microscale [5, 7]. Generally, there are two types of devices used for microdosimetry: gas-based and solid-state [8, 9]. The latter type is represented mainly by silicon-based [10] and diamond-based [11, 12] detectors, of which diamond is the superior option primarily due to its better tissue equivalence [13]. One of the mechanisms through which ionising radiation causes irreparable damage to the cell is through the introduction of double strand breaks (DSBs) into the structure of its DNA [14]. The incident ionising radiation then becomes absorbed by the cell. DSB events occur on the nanoscale since a region of missing DNA about 50 base pairs large (~15 nm) is already considered substantial damage [15]. It is obvious that the detection of such DSB events induced by ionising radiation can be realised only by using exposure sensors of comparable size. Thus, further refinement of the dosimeter resolution becomes necessary, and the new field of nanodosimetry begins to emerge [16]. A possible candidate for the role of such a sensor is nanodiamonds (NDs). The large band gap energy of diamond, the presence of nitrogen-vacancy (NV) centres that can serve as electron trap sites, the structural stability diamond exhibits when exposed to ionising radiation, and its good tissue equivalence have made this material a mainstay in the field of microdosimetry [17, 18]. These same properties are also exhibited by NDs [19, 20] with the added bonus of their nanoscale nature, with the size of the particles being in the 4–10 nm range and their much higher surface-to-volume ratio. However, while the exposure sensing side of the problem can be solved by using NDs, there is still the issue of acquiring a signal from the sensor, i.e., an effective mechanism for determining the dose delivered to the nanoscale sensor during its exposure to ionising radiation is missing. Diamond-based microdosimeters use an array of either metal-insulator-metal, diode, or field emission transistor-type structures [21] for charge detection. This approach is not practical for nanoscale detectors because of two reasons:

1. Placing the NDs into a nanoscale array like the one used at the microscale is at this point impractical since each particle would need to have a dedicated reading device attached to it, forming an array of immense size.

2. The number of trapping sites within each ND particle could be too small to cause any significant change in the electrical properties of the material, making detection impossible using microscale approaches.

While the problem of getting a reading from each individual ND particle within a practical timeframe at this point seems unsolvable, the problem of charge detection sensitivity can be solved using exoelectron emission spectrometry or dosimetry [22]. In this approach, the exoelectron emission is due to a heating process that can stimulate the electrons from the superficial atoms to raise their energy until they reach a threshold value, at which point they have enough energy to be emitted. To increase the amount of emitted electrons, the implementation of a photon source can bring additional energy to the surface without bringing any further risk to the electron detector. Therefore, in this work, we are going to take into consideration the diamond nanoparticles' photoemission and exoemission characteristics and compare them to electron radiation, which can help us understand if the diamond nanoparticles' signals maintain their characteristics or not after a radiation exposure.

2 Experiments

We considered sintetic diamond nanoparticles powder (140 nm raw DND -15 mV, Zeta 2.0% ash from Adams Nano), which were pressed with 5 bars of pressure using a modified mould and a hydraulic press to obtain a tablet with a height of less than 1 mm. In this way, the tablets' structures became more solid and easier to work with in terms of exposition. The PE and PTSE measurements were taken using an exoelectron emission spectrophotometer at the BINI lab (biomedical and nanotechnology engineering institute), which can handle both types of measurements: photostimulated and photothemalstimulated emission, working with a high pressure system of 6.58×10^{-6} atm; instead, the radiation process took place at the Oncology Centre of Latvia, in which it was used a linear accelerator "VARIAN TrueBeam", and all the samples were irradiated with an electron beam.

3 Methods

For this experiment, two types of samples were considered: common diamond nanoparticles and irradiated diamond nanoparticles, for a total of twenty samples. To easily study the diamond nanoparticles with the PE, PTSE and electron radiation techniques, it was necessary to make the diamond nanoparticles powder into solid samples. The use of a hydraulic press to obtain diamond nanoparticles tablets, which are easier to transport than powder, permits a precise measurement for the PE and PTSE techniques. Once the diamond nanoparticles tablets were ready, they were divided into two groups based on the electron radiation; the one without electron radiation was taken into consideration as a reference. The reference group did PE and PTSE irradiation using the exoelectron emission spectrophotometer present in BINI's laboratory, which can handle both PE and PTSE techniques. Starting with the PE technique, it was considered a wavelength range between 200 nm and 300 nm with a corresponding energy of 4 eV to 6 eV; subsequently, it was considered the PTSE technique, in which it took into consideration a temperature range between 50 °C and 580 °C. The second diamond nanoparticles tablets group were first irradiated with an electron beam using the instrument "VARIAN TrueBeam" at the Oncological Center of Latvia, which delivers energy of 6 meV at a dose rate of 600 MU/min, and the samples were exposed to different doses: 20 Gy, 40 Gy, 60 Gy, 80 Gy, and 100 Gy, used for clinical purposes, and each irradiation was delivered only once for each sample. The distance between samples and the source of the radiation was about 100 cm with a field of 10x10 cm, and to be sure to deliver the maximum amount of energy, a bolus with a thickness of 1.5 cm was collocated over the samples, which were covered by a paper towel to prevent any type of contamination from the bolus above them; in addiction, a plate with a thickness of 2 cm was collocated under the samples to prevent that the reflex ray from the patient bed hit the samples a second time, and after, they also did PE and PTSE measurements to compare the results to the ones obtained from the reference group.

4 Results and Discussion

4.1 Electrical Radiation Emission and PE

PE emission is based on the number of electrons emitted from the sample's surface, which will be proportional to the amount of energy delivered to it by the photons. In a certain amount of time, this will generate a photoelectron current based on the number of emitted electrons per second (Fig. 1).

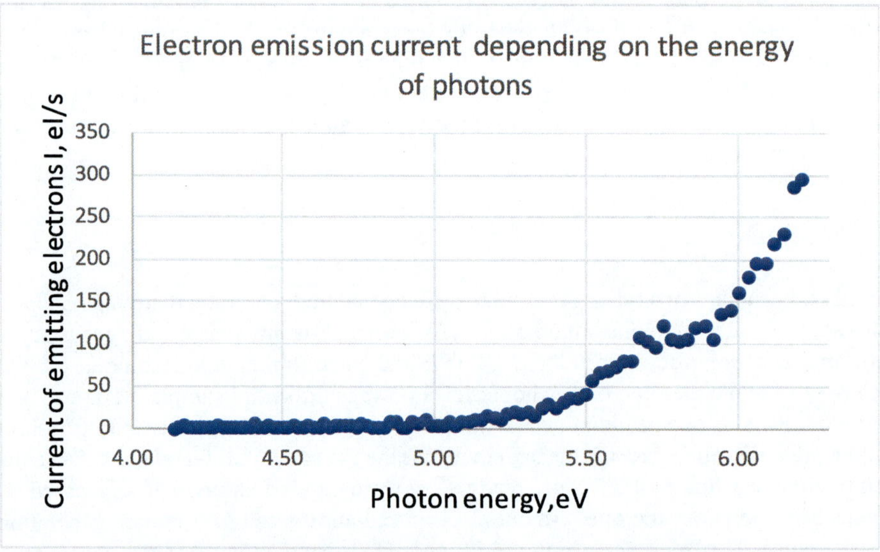

Fig. 1. Photoelectron emission current signal after correction

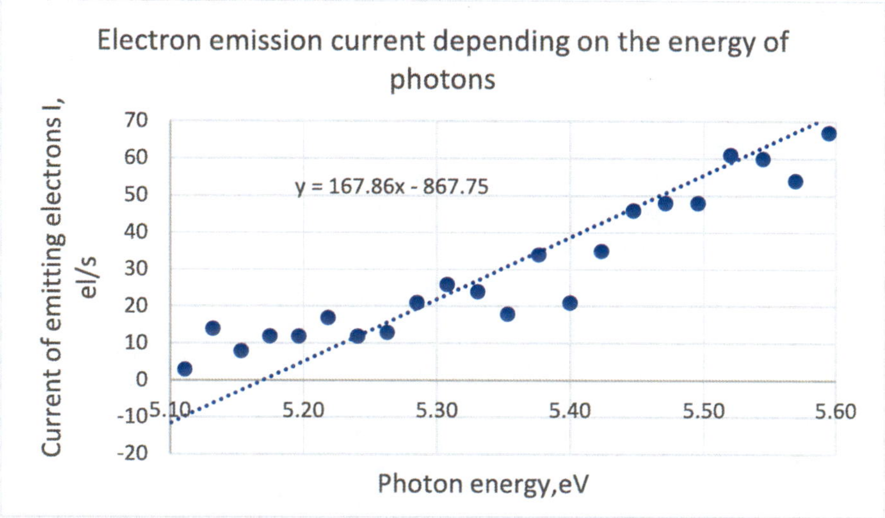

Fig. 2. Photoelectron emission spectra linear approximation for work function calculation

It was possible to study an approximation based on the signal obtained in Fig. 1, which considers a useful range between 5,1 eV and 5,6 eV to work with to understand the work function value. From Fig. 2, it is possible to see this approximation, which lets us obtain the work function of the electron emission, taking into consideration the linear approximation equation, making it possible to have the crossing value of the equation with the X-axis. The work function for the number of emitted electrons is N = f(E), and it could also be seen as a linear equation, y = ax + b [23, 24]. The aim is to obtain the x value, then it could be possible to solve the linear equation from Fig. 2 with Eq. 1:

$$\varphi = E_{N=0} = \frac{867,75}{167,86} = 5,17eV \tag{1}$$

The value obtained from Eq. 1 suggests that the electron could be expelled when its energy reaches an energy value of 5,17 eV, revealing that the emitted electron could come from a forbidden zone or from the valence zone [23].

For this experiment, the following work functions of different samples were compared (Table 1).

From Fig. 3, it's possible to highlight that in the non-irradiated samples, there's a linear increment of the work function until it reaches an absorbed dose of 80 Gy; in addition, during the analysis of the irradiated samples, it couldn't be possible to find out a regularity for their electron emission output. Analysing the data, it is possible to find out that the electron emission is strictly related to the number of doses the diamond received. To understand how the energy value increases due to the absorbed doses by the diamond, their integral value must be studied.

Table 1. Comparison of measurement value with and without electron radiation.

Dose (Gy)	1.sample (eV)	2. Sample (eV)	St.dev	Average before radiation(eV)	1. Sample (eV)	2. Sample (eV)	St.dev	Average after radiation(eV)	Pre – post average radiation
20	4,95	4,95	0	4,95	5,05	5,09	0,02828	5,07	−0,12
40	4,93	5,03	0,070711	4,98	4,99	5,07	0,05657	5,03	−0,05
60	5,01	5,03	0,014142	5,02	5,01	5,11	0,07071	5,06	−0,04
80	5,09	5,13	0,028284	5,11	5,11	5,07	0,02828	5,09	0,02
100	4,99	5,01	0,014142	5	5,03	5,11	0,05657	5,07	−0,07

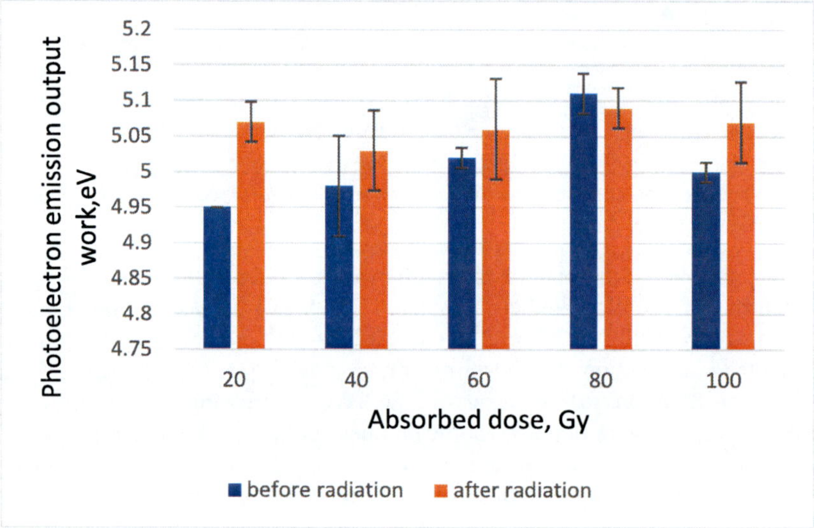

Fig. 3. Representation of the measurements took with (orange) and without (blue) electron radiation.

4.2 Electron Radiation Emission and PTSE

The exoelectron emission was measured only after the electrical radiation and compared with all samples with their different absorb doses. From Fig. 4, it is possible to highlight a specific temperature range between 530 °C and 570 °C, in which each sample had an emission signal peak.

Subsequently, the exoelectron emission signals were analysed directly between 530 °C and 570 °C. It is possible to highlight the emission peak, Fig. 5 and Fig. 6, for each sample based on the number and temperature of the exoelectron in which the emission occurred.

From Fig. 7 and Fig. 4 it is possible to highlight the diamond's exoelectrons emission, which changes along with the temperature of 200 °C, 450 °C and 550 °C. This event could be related to the crystal structures of the diamond; furthermore, from the PTSE analysis of electron irradiated diamond, only the samples' emission with a dose of 60 Gy

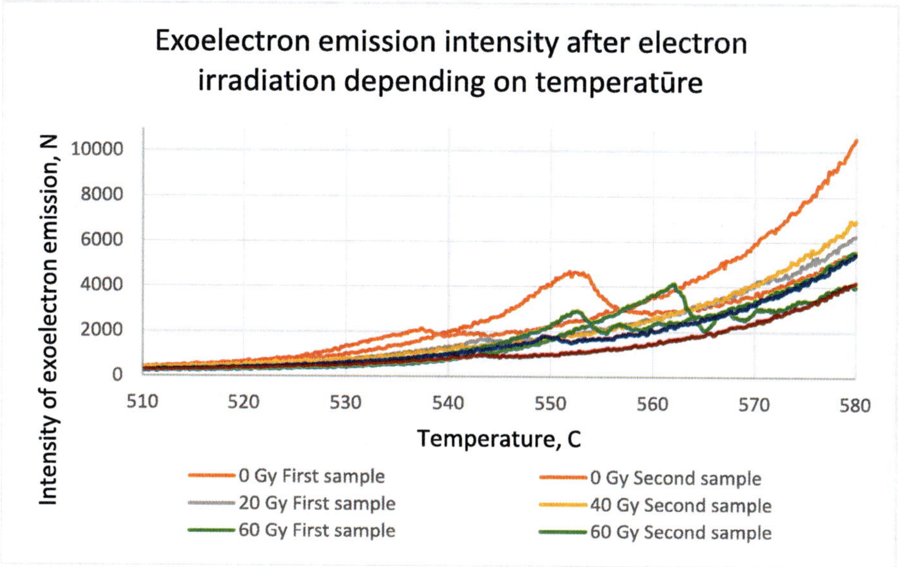

Fig. 4. Maximum value of the diamond exoelectron emission, due to the amount of absorbed doses

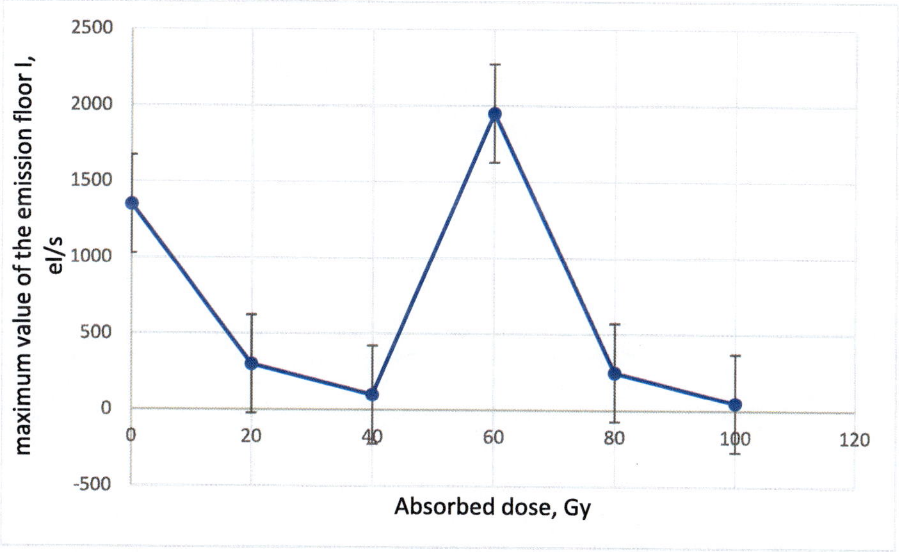

Fig. 5. Current emission due to the different absorb doses and temperature.

has a maximum value high enough to be relevant, see Fig. 5 and Fig. 6. In this experiment, diamonds are synthesised in a laboratory; they don't have a perfect chemical structure, so it is possible that the diamond could have some defects or vacancies. The most common

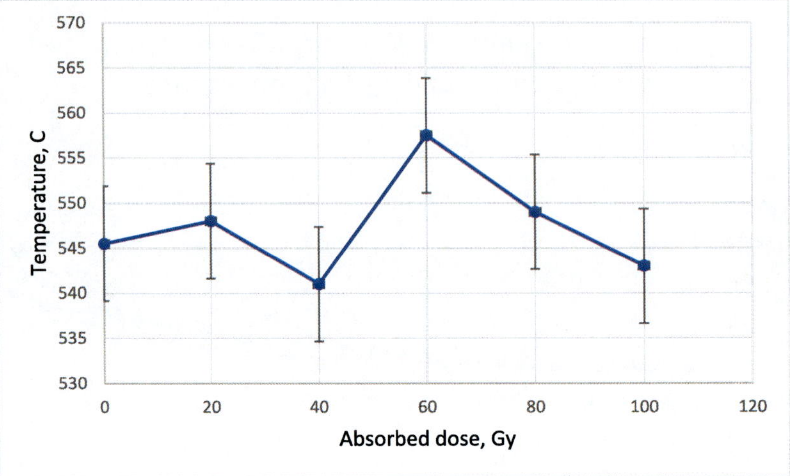

Fig. 6. Maximum value of the diamond exoelectron emission focusing on the emission temperature, due to the number of absorbed doses.

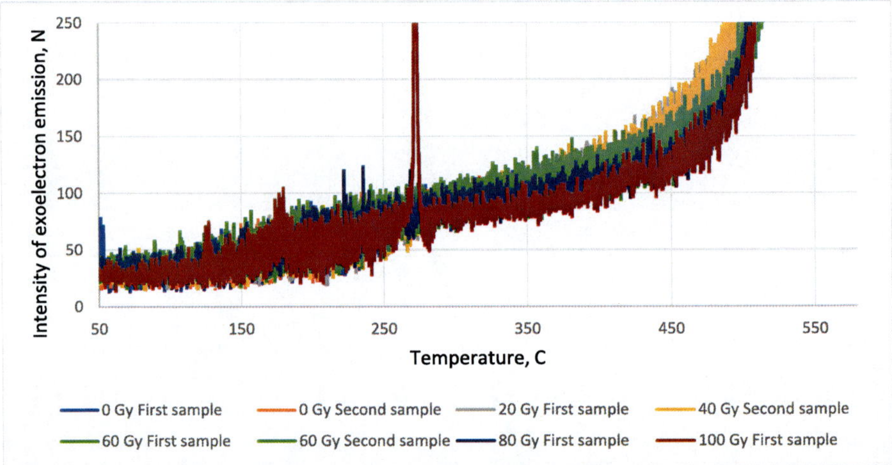

Fig. 7. Diamond's exoelectron emission with different absorbed doses, heating interval 50 °C to 500 °C

vacancy is the carbon vacancy (VC), due to the large number of carbon atoms in the structures, but this vacancy could be formed from a nitrogen vacancy (VN) or it can come from an interstitial error; in this case, an additional atom will be located between the atomic lattice. The electron irradiation promotes the formation of new vacancies, in particular VC, which leads to the formation of a new complex between a nitrogen atom and two carbon vacancies (1N-2VC). Diamond's structures start to change during the heating process, and at a temperature of 200 °C, the 1N-2VC starts to degenerate

[25]. On the other hand, this process promotes the formation of a new complex species, N-VC, with a promotion of the vacancy activity inside the forbidden zone at an energy level of 2 eV and 2,4 eV from the valence zone [25]. When the heating process reaches a temperature of 450 °C, the N-VC structure complex starts to degenerate, and at the same time, it starts to promote the formation of a new vacancy complex, VC-VC, or 2VC [25], which will collapse when the heating process reaches a temperature of 600 °C [25]. From Fig. 4 it is possible to notice each change previously described and highlight that the maximum peak of the electron emission is around 530 °C and 570 °C, which are near 600 °C. Now it's confirmed the presence of a transformation chain inside the diamond's chemical structures during the heating process; furthermore, the samples irradiated with an absorbed dose of 60 Gy clearly show this chain transformation event, like in Fig. 5 and Fig. 6 having the highest emission value. To conclude, the 60 Gy dose could be considered a limit, in which the diamond chemical structures present a transformation of VC to 2VC instead of VC to 1N-2VC. This means that with a 60 Gy irradiation, the concentration of 2VC inside the crystal lattice starts to increase, while in the meantime, the 1N-2VC concentration starts to decrease.

5 Conclusion

The aim of this work is to demonstrate the possibility of using diamond nanoparticles as dosimeters. We found out that:

- The photoemission highlights that diamond nanoparticles got a maximum peak at 5,6 eV. Studying the correlation between absorbed dose and the difference in working output, it is possible to figure out that the PE got a linear increment starting from the absorbed dose of 20 Gy to 80 Gy for the non-irradiated samples; for the irradiated samples, it is not possible to find out this correlation between dose and working output.
- The photothermal-stimulated emission spectra's maximum peak is between the range of 530 °C and 570 °C, with a rapidly increasing value at 450 °C. From the PTSE results with irradiated samples, it is possible to highlight that only the sample with an absorbed dose of 60 Gy results in a significant value; this increment could be due to some process that causes an increasing emission of electrons from the local state; instead, the other samples' results have a minimum difference in the electron emission's value, and due to their presence in the deviation standard interval, they ended up being meaningless.
- 60 Gy dose could be the limit over which it could be possible to obtain additional vacancy formation due to the high value its signal reaches with the PTSE technique; more precisely, this vacancy could be a double carbon vacancy (2VC), and the degradation of the 1N-2VC complex could end up being the event that causes the increasing amount of emitted electrons from diamond nanoparticles.

As a result, the diamond nanoparticles could be used in dosimetry techniques as nanodosimeters.

References

1. Wernil, C.: A short history and critical review of individual monitoring. In: Radiation Protection Dosimetry, pp. 1–4 (2016)
2. Liauga, P., Waker, A.J., Barthe, J.: Design of tissue-equivalentproportuonal counters. Radiat. Prot. Dosimet. **61**(4), 309–322 (1995)
3. Gibboms, J.P.: Khan's the Physics of Radiation Therapy, 6th edn (2019)
4. Santa Cruz, G.A.: Microdosimetry: Principales and applications. Rep. Pract. Oncol. Radiother. **21**(1), 135–139 (2016)
5. Waker, A.J.: Techniques for radiation measurements: Microdosimetry and dosimetry. Radiat. Prot. Dosimet. **122**(1–4), 369–373 (2007)
6. Rossi, H.H., Rosenzweig, W.: A device for the measurement of dose as a function of specific ionization. Radiation **64**, 3 (1955)
7. Goodhead, D.T.: An assessment of the role of microdosimetry in radiology. Radiat. Res. **91**(1), 45–76 (1982)
8. Agosteo, S.: Detectors for measurement of microdosimetric quantities. Radiat. Measur. **156** (2022)
9. Rosenfeld, A.B., Biasi, G., Petasecca, M., Lerch, M.L., Villani, G., Feygelman, V.: Semiconductor dosimetrhy in modern external beam radiation therapy. Phys. Med. Biol. **65**(16), 16TR01 (2020)
10. Rosenfeld, A.B.: Noveldetectors for silicon based microdosimetry, their concepts and applications. Nucl. Instrum. Methods Phys. Res. Sect. A **809**, 156–170 (2016)
11. Davis, J.A., Petasecca, M., Guatelli, S., Lerch, M.L.F., Rosenfeld, A.B.: Evolution of diamond microdosimetry. J. Phys. Conf. Ser. **1154**, 012007 (2019)
12. Talamonti, C., Kanxheri, K., Pallotta, S., Servoli, L.: Ddiamond detecors for radiotherapy X-ray small beam dosimetry. Front. Phys. Sec. Radiat. Detect. Imaging **9** (2021)
13. Yu, S., Wang, D., Zhong, X., Zhu, H., Chen, W.: A theoretical comparison of silicon and diamond in microdosimetry. JINST 18 P01032 (2023)
14. Lomax, M.E., Folkes, L.K., O'Neill, P.: Biological consequences of radiation-induction DNA damage: Relevance to radiotherapy. Clin. Oncol. **25**(10), 578–585 (2013)
15. Cannan, W.J., Pederson, D.S.: Mechanisms and consequences of double-strand DNA break formation in chromatin. J. Cell. Physiol. **231**(1), 3–14 (2016)
16. Grosswendt, B.: Nanodosimetry, the metrological tool for connecting radiation physics with radiation biology. Radiat. Prot. Dosimet. **122**(1–4), 404–414 (2006)
17. Shimaoka, T., Koizumi, S., Kaneko, J.H.: Recent progress in diamond radiation detectors. Funct. Diamond **1**(1), 205–220 (2021)
18. Marsolat, F., et al.: Why diamond dimensions and electrode geometry are crucial for small photon beam dosimetry. J. Appl. Phys. **118**, 234507 (2015)
19. Hertkorn, J., Fyta, M.: Electron features of vacancy, nitrogen, and phosphorus defects in nanodiamonds. Electrom. Struct. **1**, 025002 (2019)
20. Mochalin, V.N., Shenderova, O., Ho, D., Gogotsi, Y.: The properties and applications of nanodiamonds. Nat. Nanotechnol. **7**, 11–23 (2012)
21. Shimaoka, T., Koizumi, S., Kaneko, J.H.: Recent progress in diamond radiation detectors. Funct. Diamond **1**(1), 205–220 (2021)
22. Dekhtyar, Y.: Emission of weak electrons: Dosimetry of nanovolumes. Radiat. Meas. **55**, 34–37 (2013)
23. Dekhtyar, Y.D., Vinyarskaya, Y.A.: Exoelectron analysis of amorphous silicon. J. Appl. Phys. **75**, 8 (1994)

24. Dekhtjar, Y., Reisfeld, R., Romanova, M., Saraidarov, T., Surkov, L.: Influence of ultraviolet and electron radiation on photoelectron emission spectra of lead sulfide nanoparticles embedded in a matrix of zirconium oxide. In: First European Biomedical Engineering Conference for Young Investigator; IFMBE Proceedings, vol. 50. Springer (2015)
25. Zheng, Y., Li, C., Wei, J., Ye, H.: Diamond with nitrogen; states, control and applications. Funct. Diamond **1**(1), 63–82 (2021)

Processing of Rare Skin Disease Multispectral Images

Emilija V. Plorina[1](✉), Kristine Saulus[1], Norbert Kiss[2], Ainars Rudzitis[3], Tatjana Linova[4], Dmitrijs Bliznuks[5], Alexey Lihachev[1], and Ilze Lihacova[1]

[1] Institute of Atomic Physics and Spectroscopy, University of Latvia, Jelgavas 3, Riga, Latvia
plorina@lu.lv

[2] Department of Dermatology, Venerology and Dermatooncology, Semmelweis University, Budapest, Hungary

[3] Pauls Stradins Clinical University Hospital, Pilsonu 13, Riga, Latvia

[4] Health Center 4, Dermatology Clinic, Skanstes 50, Riga, Latvia

[5] Faculty of Computer Science and Information Technology, Riga Technical University, Kipsalas 6A, Riga, Latvia

Abstract. Rare disease has become an emerging public health priority in the European Union. The low prevalence leads to lack of knowledge and expertise in diagnosing and treating these conditions. New approaches should be found to identify, assess, and diagnose symptoms associated with rare disease. Neurofibromatosis type I, Fabry disease and Pseudoxanthoma elasticum are rare multi-organ diseases with characteristic skin manifestations. In this study a skin imaging approach previously used for skin cancer imaging was used to assess skin lesions associated with these rare diseases. The device combines diffuse reflectance imaging with 526, 663 and 964 nm illumination and autofluorescence imaging with 405 nm excitation. Image pre-processing was done using MATLAB and ImageJ. Imaging parameters were extracted using Python. The parameters were compared to similar skin lesions which are not associated with disease. A multiparameter analysis was shown to differentiate lesions characteristic of Neurofibromatosis type I from similar lesions, however, this approach was not as effective for differentiating Pseudoxanthoma elasticum plaques and Fabry disease induced angiokeratoma from similar lesions.

Keywords: Multimodal Skin Imaging · Non-invasive Skin Imaging · Rare Disease Diagnostics

1 Introduction

In the European Union there are 6172 diseases that qualify as rare – their prevalence is less than 1 in 2000 people [1]. However, in total the affected population is estimated to be from 6 to 8% which is 27 to 36 million people [2]. Due to their rarity a lot of these diseases lack adequate diagnostic and treatment options which prolongs time to diagnosis. Some of these diseases, such as Neurofibromatosis Type 1, Fabry disease and *Pseudoxanthoma elasticum* (PXE) are multi-organ diseases, however, they have characteristic cutaneous

Y. Dekhtyar and I. Saknite (Eds.): NBC 2023, IFMBE Proceedings 89, pp. 304–312, 2023.
https://doi.org/10.1007/978-3-031-37132-5_38

manifestations [3–5] which can be investigated with non-invasive skin imaging methods for potential assessment.

A multispectral imaging approach combining diffuse reflectance imaging and autofluorescence imaging was previously tested with pigmented skin lesions [6] which could also be used for assessing rare skin disease. To assess the images the imaging parameters had to be compared to a comparison group of lesions which are similar in appearance or morphology. For this purpose, a group of the most common dermatological lesions for differentiation was found for each rare skin disease.

In the case of Fabry disease, the most prominent cutaneous manifestation are angiokeratomas which are vascular lesions. Angiomas are very common benign vascular skin lesions which many people have. If there are not that many angiokeratomas present on a patient with Fabry disease, they may be misdiagnosed as angiomas [7]. Therefore, this group of lesions was chosen for comparison.

In the case of PXE, calcified plaques are formed in large patches on the skin. Such plaques cause changes in the elastic fibers of the skin's extracellular matrix. The most common type of skin change that is comparable to this kind of influence are changes in the skin's structure due to chronic sun exposure [8]. However, if PXE is still in early stages and the change is less pronounced, it may be missed completely due its similarity to normal skin. Therefore, for comparison, skin damaged by the sun's UV radiation and healthy skin was chosen.

In the case of Neurofibromatosis type I, there are two main types of cutaneous manifestations – pigmented café-au-lait macules and dermal and intradermal neurofibromas. For neurofibromas similar common lesions are dermatofibromas, intradermal nevus and sometimes, junctional nevus [9]. For café-au-lait macules the most similar are the common pigmented skin lesions – junctional melanocytic nevus and lentigo solaris [10].

Currently the main way to examine skin lesions associated with rare diseases is dermoscopy [7, 11, 12] which also requires high expertise of the dermatologist. The problem is that for all three previously described rare diseases there are usually not many lesions present at the early stages of disease. Due to their similarity to common skin lesions, they may also be missed both by the patient and a doctor unless additional symptoms are already present.

The goal of this study is to process multispectral images of rare skin disease associated lesions and common skin lesions that are similar to them, and to find multiparametric combinations where a rare skin disease can potentially be differentiated from the common skin lesions similar to it.

2 Methodology

2.1 Patient Examination

Patients with rare diseases were contacted by Rare disease cabinet of Pauls Stradins Clinical University Hospital and Department of Dermatology, Venerology and Dermatooncology of Sammelweis University, Hungary. They were examined by dermatologists at the Sammelweis University Dermatology clinic and Health Center 4, Riga. The dermatologists identified locations of lesions associated with Fabry disease, PXE and Neurofibromatosis type I and examined the lesions with dermoscopy. All involved patients were informed about the study and signed a consent form.

Table 1 summarizes the number of patients and measurements taken of each lesion group.

Table 1. Number of patients (N_{pt}) and measurements (n_m) for each lesion group

	Lesion associated with rare disease		Differential groups			
	Neurofibroma	*Café-au-lait macule*	*Dermato-fibroma*	*Intradermal nevus*	*Junctional nevus*	*Lentigo solaris*
N_{pt}	5		13	4	71	13
n_m	65	25	16	4	143	17
	Fabry disease induced angiokeratoma		*Angioma*			
N_{pt}	5		30			
n_m	52		57			
	PXE		*Healthy skin*	*Skin changes due to chronic sun exposure*		
N_{pt}	4		72	17		
n_m	55		133	29		

2.2 Experimental Setup

The multispectral imaging device used for measurements consists of four sets of narrow-band LEDs – 526 nm, 663 nm and 964 nm for diffuse reflectance imaging and 405 nm LEDs filtered through a 515 nm long-pass filter for autofluorescence imaging; an IDS 5 megapixel CMOS camera Raspberry Pi module for data processing; a 4G cellular modem for data transmission to a cloud server. A linear polarizer over LEDs as well as another one perpendicularly over the camera lens helps reduce the effects of specular reflections. The components of the device are encased in a 3D printed shell. The position of the opening of the shell and the LEDs also ensures a fixed distance between the skin being imaged and the camera objective.

The CMOS camera takes RGB images which are sent through the cloud server for further processing. A black sticker was used as a marker for later stabilization of images.

Examples of rare disease multispectral images captured with the multispectral imaging device can be seen in Fig. 1.

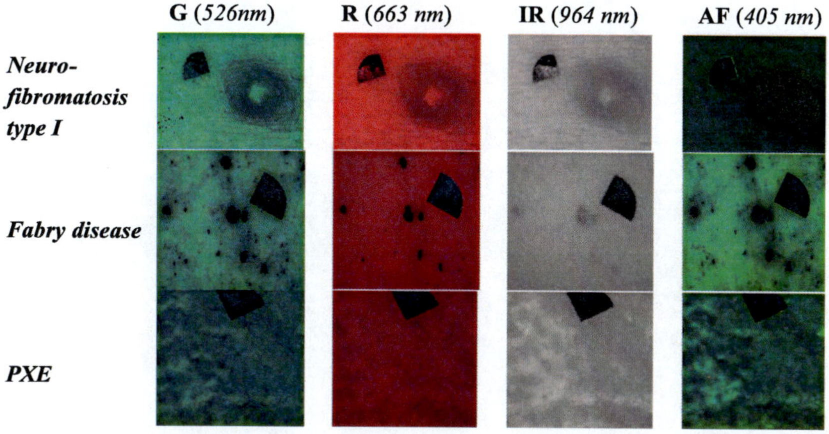

Fig. 1. Examples for rare skin disease lesion multispectral images – neurofibroma of Neurofibromatosis type I; angiokeratoma induced by Fabry disease; PXE plaque.

2.3 Image Processing

Image processing was conducted using algorithms written in MATLAB and Python programming languages. Data corrections were made using ImageJ.

Using MATLAB, the RGB images were stabilized based on the position of the black marker. If the motion artifacts were too large to compensate with stabilization, the data was not used. The thresholding segmentation method was used to extract binary arrays which contain information about the location of lesions in the image and the location of the marker. If a binary array could not be created using thresholding or if the segments were incomplete, they were created or edited using ImageJ.

Image parameter extraction was done using Python. The mean, minimum, maximum, standard deviation and variance of lesion area's intensity could be extracted. Then, the parameters were compared in groups to find the optimal combination for differentiating relevant lesion groups.

3 Results

A large number of graphs were produced for analysis, however, not all of them were informative so only the most relevant will be presented here.

The combinations of following parameters were shown to have the best differentiation ability – the minimum 5% of the mean lesion area intensity in the 526 nm (G) illumination (G(lesions) min 5% mean); the minimum 5% of the mean lesion area intensity in the 663 nm (R) illumination (R(lesion) min 5% mean); the mean intensity of the lesion area in the 663 nm (R) illumination (R(lesion) mean); the mean intensity of the lesion area in the 964 nm (IR) illumination (IR(lesion) mean); This will be demonstrated in the following graphs.

A comparison of Neurofibromatosis type I lesions café-au-lait macules and neurofibromas and their differentiation groups are shown in Fig. 2. Included in the 2D graphs are 95% confidence ellipses for each group of lesions.

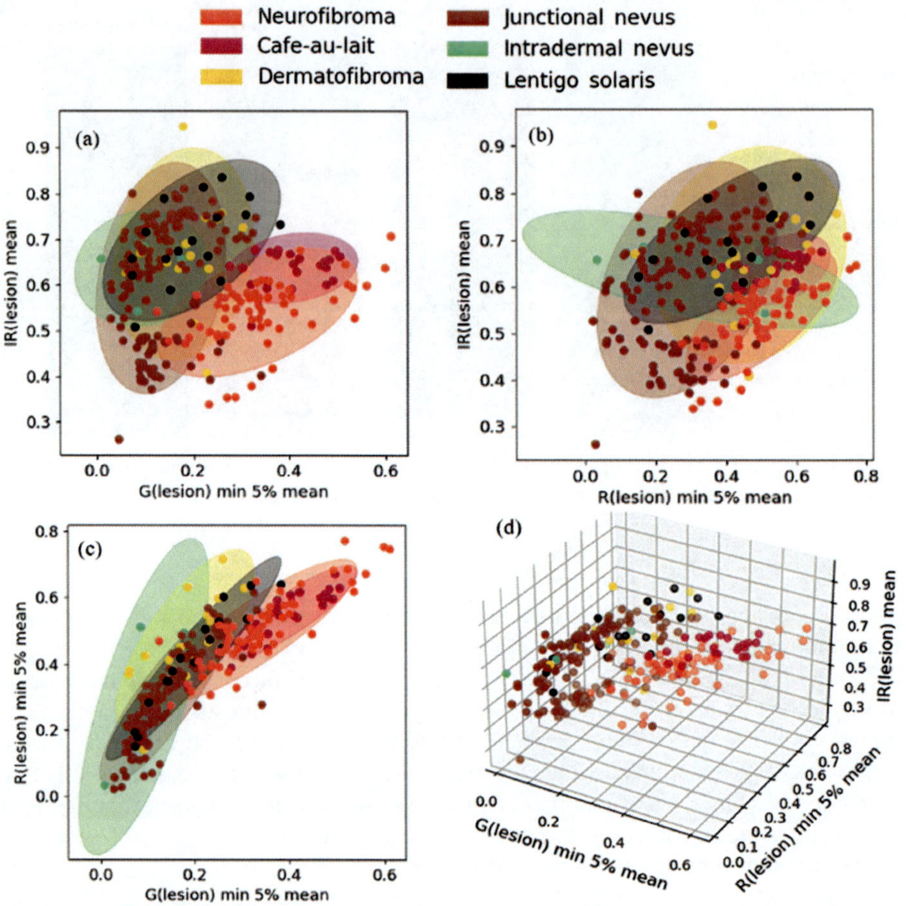

Fig. 2. 2D (a-c) and 3D (d) graphs showing Neurofibromatosis type I lesions and similar common lesions with parameters G(lesion) minimum 5% mean, R(lesion) minimum 5% mean and IR(lesion) mean.

Neurofibromas show a larger dispersion of data points compared to café-au-lait macules. With the multiparametric approach café-au-lait macules are visually very well differentiated as can be seen in graphs (a) and (c). Neurofibromas, on the other hand, showed some overlap with junctional nevus and small overlap with the other lesion groups in all graphs.

A comparison of Fabry disease induced angiokeratoma and angiomas are shown in Fig. 3. Included in the 2D graphs are 95% confidence ellipses for each group of lesions.

Compared to neurofibromatosis type I lesions there was no combination of parameters that could differentiate angiokeratomas and angiomas completely. A part of angiokeratomas could be differentiated from angiomas in graphs (b) and (c). The vertical line for angioma data at 0.00 seen in graphs (a) and (c) may also indicate that part of the angioma images were underexposed when taking images at 663 nm illumination.

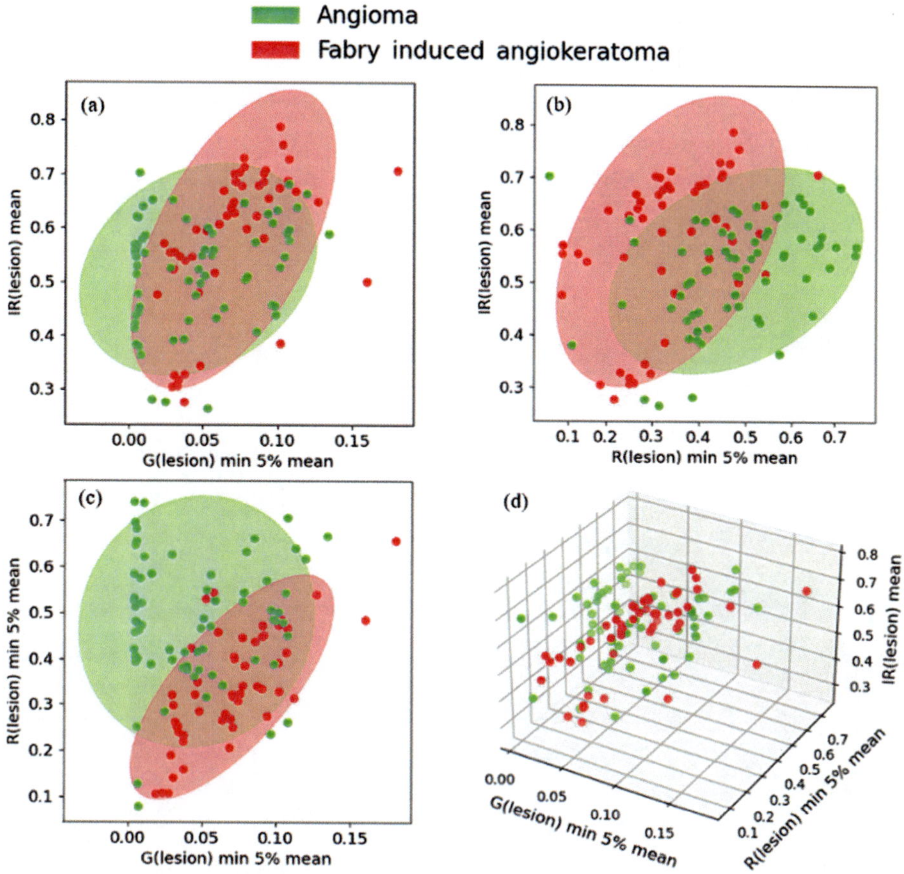

Fig. 3. 2D (a-c) and 3D (d) graphs showing Fabry disease induced angiokeratoma and angiomas with imaging parameters G(lesion) minimum 5% mean, R(lesion) minimum 5% mean and IR(lesion) mean.

A comparison of PXE lesions and their differentiation groups are shown in Fig. 4. Partial differentiation was visible in the combination between parameters G(lesion) minimum 5% mean, R(lesion) mean and R(lesion) minimum 5% mean. Included in the 2D graphs are 95% confidence ellipses for each group.

While healthy skin and PXE are mostly differentiated, a part of sun damaged skin data overlaps with PXE data in all planes.

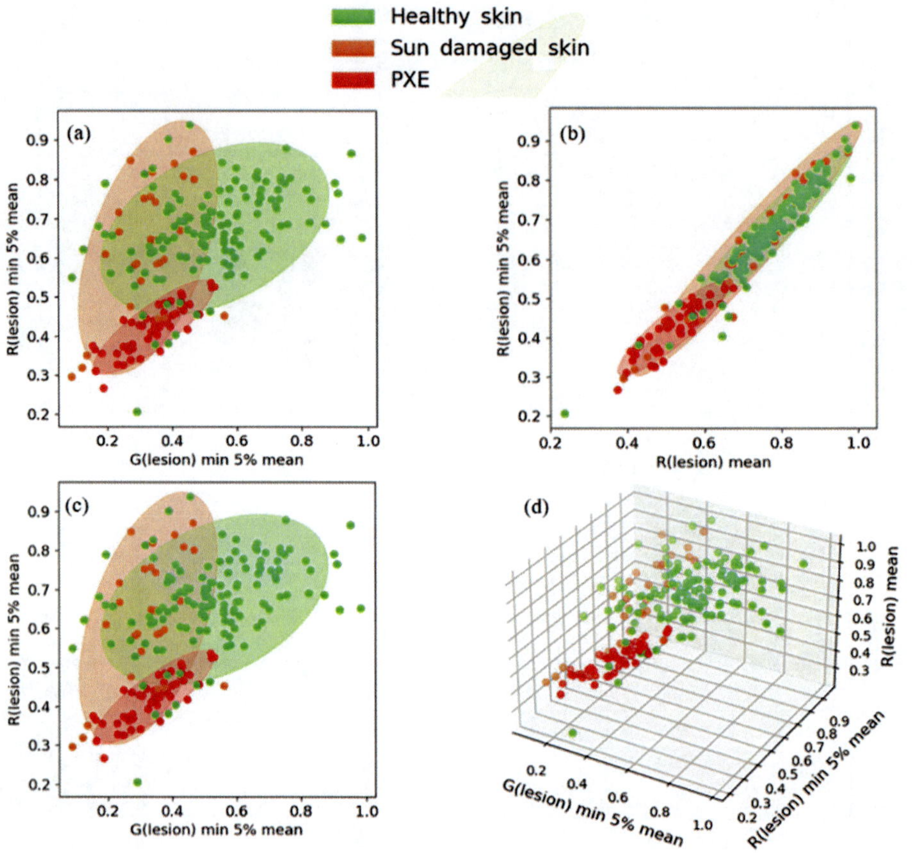

Fig. 4. 2D (a-c) and 3D (d) graphs showing PXE, skin damage due to chronic sun exposure (sun damaged skin) and healthy skin with imaging parameters G(lesion) minimum 5% mean, R(lesion) minimum 5% mean and IR(lesion) mean.

4 Discussion

Previous studies that have used the multispectral imaging method described here, have focused on differentiating skin cancer from common skin lesions [13, 14]. The main way this was done was using quantitative analysis of imaging parameters. To differentiate rare skin disease images, a similar approach using only single parameters was used but this did not yield significant results.

The multiparametric approach to analyze multispectral images described in this study had not been used before by our group when working with multispectral images of neither common lesions, nor malignant skin lesions.

5 Conclusions

The described study shows that there is potential in using a multiparametric approach to differentiate rare skin disease lesions for similar common dermatological lesions. The combination of parameters - minimum 5% of the mean lesion area intensity in the 526 nm (G) illumination (G(lesions) min 5% mean), the minimum 5% of the mean lesion area intensity in the 663 nm (R) illumination (R(lesion) min 5% mean) and the mean intensity of the lesion area in the 964 nm (IR) illumination (IR(lesion) mean) - was shown differentiate neurofibromatosis type I lesions café-au-lait macules and neurofibromas from similar common dermatological lesions – dermatofibromas, junctional nevus, intradermal nevus and lentigo solaris.

For Fabry disease induced angiokeratomas more or different parameters could be necessary for differentiation. Alternatively, they may have very similar optical properties with their comparison groups and therefore this may not be the appropriate method for their differentiation. The 0.00 values that appear for angioma data may indicate that part of the images were underexposed in 663 nm illumination or it might. For further analysis these undexposed values have to be filtered out or if a large part of the imaged lesion is affected, the whole measurement should be filtered out and new ones acquired instead. If this is done, it could also potentially reveal a multiparametric combination that can differentiate these two lesion groups.

For PXE the combination - minimum 5% of the mean lesion area intensity in the 526 nm (G) illumination (G(lesions) min 5% mean), the minimum 5% of the mean lesion area intensity in the 663 nm (R) illumination (R(lesion) min 5% mean) and the mean intensity of the lesion area in the 663 nm (R) illumination (R(lesion) mean) showed potential of differentiation, however, there is some overlap with healthy skin values and sun damaged skin values. This can be explained with the fact that for both sun damaged skin images and PXE plaque images the measurement includes some healthy skin areas as well which may not have been segmented from the affected skin. Therefore, for differentiating this rare disease a mapping approach may be better.

Acknowledgements. This research is funded by project "Rare skin diseases efficient identification and multi-modal diagnostic system" (agreement No. 1.1.1.1/20/A/072).

References

1. Nguengang Wakap, S., Lambert, D.M., Olry, A., et al.: Eur. J. Hum. Genet. **28**, 165–173 (2020). https://doi.org/10.1038/s41431-019-0508-0
2. Council of the European Union. Council recommendation on action in the field of rare diseases—2947th employment, social policy, health and consumer affairs—council meeting (2009). http://www.europlanproject.eu/Resources/docs/CouncilRecommendation_2009-C151-02.pdf. Accessed 15 Jan 2019
3. Luna, P.C., Boggio, P., Larralde, M.: Dermatologic aspects of fabry disease. J. Inborn Errors Metabol. Screen. (2016). https://doi.org/10.1177/2326409816661353
4. Boyd, K.P., Korf, B.R., Theos, A.: J. Am. Acad. Dermatol. **61**(1), 1–16 (2009). https://doi.org/10.1016/j.jaad.2008.12.051

5. Marconi, B., et al.: Pseudoxanthoma elasticum and skin: Clinical manifestations, histopathology, pathomechanism, perspectives of treatment. Intractable Rare Dis. Res. **4**(3), 113–122 (2015). https://doi.org/10.5582/irdr.2015.01014. PMID: 26361562; PMCID: PMC4561240
6. Bolochko, K., Bliznuks, D., Lihacova, I., Lihachov, A.: Quality enhancement of multispectral images for skin cancer optical diagnostics. In: Proceedings of the SPIE 10679, Optics, Photonics, and Digital Technologies for Imaging Applications V, p. 1067903 (2018)
7. Piccolo, V., Russo, T., Moscarella, E., Brancaccio, G., Alfano, R., Argenziano, G.: Dermatoscopy of vascular lesions. Dermatol. Clin. **36**(4), 389–395 (2018). https://doi.org/10.1016/j.det.2018.05.006. Epub 2018 Jul 31. PMID: 30201148
8. Lewis, K.G., Bercovitch, L., Dill, S.W., Robinson-Bostom, L.: Acquired disorders of elastic tissue: part I. Increased elastic tissue and solar elastotic syndromes. J. Am. Acad. Dermatol. **51**(1), 1–21; quiz 22-4 (2004). https://doi.org/10.1016/j.jaad.2004.03.013. PMID: 15243519
9. Messersmith, L., Krauland, K.: Neurofibroma. StatPearls Publishing (2022). https://www.ncbi.nlm.nih.gov/books/NBK539707/
10. Jha, S.K., Mendez, M.D.: Cafe Au Lait Macules. [Updated 2023 Jan 31]. In: StatPearls [Internet]. Treasure Island (FL): StatPearls Publishing; 2023 Jan-. https://www.ncbi.nlm.nih.gov/books/NBK557492/?report=classic
11. Kawashima, S., Togawa, Y., Miyachi, H., Matsue, H.: Dermoscopic features of pseudoxanthoma elasticum. Clin. Exp. Dermatol. **43**(2), 175–179 (2018). https://doi.org/10.1111/ced.13308. Epub 2017 Dec 22 PMID: 29271496
12. Behera, B., Kumari, R., Thappa, D.M., Gochhait, D., Srinivas, B.H., Ayyanar, P.: Dermoscopic features of solitary neurofibroma: A retrospective analysis of 32 cases. Australas. J. Dermatol. **61**(4), e406–e409 (2020). https://doi.org/10.1111/ajd.13338. Epub 2020 May 29 PMID: 32469460
13. Bozsányi, S., et al.: Quantitative multispectral imaging differentiates melanoma from seborrheic keratosis. Diagnostics (Basel). **11**(8), 1315 (2021). https://doi.org/10.3390/diagnostics11081315. PMID: 34441250; PMCID: PMC8392390
14. Lihachev, A., Lihacova, I., Plorina, E.V., Lange, M., Derjabo, A., Spigulis, J.: Differentiation of seborrheic keratosis from basal cell carcinoma, nevi and melanoma by RGB autofluorescence imaging. Biomed. Opt. Exp. **9**(4), 1852–1858 (2018). https://doi.org/10.1364/BOE.9.001852. PMID: 29675324; PMCID: PMC5905929

Dynamic Laser Speckle Imaging for Fast Evaluation of the Antibacterial Susceptibility by the Disc Diffusion Method

Ilze Lihacova[1(✉)], Ilya Balmages[1,2], Aigars Reinis[3,4], Svjatoslavs Kistkins[1,3], Dmitrijs Bliznuks[2], Emilija Vija Plorina[1], and Alexey Lihachev[1]

[1] Institute of Atomic Physics and Spectroscopy, University of Latvia, Riga, Latvia
ilze.lihacova@lu.lv

[2] Riga Technical University, Zunda Krastmala 10, Riga, Latvia

[3] Pauls Stradins Clinical University Hospital, Pilsoņu 13, Riga, Latvia

[4] Department of Biology and Microbiology, Riga Stradins University, Dzirciema 16, Riga, Latvia

Abstract. Phenotypic resistance tests e.g., disc diffusion method require 16–24 h to obtain the results, while the PCR tests provide only genotypic type of antibacterial resistance. New methods are needed to assess antibacterial resistance faster than existing methods, thus providing targeted pharmacological intervention at the early stage of the disease, increasing a patient's survival chances. Laser speckle imaging technique allows tracking moving particles in optically inhomogeneous media, allowing to observe the processes of changing the behavior of the sterile zone. The choice and adaptation of a mathematical model describing changes in the growth diameter of the sterile zone as a function of time and comparison with experimental data will help to understand and describe the formation of the sterile zone. The aim is to create processing algorithms for the obtained images to predict the diameter at the end of the disc diffusion test from the diameter of the earliest visible sterile zone.

Keywords: phenotypic antibacterial resistance · antibacterial resistance estimation · laser speckle imaging · sub-pixel correlation analysis · image processing · disc diffusion method

1 Introduction

Early identification of pathogen and its susceptibility to antibiotics can provide targeted pharmacological intervention at the early stage of the disease, increasing a patient's survival chances. The issue has become relevant in the treatment of elderly patients, immunosuppressed patients, and patients with secondary bacterial complications. Antibacterial therapy based on the results of laboratory tests, on the one hand, helps to reduce patient mortality, and on the other hand, the spread of resistant bacteria in the human and animal population. Phenotypic resistance tests such as disk diffusion

Y. Dekhtyar and I. Saknite (Eds.): NBC 2023, IFMBE Proceedings 89, pp. 313–319, 2023.
https://doi.org/10.1007/978-3-031-37132-5_39

and E-test require 16-24 hours to obtain the results [1], while the PCR tests provide only genotypic type of antibacterial resistance. The development of new cost-effective methods for evaluation of microbial activity to reduce detection time is a potential field of research. The laser speckle imaging technique with sub-pixel correlation analysis demonstrated promising results for the detection of sub - microbial activity [2, 3].

Laser speckle is an interference pattern produced by coherent light reflected from an illuminated rough surface. If changes occur on the surface, a "boiling" effect is created and will be detected. This method allows tracking moving particles in optically inhomogeneous media [4]. In our study of the laser speckle imaging method, a laser beam is scattered on a Petri dish where the test bacteria and antibiotic discs are located. A sub-pixel correlation analysis allows detection of small changes in bacterial activity.

Our previous study has shown that this analysis allows us to determine the bacteria growth after 3 hours from the beginning of activity [5]. Also, this analysis allowed us to identify changes in the sterile zone radius and will allow predicting these changes significantly earlier than the disk diffusion method which is recommended by the European Committee on Antimicrobial susceptibility testing (EUCAST) [6].

Based on the results from experiments with different bacteria that respond to different antibiotics, using the laser speckle imaging technique with sub-pixel correlation analysis it is possible to construct experimental curves for the growth of the diameter of the sterile zone around the antibiotic as a function of time. This study compares the mathematical model of growth of the diameter of the sterile zone with experimental results. This will help to understand the behavior of the growth curves of the sterile zone and in the future will allow us to predict the subsequent change in the diameter of the sterile zone based on short-term experiments.

Experiments were performed at the Pauls Stradins Clinical University Hospital Joint Laboratory on different bacteria and their corresponding antibiotics [7].

2 Methodology

2.1 The Experimental Setup

The laser speckles were generated by a 658 nm diode pumped solid state laser (output power 60 mW). Images were captured with 20-s intervals by a CMOS camera (Fig. 1). To read the inscriptions on the antibiotic discs, an RGB image was captured under white LED illumination.

2.2 Microbial Strains and Cultivation Conditions

The clinical isolate of E. coli was chosen for the experiment. As an antibiotic, we chose Ciprofloxacin disc (5 μg). To avoid artifacts, the experiments were performed in a separate room, in an incubator at 37 °C. The suspension of the E. coli was made in saline to the density of a 0.5 McFarland turbidity standard. Culturing on Petri dishes was prepared according to an EUCAST standard procedure - bacteria were inoculated on Mueller-Hinton agar and then antibiotic discs were placed on the surface [8].

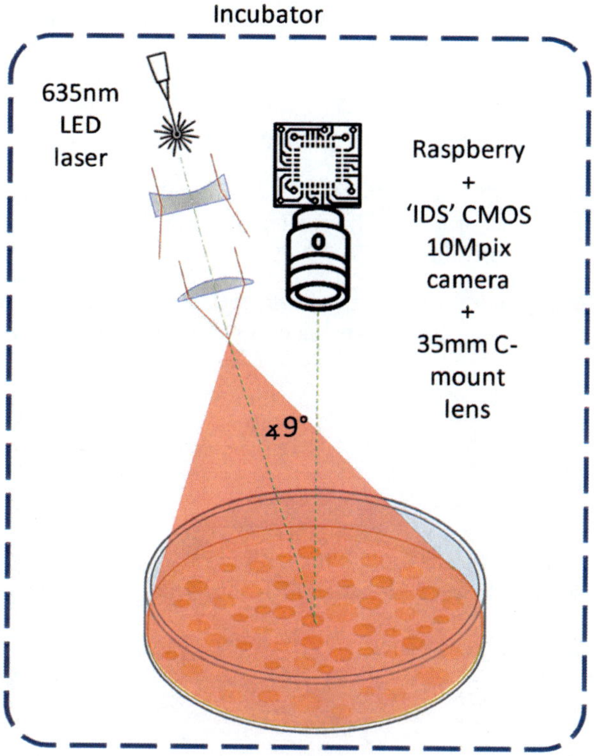

Fig. 1. Setup scheme for burst image capturing of bacteria growing process under 658 nm laser and white LED illumination.

2.3 Laser Speckle Image Conversion to Time Signals

The images captured at 20 s intervals were processed by dividing the experimental field into small sections of NxN pixels.

1) A two-dimensional normalized cross-correlation was performed between consecutive NxN image fragments throughout the experiment [9].

$$NCC(u, v) = \frac{\sum_x \sum_y ((a(x, y) - \overline{a}) * (b(x - u, y - v) - \overline{b}))}{\sqrt{\sum_x \sum_y (a(x, y) - \overline{a})^2 * \sum_x \sum_y (b(x - u, y - v) - \overline{b})^2}} \tag{1}$$

where a(x,y) and b(x,y) are two adjacent frames, $\overline{a}$ and $\overline{b}$ are the average values of these two frames, u and v are spatial displacements between frames a(x,y) and b(x,y) towards x and y, respectively.

2) The changes that occur between consecutive frames were found by the offset at the location of the correlation peak. To find a more accurate value of the offset, the interpolation was performed.
3) Offsets obtained between each pair of adjacent frames were accumulated, converting into a time signal that can be analyzed.

4) To avoid the influence of local transient spikes, a signal envelope within a certain window was used [10].

$$Env[n] = \sqrt{\frac{1}{N}\sum\nolimits_{k=n-N+1}^{n} sig[k]^2} \tag{2}$$

where N is the length of the window, n is the current sample, k is the index running inside the window. Accordingly, N – the length of the window – is responsible for the degree of signal smoothing.

An increase of signal values will occur when there is an activity increase (due to bacterial growth), and a decrease in signal occurs either due to nutrient depletion or due to the antibiotics action.

3 Results

3.1 The Sterile Zone Analysis

A disc of the sterile zone forms around the antibiotic, changing over time. The signal envelopes over each radius were averaged (Eq. 3), to avoid the influence of noise.

$$\bar{Env}[r, n] = \left[\frac{1}{M_{R1}}\sum_{m_{R1}=1}^{M_{R1}} Env_{m_{R1}}[n], \ldots, \frac{1}{M_{Rk}}\sum_{m_{Rk}=1}^{M_{Rk}} Env_{m_{Rk}}[n]\right] \tag{3}$$

where M_{Rk} is the number of envelope signals at a given radius from the center.

A spatial-temporal behavior of the sterile zone can be obtained by putting all the averaged signal envelopes for all radii together (Fig. 2). Consider the same bacterium: E. coli, and the same antibiotic: Ciprofloxacin, but for two different cases: 1) the antibiotic is placed immediately after the bacteria (almost simultaneously); 2) the antibiotic is placed 4.3 h after the bacteria. Our previous work describes this situation [11]. The sterile zone can be observed only at certain bacterial activity. 4 h after inoculation the bacteria have grown to a sufficient concentration to demonstrate a signal from bacterial activity. Putting of antibiotic discs 4–5 h after bacterial inoculation allows us to immediately observe the formation of the sterile zone (case 2). Otherwise, the sterile zone will be observed only 3–5 h after the start of the experiment (case 1).

3.2 Mathematical Model of the Sterile Zone Size Changing

The agar diffusion analysis - a method for assessing the ability of antibiotics to inhibit bacterial growth. The minimum inhibitory concentration (MIC) in solid media is usually determined using the agar diffusion technique [12]. Antibiotic diffusion into the agarose medium leads to inhibition of bacterial growth and to the formation of sterile zones. The diameter of sterile zones increases with increasing antibiotic concentration. It is possible to obtain a model of sterile zone radius growth as a function of time [13]. The radius of sterile zone is proportional to the root square of the difference between natural logarithm

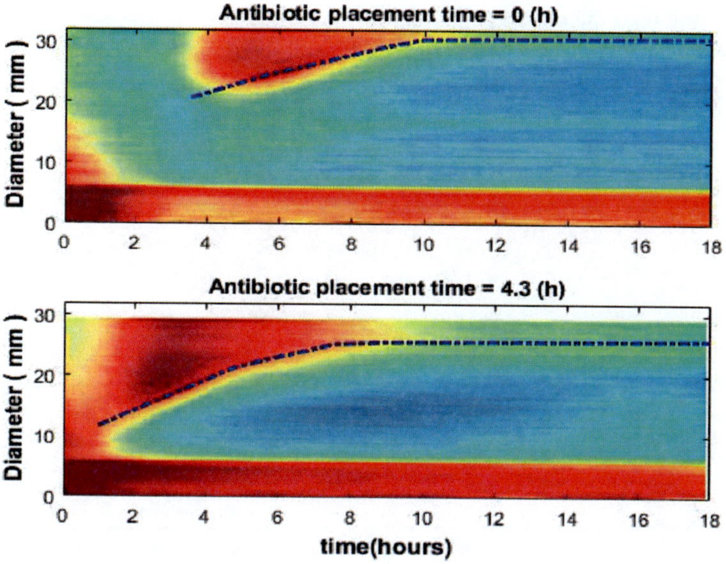

Fig. 2. The growth of E. coli around the Ciprofloxacin disc (5 μg). Spatial-temporal changes of the sterile zone. 1) The antibiotic is placed immediately after the bacteria (top). 2) The antibiotic is placed 4.3 h after the bacteria (bottom).

of the antibiotic concentration c, and natural logarithm of the MIC, multiplied by the diffusion coefficient and the time of antibiotic diffusion (Eq. 4).

$$r = \sqrt[2]{4D(\ln(c) - \ln(\mathrm{MIC})) \cdot \mathrm{t}} \tag{4}$$

where c is antibiotic concentration, and D is the diffusion coefficient received from Fick's second law of diffusion [14] (Eq. 5).

$$\frac{\partial c(x, t)}{\partial t} = D\frac{\partial^2 c(x, t)}{\partial x^2} \tag{5}$$

Fick called this coefficient: "a constant dependent upon the nature of the substances." Parameter c(x,t) describes the dependence of antibiotic concentration on time and on distance from the source. This approach has been applied to studies of two-dimensional diffusion of dyes and antibiotics, most notably penicillin [15], where it was assumed that D is independent of concentration. A newer study, such as [13] for agar diffusion assay of antibiotics to inhibit bacterial growth, also based on these formulas and on the assumption that D is independent of concentration. Then the authors consider the loss of antibiotics during diffusion and provide higher accuracy of the MIC which was determined by the proposed new dissipative diffusion models for several antibiotics.

Thus, using this model, we have three unknowns: MIC, D, and c. Their total influence in the formula can be replaced by one constant (in which it is also worth adding the number 4). Thus, a simple relationship is obtained between the radius of the sterile zone and time (Eq. 6).

$$r = \sqrt[2]{\mathrm{const} \cdot \mathrm{t}} \tag{6}$$

Having obtained the experimental curves, it is possible to choose the parameter of this constant, at which it will be as close as possible to the model. Then compare how much in this case the shape of the experimental curves corresponds to the shape of the curves from the model.

To compare the change in the theoretical and experimental growth curves of the sterile zone, several different bacterial species were used that reacted to different types of antibiotics. The results are presented in Fig. 3.

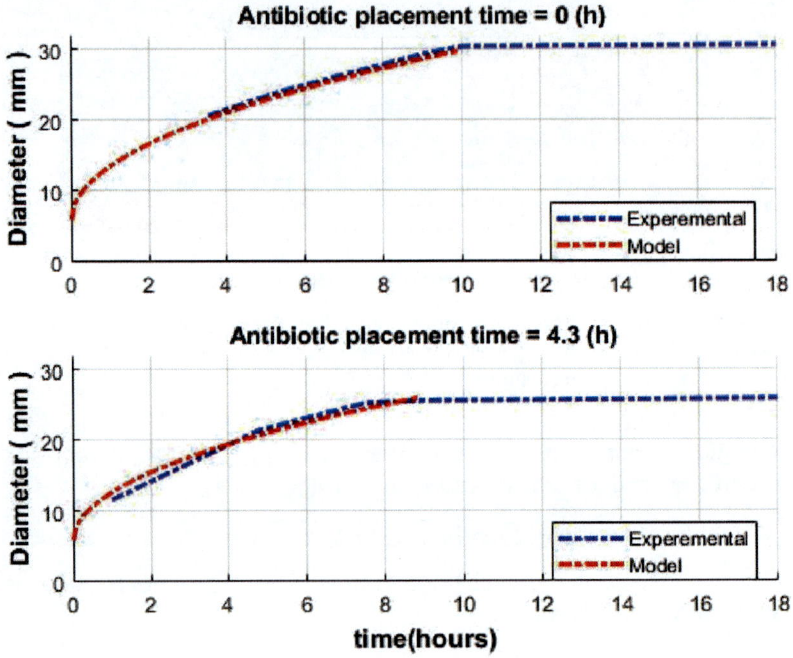

Fig. 3. Spatio-temporal curves for experimental data and model for E. coli around the Ciprofloxacin disc (5 μg).

4 Conclusions

In the current study, the ability of finding experimental growth curves of sterile zones using sub-pixel correlation analysis of laser speckle images was demonstrated.

The accuracy of the proposed mathematical model of growth curves for sterile zones has been verified in comparison with experimentally obtained curves (using sub-pixel correlation analysis of laser speckle images). The approximation, fitted to our case, describes well the experimental curves obtained. For each sensitive bacterium, to determine the radius of the sterile zone early, these curves will be different, and it is necessary to find the set of parameters for each bacterium against each antibiotic. In the case of resistant bacteria, these curves might be unpredictable and then approximations will not work here.

Thus, the effort put into the determination of these parameters will contribute to the development of an algorithm that allows predicting the change in diameter of the sterile zone around the antibiotic disk much earlier than using the standard disk diffusion method.

Acknowledgements. This work has been supported by the European Regional Development Fund project "Rapid assessment system of antibacterial resistance for patients with secondary bacterial infections" (No. 1.1.1.1/21/A/034).

References

1. Tenover, F.C.: Antimicrobial Susceptibility Testing, 4th ed., pp. 166–175. Academic Press (2019)
2. Balmages, I., et al.: Use of the speckle imaging sub-pixel correlation analysis in revealing a mechanism of microbial colony growth. Sci. Rep. **13**(1), 2613 (2023)
3. Balmages, I., Liepins, J., Bliznuks, D., Zolins, S., Lihacova, I., Lihachev, A.: Laser speckle imaging reveals bacterial activity within colony, In: Proceedings of SPIE 11920, Diffuse Optical Spectroscopy and Imaging VIII, p. 1192024 (2021)
4. Goodman, J.W.: Speckle Phenomena in Optics. Theory and Application, Roberts and Company, Englewood, Colorado (2007)
5. Balmages, I., Liepins, J., Zolins, S., Bliznuks, D., Lihacova, I., Lihachev, A.: Laser speckle imaging for early detection of microbial colony forming units. Biomed. Opt. Express **12**(3), 1609–1620 (2021)
6. The European Committee on Antimicrobial Susceptibility Testing, available online. https://www.eucast.org/ast_of_bacteria/disk_diffusion_methodology
7. Karlowsky, J.A., Weaver, M.K., Thornsberry, C., Dowzicky, M.J., Jones, M.E., Sahm, D.F.: Comparison of four antimicrobial susceptibility testing methods to determine the in vitro activities of piperacillin and piperacillin-tazobactam against clinical isolates of Enterobacteriaceae and Pseudomonas aeruginosa. J. Clin. Microbiol. **41**(7), 3339–3343 (2003)
8. The European Committee on Antimicrobial Susceptibility Testing, Clinical breakpoints-breakpoints and guidance. https://www.eucast.org/clinical_breakpoints/
9. Gåsvik, K.J.: Optical Metrology, 3rd edn. John Wiley & Sons, New York (2002)
10. Tukey, J.W.: Nonlinear (nonsuperimposable) methods for smoothing data. In: Proceedings of Con. Rec. EASCON, vol. 673 (1974)
11. Balmages, I.: Use of a laser speckle system in the determination of antibacterial susceptibility by the disc diffusion method (in press)
12. Abraham, E.P., Gardner, A.D., Chain, E., et al.: Further observations on penicillin. Lancet, 177–89 (1941)
13. Bonev, B., Hooper, J., Parisot, J.: Principles of assessing bacterial susceptibility to antibiotics using the agar diffusion method. J. Antimicrob. Chemother. **61**, 1295–1301 (2008)
14. Fick, A.: Poggendorff's Annalen. 94, 59–86. and (in English). Phil.Mag. S.4(10), 30–39 (1855)
15. Cooper, K.E., Woodman, D.: The diffusion of antiseptics through agar gels, with special reference to the agar cup assay method of estimating the activity of penicillin. J. Pathol. Bacteriol. **58**, 75–84 (1946)

Test for the Assessment of Crossed and Uncrossed Stereovision Acuity

Liva Volberga[1], Carlo Adami[2], Elizabete Strauta[3], Vsevolod Lyakhovetskii[1], and Gunta Krumina[1](✉)

[1] Department of Optometry and Vision Science, University of Latvia, Riga, Latvia
Gunta.Krumina@lu.lv
[2] Ottica Adami Ferdinando, Verona, Italy
[3] Lukin's Eye Clinic, Riga, Latvia

Abstract. Stereovision, the ability to perceive depth and spatial relationships using both eyes, is critical for human visual perception and navigation. However, some individuals have trouble detecting disparities in the images received by each eye, leading to stereoanomaly. Current methods for assessing stereovision disparities have limitations and may not provide precise and dependable measures. To address this need, we developed a stereovision test based on global random dot and anaglyph principles, incorporating a staircase method to evaluate stereovision thresholds quickly. The test can be repeated as many times as necessary, with each subsequent set being distinct from the preceding set, allowing for simultaneous assessment of crossed and uncrossed disparities. In a pilot study of 55 participants, we observed that subjects encountered difficulty in perceiving the stimuli with a stimulus presentation time of 200 ms, and 42% experienced single disparity stereoanomaly. Our newly developed stereovision test overcomes some of the limitations of current methods and provides a more precise and dependable approach for assessing crossed and uncrossed disparities. By delivering more accurate and reliable measures of stereovision, this test could have significant implications for various fields, including visual neuroscience, clinical assessment of vision disorders, and the development of new visual technologies. The results of this study contribute to the advancement of stereovision assessment and pave the way for future research in this area.

Keywords: Anaglyph · Crossed Disparity · Diagnostics · Stereoacuity · Stereoanomaly · Uncrossed Disparity

1 Introduction

Stereovision, which is the ability to neurologically integrate and fuse two horizontally disparate retinal images to produce a sense of depth, is critical for navigating the world, judging distances, and interacting with objects in our environment with precision and accuracy. However, poor stereoacuity can pose significant challenges in various aspects of life, such as hindering learning in children [1], impairing a driver's ability to gauge distances between their car and other objects [2, 3], and compromising accurate depth

Y. Dekhtyar and I. Saknite (Eds.): NBC 2023, IFMBE Proceedings 89, pp. 320–327, 2023.
https://doi.org/10.1007/978-3-031-37132-5_40

perception in sports [4], such as determining the distance to a basketball hoop or home plate, judging the flight of a ball, or assessing the movement of players. Moreover, stereovision is essential for performing everyday tasks safely [5], such as crossing a busy street or walking downstairs without the risk of misjudgement.

When binocularity is disrupted during early life due to conditions such as strabismus [6], anisometropia [7, 8], or monocular opacity [9, 10], it can result in compromised vision in the affected eye and loss of stereopsis, which can disrupt hand-eye coordination and greatly impact self-esteem, lifestyle, and career choices. However, some individuals may have trouble in detecting either crossed or uncrossed disparities, which are differences in the images received by each eye, a problem referred to as stereoanomaly. Richards was among the first scientists to research stereoanomaly and found that 30% of people have this condition [11]. Patterson and Fox [12] found that anomalies in stereopsis could be influenced by brief exposure time (<100 ms) rather than a real lack of neural mechanisms responsible for stereopsis. They reported an incidence of 4% for stereo-blindness.

Stereoscopic depth perception can be achieved through small retinal disparities that can be fused to create a single image (fine stereopsis), but more reliable depth information is often obtained through larger disparities that result in double vision (coarse stereopsis) [13]. The precise function of coarse stereopsis is not yet fully understood, but it may play a critical role in guiding vergence eye movements that are necessary for the development of fine stereopsis, particularly during early stages of visual development. It can be assessed through various clinical stereotests, such as polarized stimuli, anaglyph stimuli, random dot stimuli, and real-world stimuli [14], with most of them used solely for assessing crossed disparity – the perceived stereoscopic image appear before reference plane.

To accurately assess stereoacuity, it is important to use a test that does not include any monocular cues, which could enable observers to complete the test without relying on stereovision. Examples of monocular cues include occlusion, texture gradient, relative size, and motion parallax [15]. There are several clinical stereotests, such as the Randot, Frisby, TNO, Lang, and Random-Dot E tests, which aim to eliminate these monocular depth cues by using global random-element stereograms developed by Julesz [16]. These tests require the subject to detect a target that is perfectly camouflaged in the monocular images and can only be detected through stereopsis.

Despite efforts to eliminate monocular cues in these tests, the presence of monocular artifacts remains a potential issue, and caution must be taken during the administration of the test to ensure that only stereovision is used. Additionally, accurately measuring stereoacuity is still a complex and challenging task, and more research is needed to develop reliable and precise methods for assessing stereovision disparities. By addressing these challenges, the development of new and improved stereovision tests could have significant implications for various fields, including visual neuroscience, clinical assessment of vision disorders, and the development of new visual technologies.

The aim of this study is to develop a new test that can accurately assess crossed and uncrossed disparity for global stereopsis by varying parameters, such as random dot size, exposure time, random dot density, coarse and fine disparities.

2 Method

Participants
In this study, we recruited 55 adult participants with normal or corrected-to-normal (0.00 logMAR or better) binocular vision. Prior to the experimental measurements, all participants underwent a stereo vision assessment using the clinical TNO test, which assesses crossed disparity. The participants consisted of both students and staff members, with an age range of 19 to 42 years old. The research protocol adhered to the principles outlined in the Declaration of Helsinki and was approved by the ethics committee of the University of Latvia (No. 13/03/2020).

Stimuli
The global stereovision test (GST) was developed using the principles of anaglyph and random dot stereograms, and was presented on a computer screen. The background of the stereotest was grey with an RGB value of 170, 170, 170, and a brightness of 23.5 cd/m^2 ($SD = 0.6$). The average brightness of the red dots (with an RGB range of 180–183 (varied by subjects) 0, 0) was up to 6.7 cd/m^2 ($SD = 0.3$), while the cyan dots (with an RGB range of 0, 172–175, 171–175) had a brightness of up to 19.8 cd/m^2 ($SD = 0.9$). The room illumination was 456 lx ($SD = 29$). The stereo threshold evaluation was affected by the pixel size of the computer screen, which was 0.21 mm. To ensure a minimum stereo threshold of approximately 43 arcseconds, the screen was positioned at a distance of 1 m from the participant. The stimulus presented on the screen consisted of a square with a circle inside, with a diameter of 2.77° and a background size of 3.34° (see Fig. 1 a,b). The stimulus size was similar to that of the TNO circle with the cut-out. The density of the random dots used was 25%, and the size of each random dot was a 2x2 pixel matrix to ensure that the stereogram could be viewed from a distance of 1 m. However, it should be noted that the size of the random dot used in this test was smaller than that used in the original TNO test.

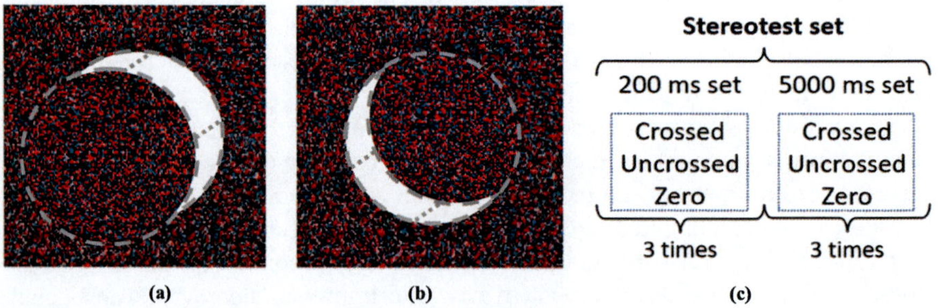

Fig. 1. Stimuli of the global stereovision test (GST) – appeared ring in front of the screen **(a)**, behind the screen **(b)**. The presentation of stereovision test sets **(c)**.

The GST employed a staircase method for presenting the stereo stimulus, which means that the level of difficulty of the test is adapted to the subject's responses. Specifically, after three correct responses, the offset or depth of the stereo stimulus was reduced, making it more challenging to discriminate between the stimuli. Conversely, after two incorrect responses, the offset or depth was increased, making it easier to distinguish between the stimuli. The maximum stereo depth was limited to 1078 arcseconds to reduce test time. Two sets of stimuli were presented with different time limits, 200 ms and 5000 ms or 5 s, respectively. If the subject was unable to view the stereogram within 5 s, the test result was recorded as an incorrect response. Each time set was repeated three times to obtain a more reliable measure of the subject's stereo threshold. The order of presentation of crossed, uncrossed, and zero disparity stimuli was randomized to avoid any learning effects or biases (see Fig. 1 c).

Data Analysis
The statistical analysis of the results was performed using *GraphPad Prism 7.0*. The *D'Agostino-Pearson* was applied, and it was determined that the data did not follow a normal distribution. Consequently, the non-parametric *Mann-Whitney* and *Wilcoxon matched pairs signed rank tests* were used to compare the stereoacuity determined by TNO and GST, the means of interdependent samples in both the limited and unlimited test group, as well as the crossed and uncrossed disparity group. *Pearson correlation analysis* and its associated confidence test were employed to investigate the correlation between response time and both crossed and uncrossed disparity. These statistical methods were chosen to ensure a rigorous and appropriate analysis of the obtained data.

3 Results

The Comparison Between the Clinical TNO Test and the GST
When administering the global stereovision test in two sets of tests, namely the "Limited" (200 ms) and "Unlimited" (5000 ms), with each set repeated three times, it was discovered that 42% of participants experienced difficulty with the Limited test at the uncrossed disparity. Specifically, they were unable to perceive the stereovision stimulus in all three repetitions or only saw it in two out of three repetitions. However, all participants were found to possess stereovision based on clinical assessment with the TNO test, which tests for crossed disparity, with eight of them exhibiting at least coarse stereovision (>480 arcsec). Only seven participants, or 13%, were unable to perceive the uncrossed stereovision stimulus during the Unlimited test. Furthermore, the mean crossed stereoacuity determined by the TNO and GST tests did not exhibit any significant difference ($U = 1214$, $p = 0.48$), being 96 arc sec ($SD = 65$) and 132 arc sec ($SD = 202$), respectively.

The Assessment of Crossed and Uncrossed Disparity Stereovision Threshold
To evaluate the stereo threshold, we included data only from participants who completed both crossed and uncrossed disparity measurements using the global stereovision test (GST) (*see Fig. 2*). Therefore, the number of measurements analyzed varied, as some participants struggled to detect the stereostimulus under the conditions of limited test time and uncrossed disparity. In the limited time condition, not all participants could identify the stereostimulus. However, with more time, participants' performance improved, resulting in better stereoacuity.

The Assessment of Response Time

The stereovision test affords the manipulation of stimulus timing, and records the

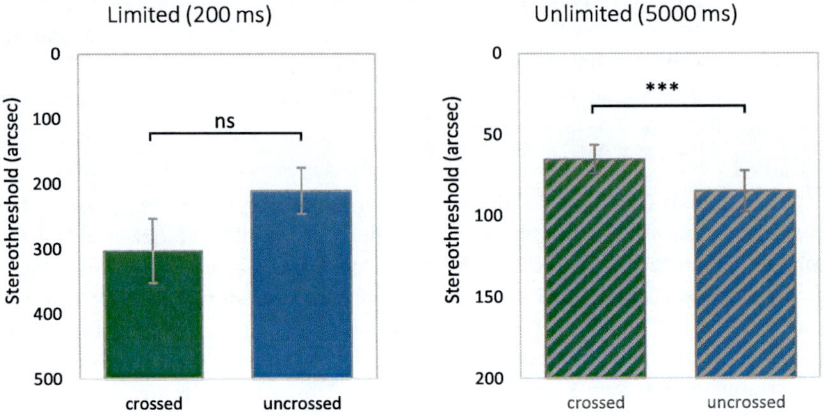

(a) (b)

Fig. 2. In case of the limited presentation time of stereovision stimuli, the stereo thresholds are statistically different **(a)**, n = 32. When the stimuli are presented longer time and subject can look at it longer, then there is no see statistically different stereo thresholds values between crossed and uncrossed disparity **(b)**, n = 48. The average value and standard errors are shown, ns – p-value is more than 0.05 and *** – $p < 0.0001$.

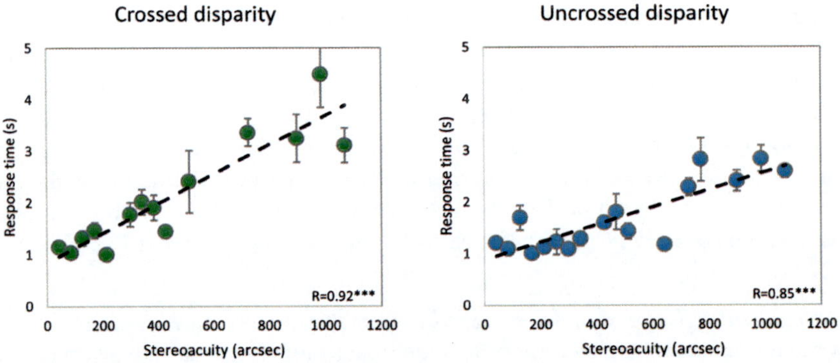

(a) (b)

Fig. 3. The correlations between crossed **(a)** or uncrossed disparity **(b)** and response time were calculated only for correct answers. The average value and standard errors are shown. The Person correlation coefficient was found to be statistically significant, *** – $p < 0.0001$.

response time from stimulus presentation to participant key-press. Response time enables comparison of the perception of crossed and uncrossed disparity images, determining their speed of perception and their relative position in relation to the screen plane (*see Fig. 3*). The presentation of stimuli for only 200 ms allowed assessment of the detectability of stereoanomalies using this test. In our study, we observed a 42% rate of detection for uncrossed disparity in stereo image processing and perception using the limited test. Conversely, the unlimited test enabled us to evaluate the speed of processes occurring in each of the stereoacoustic processing channels.

On average, each administration of the stereovision test, which assessed both crossed and uncrossed disparity conditions, took approximately 55 s ($SD = 26$) for participants to complete. Interestingly, the data indicated that participants were able to detect the uncrossed disparity stimuli faster than the crossed disparity images.

4 Discussion and Conclusion

The use of stereovision tests is crucial for assessing stereoacuity, but these tests can be categorized into local and global stereovision tests. Local stereovision tests incorporate monocular cues that may lead to incorrect conclusions by vision specialists if the subject is not seeing the stereo image and instead focusing on the divergent image. In contrast, global stereovision tests lack monocular cues and can be more challenging to see, especially when viewed for the first time [17]. It is believed that global stereovision tests do not solely assess global stereovision, as the subject must also integrate the conditional contour of the image by matching similar dots.

The global stereovision test developed by the authors allows for the adjustment of various parameters, such as the size of the random dot, the dot density, the algorithm of the staircase method, the stimulus display time, the stimulus size, and the stereogram size. Additionally, the colours of the respective filters can be adjusted to separate the images of the two eyes. As technology advances and pixel sizes of screens decrease, computer-based tests will become increasingly prevalent for assessing the quality of stereovision.

Different principles of stereoacuity assessment may yield varying results for the same individual [18, 19], highlighting the importance of understanding the advantages and disadvantages of each method. It is also essential to recognize the factors that may impact the quality of stereo vision, including test parameters, internal factors in the individual, and external factors.

In clinical practice, it is important to consider that stereo acuity can decline when one eye's defocus is 0.50 D, while visual acuity may remain intact or only slightly diminished. Monocular defocus or blurring can have a more significant impact on stereo vision quality than binocular defocus [7, 10, 20–23]. This knowledge is valuable in prescribing vision correction and recommending monocorrection or presbyopia-specific contact lenses to patients.

An unanswered question is why stereoacuity assessments performed by clinical stereoacuity tests using test plates yields better results than on computer screens [24]. In our study, the use of colour filters in the anaglyph method may have hindered accurate stereoacuity assessment [25]. One possible explanation for this discrepancy could be the absence of blue whiskers in the central retina, suggesting that stereoacuity may be worse with red-green stimuli [26, 27]. However, it is also plausible that the size of the random dot may influence stereopsis quality [28]. In our pilot study, the random dot's angular size was smaller than that of the TNO test due to the test's greater distance. Future studies will examine how test distance, dot size, dot density, stimulus size, and stimulus shape affect stereopsis quality. Since stereopsis quality can improve with repeated testing [29], it is crucial that tests can modify the disparity, shape parameters, and stimulus appearance. Our developed test allows for such modifications, making it useful not only for stereopsis assessment but also for training.

To assess stereovision threshold accurately, it is necessary to use computerized tests that can adjust the parameters of the stimuli and present them multiple times while ensuring that the stimuli are not remembered by the subject. Additionally, it is crucial to consider the principles of test design and the factors that affect the assessment of stereovision quality. In the near future, it is expected that tests based on smart devices will become more prevalent, and specialists will be able to use advanced algorithms to adjust the stimulus parameters and minimize their impact on the assessment of stereoacuity. By leveraging technology and advanced methodologies, we can improve the accuracy and reliability of stereo vision tests and pave the way for better diagnosis and treatment of vision disorders.

Acknowledgements. The study is supported by the UL Projects No Y5-AZ77 and No Y9-B003, and the UL Foundation Project No 2260. We would like to express our gratitude to two anonymous reviewers for their helpful comments.

References

1. Abu Bakar, N.F., Chen, A.H.: Comparison on testability of visual acuity, stereo acuity and colour vision tests between children with learning disabilities and children without learning disabilities in government primary schools. Indian J. Ophthalmol. **62**(2), 141–144 (2014)
2. Bauer, A., Kolling, G., Dietz, K., Zrenner, E., Schiefer, U.: Sind schielende Personen schlechtere Autofahrer? Einfluss des querdisparaten Stereosehens auf das Fahrvermögen. Klinische Monatsblätter Augenheilkunde **217**, 183–189 (2000)
3. Bauer, A., Dietz, K., Kolling, G., Hart, W., Schiefer, U.: The relevance of stereopsis for motorists: a pilot study. Graefes Arch. Clin. Exp. Ophthalmol. **239**(6), 400–406 (2001). https://doi.org/10.1007/s004170100273
4. Presta, V., Vitale, C., Ambrosini, L., Gobbi, G.: Stereopsis in sports: visual skills and visuomotor integration models in professional and non-professional athletes. Int. J. Environ. Res. Public Health **18**(21), 11281, 1–10 (2021)
5. O'Connor, A.R., Birch, E.E., Anderson, S., Draper, H.: FSOS Research Group. The functional significance of stereopsis. Investigative Ophthal. Vis. Sci. **51**(4), 2019–2023 (2010)
6. Read, J.C.: Stereo vision and strabismus. Eye **29**(2), 214–224 (2015)
7. Papelba, G., Cipane, I., Ozolinsh, M.: Stereovision studies by disbalanced images. Proc. SPIE **5123**, 323–329 (2002)
8. Khan, N., Zaka-ur-Rab, S., Ashraf, M., Mishra, A.: Comparison of stereoacuity in patients of anisometropia, isometropia and emmetropia. Indian J. Ophthalmol. **70**(12), 4405–4409 (2022)
9. Krumina, G.: Correlation between stereothreshold and quality of monocular images. Doctoral Thesis, University of Latvia, Riga, p. 43 (2004), https://www.cfi.lu.lv/fileadmin/user_upload/lu_portal/projekti/cfi/PhD_thesis/Krumina_PhD_Thesis_Eng.pdf, (Accessed 5 April 2023)

10. Krumina, G., Ozolinsh, M., Lacis, I., Lyakhovetskii, V.: Estimation of stereovision in conditions of blurring simulation. Proc. SPIE **5946**(59461N), 1–10 (2005)
11. Richards, W.: Stereopsis and stereoblindness. Exp. Brain Res. **10**(4), 380–388 (1970)
12. Patterson, R., Fox, R.: The effect of testing method on stereoanomaly. Vision. Res. **24**(5), 403–408 (1984)
13. Giaschi, D., Narasimhan, S., Solski, A., Harrison, E., Wilcox, L.M.: On the typical development of stereopsis: fine and coarse processing. Vision. Res. **89**, 65–71 (2013)
14. Pateras, E., Plakitsi, A., Chatzipantelis, A.: Stereoscopic vision and testing techniques – overview. Biomed. J. Scient. Tech. Res. **26**(5), 1–9 (2020)
15. Chopin, A., Chan, S.W., Guellai, B, Bavelier, D., Levi, D.M.: Binocular non-stereoscopic cues can deceive clinical tests of stereopsis. Scient. Report **9**(5789), 1–11 (2019)
16. Julesz, B.: Binocular depth perception of computer-generated patterns. Bell Labs Tech. J. **39**(5), 1125–1162 (1960)
17. Westheimer, G.: Clinical evaluation of stereopsis. Vision. Res. **90**, 38–42 (2013)
18. Piano, M.E., Tidbury, L.P., O'Connor, A.R.: Normative values for near and distance clinical tests of stereoacuity. Strabismus **24**(4), 169–172 (2016)
19. Zhao, L., Wu, H.: The difference in stereoacuity testing: contour-based and random dot-based graphs at far and near distances. Annal. Trans. Med. **7**(9), 193, 1–8 (2019)
20. Legge, G.E.: Spatial frequency masking in human vision: binocular interactions. J. Opt. Soc. Am. **69**(6), 838–847 (1979)
21. Goodwin, R.T., Romano, P.E.: Stereoacuity degradation by experimental and real monocular and binocular amblyopia. Invest. Ophthalmol. Vis. Sci. **26**(7), 917–923 (1985)
22. Ozolinsh, M., Papelba, G.: Eye cataract simulation using polymer dispersed liquid crystal scattering obstacles. Ferroelectrics **304**, 207–212 (2004)
23. Odell, N.V., Hatt, S.R., Leske, D.A., Adams, W.E., Holmes, J.M.: The effect of induced monocular blur on measures of stereoacuity. J. AAPOS **13**(2), 136–141 (2009)
24. Gadia, D., Garipoli, G., Bonanomi, C., Albani, L., Rizzi, A.: Assessing stereo blindness and stereo acuity on digital displays. Displays **35**(4), 206–212 (2014)
25. Papelba, G., Ozolinsh, M., Petrova, J., Cipane, I.: Stereoacuity determination at changing contrast of colored stereostimuli. Proc. SPIE **5123**, 330–338 (2002)
26. Pennington, J.: The effect of wavelength on stereoacuity. Am. J. Optom. Arch. Am. Acad. Optom. **47**(4), 288–294 (1970)
27. Kingdom, F.A., Simmons, D.R.: Stereoacuity and colour contrast. Vision. Res. **36**(9), 1311–1319 (1996)
28. Zhao, L., Wu, H.: The effect of dot size in random-dot stereograms on the results of stereoacuity measurements. BMC Ophthal. **20**(1), 253, 1–6 (2020)
29. Clayton, R., Siderov, J.: Differences in stereoacuity between crossed and uncrossed disparities reduce with practice. Ophthalmic Physiol. Opt. **42**(6), 1353–1362 (2022)

Macular Thickness in Latvian Children with Refractive and Anisometropic Amblyopia

Kristine Kalnica-Dorosenko[1](✉), Elina Karelska[2], Aiga Svede[2], and Sandra Valeina[1]

[1] Eye Diseases Clinic, Children's Clinical University Hospital, Vienibas gatve 45, Riga LV-1004, Latvia
kristinekalnica@gmail.com

[2] Department of Optometry and Vision Science, University of Latvia, Jelgavas iela 1, Riga LV-1004, Latvia

Abstract. This work aimed to measure and compare macular thickness in Latvian children with refractive and anisometropic amblyopia and control group eyes using optical coherence tomography (OCT) and to define OCT as a diagnostic tool. In Latvia, it's still a problem that amblyopia is diagnosed later when the part of the brain responsible for visual perception is already mature and cannot be treated completely. This is due to the lack of strict monitoring of mandatory eye examinations and the fact that assessing the posterior part of the eye (especially the macula) isn't part of every child's visit. A total of 85 patients aged 3–9 years (51.8% female subjects) were studied, of whom 20 had refractive amblyopia, 24 had anisometropic amblyopia, and 41 children had normal visual acuity in both eyes and were considered the control group. The macular thickness was measured by OCT in the centre (foveola), 1 mm ring (fovea), 3 mm and 6 mm rings and compared. Comparing the amblyopic eyes with healthy eyes of the same children, there were no statistically significant differences in macular thickness in the central part, 1 mm ring and 3 mm ring. In comparison, in the inferior part of the 6 mm ring, the macular thickness was lower in the amblyopic than in the healthy eyes. Comparison of macular thickness in the centre of 1 mm, 3 mm, and 6 mm ring amblyopic and control eyes showed no significant difference. Further studies with larger numbers of refractive and anisometropic amblyopia cases are recommended.

Keywords: Amblyopia · Optical coherent tomography · Macular thickness · Refractive amblyopia · Anisometropic amblyopia

1 Introduction

1.1 Amblyopia

Amblyopia is defined clinically as a decrease in visual acuity in one, or more often in both eyes, in the absence of an apparent disorder or ocular disease. Amblyopia is the leading cause of monocular vision loss in young and middle-aged people. If left untreated, amblyopia can increase the risk of vision loss in the other eye [1]. Diagnosing

Y. Dekhtyar and I. Saknite (Eds.): NBC 2023, IFMBE Proceedings 89, pp. 328–335, 2023.
https://doi.org/10.1007/978-3-031-37132-5_41

amblyopia at a young age is very important because its treatment is more effective early [2]. Children in Latvia have three mandatory vision examinations: at 13–24 months, three years and before school at 6–7 years. These examinations are defined by the Cabinet of Ministers Regulation No. 555 (Minister of Cabinet Regulation No 555, 2019). The assessment of the posterior part of the eye is often overlooked during a mandatory eye examination. As part of this examination, optical coherence tomography is used.

1.2 Optical Coherence Tomography

Optical coherence tomography (OCT) is used to perform a non-contact, non-invasive and precise examination of the posterior part of the eye assessment. This technique allows eye care professionals to assess the fine details of the posterior segment of the eye at the micron (μ) scale and to quantify subtle changes in macular thickness, thus enabling objective monitoring of disease progression and the effectiveness of different therapies for a variety of eye diseases, including in people with established amblyopia [3]. The study of macular and RNFL changes in people with amblyopia remains an interesting research topic due to the variability of published results [4–8] and the lack of an apparent correlation between these changes and patient characteristics. In the present study, we aimed to compare macular thickness in refractive and anisometropic amblyopic eyes of Latvian children with fellow eyes and with a control group.

2 Materials and Methods

2.1 Ethical Consideration

All procedures followed the Declaration of Helsinki and were approved by the Ethics Committee of the Experimental and Clinical Medicine Institute of the University of Latvia.

2.2 Participants

A total of 85 potential study participants from the Eye Diseases Clinic of Children's Clinical University Hospital in Riga, Latvia, aged 3–9 years (the average age was 6 $\pm$ 1 years; 51.8% female subjects) were recruited from 385 patients, of whom 20 had refractive amblyopia, 24 had anisometropic amblyopia, and 41 children had normal visual acuity in both eyes and were considered the control group. All participants were generally and neurologically healthy. Visual acuity was converted from decimal units to logMAR (logarithm of the minimum angle of resolution). A participant was classified as having anisometropic amblyopia if there was a difference of at least 1.00 D in one of the meridians. The spherical equivalent was used to analyse the refractive error, defined as the sphere power plus half of the cylinder power. Table 1 shows the participants' demographics and essential data for all patients.

Table 1. Participants' demographic data.

Characteristics	Amblyopic children		Control
	Refractive	Anisometropic	
Number	20	24	41
Age (year) %			
Mean ± SD	6 ± 1	6 ± 2	7 ± 1
Sex (%)			
Male	11 (55.0)	9 (37.5)	21 (51.2)
Female	9 (45.0)	15 (62.5)	20 (48.8)
BCVA (logMAR)			
AE X ± SD	0.40 ± 0.23	0.49 ± 0.32	
FE X ± SD	−0.10 ± 0.07	−0.12 ± 0.06	−0.12 ± 0.06
SE (D)			
AE X ± SD	4.56 ± 3.06	3.67 ± 3.02	
FE X ± SD	2.13 ± 2.98	2.12 ± 2.34	0.50 ± 2.12

Results are presented as mean ± SD; SD, standart deviation; BCVA, best corrected visual acuity; logMAR, logarithm of the minimum angle of resolution; AE, amblyopic eye; FE, fellow eye; SE, spherical equivalent; D, diopter

2.3 Method

The OCT SPECTRALIS laser scanning camera (s/n Spec-CAM-14866-S2610, Heidelberg Engineering, Germany) was used in the experiment. Ophthalmologists and optometrists previously measured macular thickness at the Children's Clinical University Hospital (Table 2). Usually, five measurements are taken for each eye to obtain accurate results. The OCT data for each selected patient was opened in Heidelberg Eye Explorer to get the macular thickness data. The resulting macular thickness image is then opened as a thickness map. A 6 mm ETDRS ring with nine regions assessed the macular thickness. Three concentric rings of 1-, 3- and 6-mm diameter, internal and external macular diameter, with outer concentric rings, each divided into four quadrants.

3 Results

Statistical analysis using paired T-test showed no statistically significant differences in macular thickness in the central part, 1 mm ring and 3 mm ring ($P > 0.05$), while in the inferior part of the 6 mm ring, the macular thickness was lower in the amblyopic eyes than in the healthy eyes ($P = 0.01$) for the same participant. Comparison of macular thickness in the center, 1-, 3-, and 6-mm ring in amblyopic and control eyes showed no significant differences ($P > 0.05$) (Table 2). Central macular thickness was compared in refractive and anisometropic amblyopic eyes. The data showed no significant differences in macular thickness and were not statistically significant ($P > 0.05$) (Fig. 1). The 1 mm

Table 2. Central, 1, 3, 6 mm ring macular thickness in children with amblyopia.

	Amblyopic children		Control	P^a	P^b
	AE	FE			
CT (foveola) (μm)	225 ± 30	229 ± 28	222 ± 13	0.079	0.593
1 mm ring (μm)	263 ± 36	265 ± 34	269 ± 21	0.522	0.163
3 mm ring (μm)					
Nasal	347 ± 20	352 ± 18	348 ± 18	0.090	0.982
Temporal	342 ± 19	342 ± 18	337 ± 18	0.990	0.652
Superior	348 ± 22	345 ± 20	353 ± 17	0.183	0.124
Inferior	332 ± 24	332 ± 21	352 ± 16	0.836	0.146
6 mm ring (μm)					
Nasal	320 ± 24	319 ± 22	338 ± 21	0.898	0.540
Temporal	303 ± 29	314 ± 23	308 ± 17	0.409	0.218
Superior	335 ± 22	335 ± 18	324 ± 20	0.831	0.582
Inferior	301 ± 24	306 ± 24	309 ± 18	0.011*	0.310

Results are presented as mean ± SD; SD, standart deviation; CT, central thickness; AE, amblyopic eye; FE, fellow eye; P, probability; P^a, amblyopic eye and healthy eye of the same participant; P^b, amblyopic eyes and control group eyes; *statistically significant

macular ring thickness is significantly greater in eyes with refractive amblyopia than in eyes with anisometropic amblyopia. This difference is also statistically significant ($P < 0.05$) (Fig. 1).

In refractive amblyopia, there was no statistically significant difference in the central macular thickness ($P > 0.05$) when comparing macular thickness between amblyopic eyes and healthy eyes of the same children or when comparing macular thickness between eyes with amblyopia and control eyes (Fig. 2).

Reaching the central macular thickness in amblyopic and healthy eyes of children with anisometropic amblyopia showed a statistically significant difference between macular thicknesses ($P = 0.05$). A comparison of macular thickness in amblyopic and control eyes showed that the difference was not statistically significant ($P > 0.05$) (Fig. 3).

It was interesting to compare the data between girls and boys. Hence, the same children's amblyopic eyes, the healthy ones, and the eyes of a control group were compared

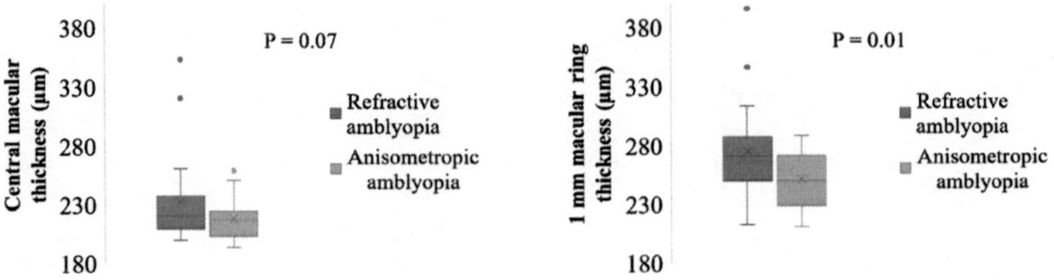

Fig. 1. Central and 1-mm ring macular thickness in eyes with refractive and anisometropic amblyopia.

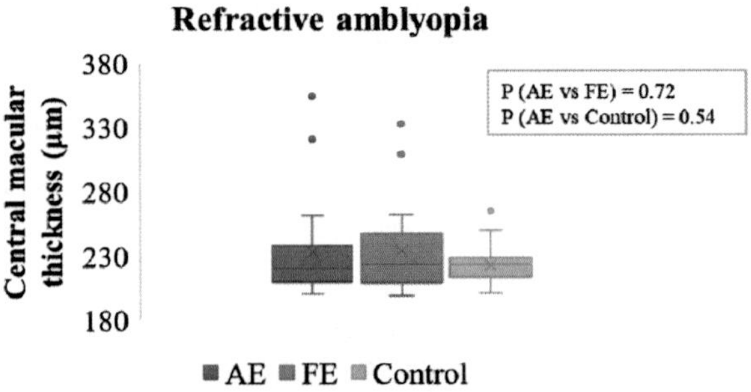

Fig. 2. Central macular thickness (foveola) in refractive amblyopia.

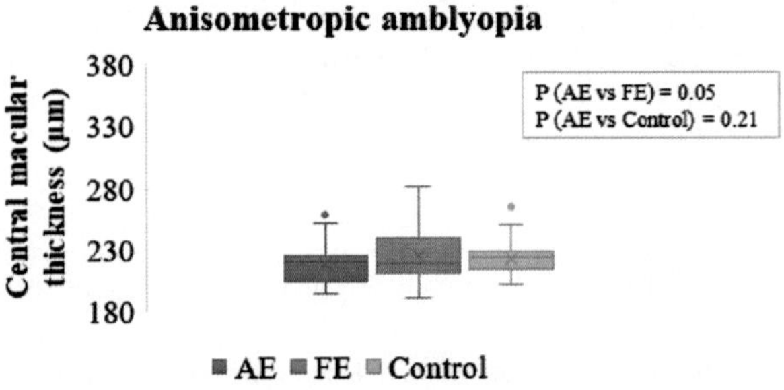

Fig. 3. Central macular thickness (foveola) in anisometropic amblyopia.

between boys and girls. Based on the statistical analysis, the macular thickness is significantly lower in girls than in boys when comparing amblyopic, healthy and control eyes ($P < 0.05$).

Five of the 44 amblyopic patients had OCT examinations before and after amblyopia treatment. The mean age at diagnosis of amblyopia was 4 ± 1 years. All patients had a refractive correction and occlusion of the healthy eye. The mean visual acuity in amblyopic eyes before the treatment was 0.50 ± 0.23 logMAR. Visual acuity after amblyopia therapy for amblyopic eyes was 0.15 ± 0.10 logMAR ($P < 0.05$). Macular thickness before and after treatment was compared in amblyopic and healthy eyes. Statistical data analysis showed that macular thickness in amblyopic and healthy eyes before treatment was not statistically different in all parts of the retina (P > 0.05). A comparison of macular thickness in amblyopic eyes before and after treatment showed that the thickness in the centre, 1 mm in the ring, 3 mm in the superior part of the ring and 6 mm in the inferior part of the ring decreased after treatment. This difference was considered statistically significant. A Bonferroni correction test showed that the difference was statistically significant only in the macula's centre and the 1 mm ring (P < 0.01). Figure 4 shows the correlation between changes in visual acuity (logMAR units) against changes in central macular thickness and 1 mm ring after treatment in patients with amblyopia.

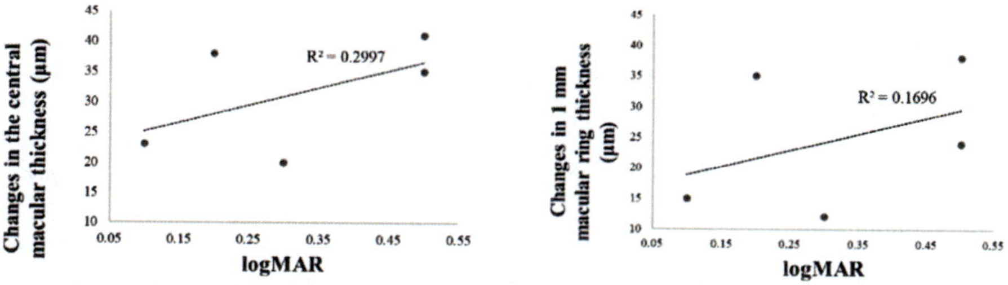

Fig. 4. Improvement in visual acuity (logMAR) and central macular thickness (left), 1 mm macular ring thickness change (right).

There was no significant correlation between these two parameters after amblyopia treatment (R2 = 0.2997; P = 0.34; R2 = 0.1696; P = 0.50).

4 Discussion

The presence of amblyopia in children can affect different levels of the visual pathway. Some studies have found a correlation between amblyopic eye and cell apoptosis in the lateral geniculate nucleus [9, 10]. However, the involvement of the retina, especially the macula, in the development of amblyopia is still controversial.

In the current study, significant differences in macular thickness between amblyopic and healthy eyes of the same children showed that the macula's central thickness, 1-mm ring and the 3-mm ring did not differ (P > 0.05), while in the inferior part of the 6-mm ring macular thickness was significantly lower in amblyopic eyes. No significant difference in macular thickness between amblyopic and control eyes (P > 0.05).

This was consistent with previous studies [8, 11–14], but other authors reported significant differences in macular thickness [6, 15, 16] or retinal nerve fibre layer thickness (RNFLT) [6] in amblyopic eyes.

As amblyopia can be caused by refraction or anisometropia but also by strabismus and deprivation, it is interesting to know whether the thickness of the macula differs in these types of amblyopia. For instance, in the case of deprivation amblyopia, Kasem and Badawi [6] observed a significant increase in the foveal thickness and the superior and inferior RNFLT. Meanwhile, Kim and colleagues [17] found no significant difference in deprived amblyopic eyes compared to normal eyes regarding macular thickness. They found a thicker nasal RNFL in amblyopic eyes. Pal and Malik's [18] 2020 work also showed no statistically significant difference between amblyopic and normal eyes in patients with deprivation amblyopia regarding peripapillary RFNL thickness. When comparing foveal thickness in strabismic amblyopic eyes, some studies show a difference between the better-seeing eyes or controls [6, 19, 20], while others, on the contrary, show no difference [18, 21, 22]. Regarding anisometropic amblyopia, Wang and colleagues [23] argue that the average central macular thickness is thicker than normal eyes, but other areas have no significant difference. In the study by Xia and colleagues [24], there was no significant difference in total macular thickness in the five sectors between amblyopic and control eyes. As well as, the macular RNFLT did not differ significantly in the five sectors between amblyopic and control eyes.

Furthermore, although the number of patients was not large, a study of amblyopia treatment's effect on macular thickness changes found that the central macular thickness and 1 mm macular ring in eyes with amblyopia decreased after treatment. A 2015 study by Pang and colleagues [25] found similar changes in macular thickness. The researchers concluded that macular thickness in the centre decreased significantly after amblyopia treatment in eyes with amblyopia.

In conclusion, foveola thickness was significantly higher in eyes with amblyopia than in healthy eyes with anisometropic amblyopia. The central macular thickness was not significantly higher in amblyopic eyes compared to control eyes. There is no basis for the assumption that macular thickness changes correlate with visual acuity changes. Further studies with more significant numbers of refractive and anisometropic amblyopia cases are recommended.

References

1. Donahue, S.P.: Relationship between anisometropia, patient age, and the development of amblyopia. Am. J. Ophthalmol. **142**(1), 132–140 (2006)
2. Bradfield, Y.S.: Identification and treatment of amblyopia. Am. Fam. Physician **87**(5), 348–349 (2013)
3. Shrestha, C., Rajbhandari, R., Poudel, M.: Macular and peripapillary retinal nerve fiber thickness in unilateral amblyopic eye. Nepal Med. Coll. J. **23**(2), 132–138 (2021)
4. Rajavi, Z., Moghadasifar, H., Feizi, M., et al.: Macular thickness and amblyopia. J. Ophthalmic Vis. Res. **9**(4), 478–483 (2015)
5. Altindağ, S., Şahin, M.: Evaluation of the macular thickness by optical coherence tomography in amblyopia. J. Clin. Exp. Investig. **7**(2), 178–183 (2016)
6. Kasem, M.A., Badawi, A.E.: Changes in macular parameters in different types of amblyopia: optical coherence tomography study. Clin. Ophthalmol. **11**, 1407–1416 (2017)
7. Rajavi, Z., Sabbaghi, H., Behradfar, N., et al.: Macular thickness in moderate to severe amblyopia. Korean J. Ophthalmol. **32**(4), 312–318 (2018)
8. Kurt, R.A., Bayar, S.A., Ercan, Z.E., et al.: Choroidal and macular thickness in eyes with amblyopia. Beyoglu Eye J. **6**(4), 320–327 (2021)
9. Hess, R.F., Thompson, B., Gole, G., et al.: Deficient responses from the lateral geniculate nucleus in humans with amblyopia. Eur. J. Neurosci. **29**(5), 1064–1070 (2009)

10. Von Norden, G.K., Crawford, M.L.: The lateral geniculate nucleus in human strabismic amblyopia. Invest. Ophthalmol. Vis. Sci. **33**(9), 2729–2732 (1992)
11. Kee, S.Y., Lee, S.Y., Lee, Y.C.: Thicknesses of the fovea and retinal nerve fiber layer in amblyopic and normal eyes in children. Korean J. Ophthalmol. **20**(3), 177–181 (2006)
12. Walker, R.A., Rubab, S., Voll, A.R., et al.: Macular and peripapillary retinal nerve fibre layer thickness in adults with amblyopia. Can. J. Ophthalmol. **46**(5), 425–442 (2011)
13. Yassin, S.A., Al-Tamimi, E.R., Al-Hassan, S.: Macular and retinal nerve fiber thickness in recovered and persistent amblyopia. Int. Ophthalmol. **35**(6), 833–842 (2015). https://doi.org/10.1007/s10792-015-0055-9
14. Repka, M.X., Kraker, R.T., Tamkins, S.M., et al.: Retinal nerve fiber layer thickness in amblyopic eyes. Am. J. Ophthalmol. **148**(1), 143–147 (2009)
15. Huynh, S.C., Samarawickrama, C., Wang, X.Y., et al.: Macular and nerve fiber layer thickness in amblyopia: the Sydney Childhood Eye Study. Ophthalmology **116**(9), 1604–1609 (2009)
16. Subedi, S., Shrestha, J.B., Sharma, A.K., et al.: Evaluation of retinal nerve fibre layer and macular thickness in amblyopia. Nepal Med. J. **05**(01), 45–49 (2022)
17. Kim, Y.W., Kim, S.-J., Yu, Y.S.: Spectral-domain optical coherence tomography analysis in deprivational amblyopia: a pilot study with unilateral pediatric cataract patients. Graefes Arch. Clin. Exp. Ophthalmol. **251**(12), 2811–2819 (2013). https://doi.org/10.1007/s00417-013-2494-1
18. Pal, S., Malik, P.: Comparison of macular and peripapillary RNFL thickness between normal and abnormal eye in children with unilateral amblyopia. Eur. J. Mol. Clin. Med. **7**(2), 6915–6918 (2022)
19. Dickmann, A., Petroni, S., Perrotta, V., et al.: A morpho-functional study of amblyopic eyes with the use of optical coherence tomography and microperimetry. J. Am. Assoc. Pediatric Ophthalmol. Strabismus **15**(4), 338–341 (2011)
20. Agrawal, S., Singh, V., Singhal, V.: Cross-sectional study of macular thickness variations in unilateral amblyopia. J. Clin. Ophthalmol. Res. **2**(1), 15–17 (2014)
21. Altintas, O., Yüksel, N., Ozkan, B., et al.: Thickness of the retinal nerve fiber layer, macular thickness, and macular volume in patients with strabismic amblyopia. J. Pediatr. Ophthalmol. Strabismus **42**(4), 216–221 (2005)
22. Alotaibi, A.G., Al Enazi, B.: Unilateral amblyopia: optical coherence tomography findings. Saudi J. Ophthalmol. **25**(4), 405–409 (2011)
23. Wang, X.M., Cui, D.M., Zhen, L., et al.: Characteristics of the macula in amblyopic eyes by optical coherence tomography. Int. J. Ophthalmol. **5**(2), 172–176 (2012)
24. Xia, Z., Chen, H., Zheng, S.: Thicknesses of macular inner retinal layers in children with anisometropic amblyopia. Biomed. Res. Int. 6853258 (2020)
25. Pang, Y., Frantz, K.A., Block, S., et al.: Effect of amblyopia treatment on macular thickness in eyes with myopic anisometropic amblyopia. Invest. Ophthalmol. Vis. Sci. **56**(4), 2677–2683 (2015)

Coherent Analysis of the EEG to Study the Correlation of Neural Connections of the Cerebral Cortex with the Cognitive Functions of 6–8 Y.O. Children Having Difficulties in Learning and Behavioral Problems

Konstantins Pudovskis[1](✉), Nelli Tolmača[1], Andrejs Bondarenko[2,3], Viesturs Larins[4], and Janis Vandans[5]

[1] Centre of Psychoneurophysiological and Bioregulation Research, Kr. Valdemara 17A, 2nd Floor, Riga 1010, Latvia
neurofiziol@inbox.lv
[2] CTCo, Meistaru Street 33, Valdlauci, Kekavas Parish 1076, Latvia
[3] Riga Technical University, 10 Zunda Embankment, Riga 1048, Latvia
[4] Latvian Academy of Sport Education, Brivibas Gatve 333, Riga 1006, Latvia
[5] Latvian Association of Functional Bioregulation Specialists, Kr. Valdemara 17A, 2nd Floor, Riga 1010, Latvia

Abstract. Current research is done in groups of children within the norm and from those having learning difficulties from preschool (6–7) and primary school ages (7–8). A comparative analysis of cognitive brain functions during orientation in a maze has been performed using coherent EEG analysis with respect to age-related features of the formation of neural connections in the brain cortex. Identified differences in the quantitative and spatial organization of neural connections in the cerebral cortex in both hemispheres of the brain. The dynamics of neural connections in children with learning difficulties indicate a pronounced age-related functional immaturity of the cortical regions of the brain, as well as disruptions in the regulatory brain structures – an imbalance between inhibitory and activating influences. The data from the coherent EEG analysis correlated with a high level of motor activity (often chaotic) in the maze, fragmented perception, unstable attention, low working memory capacity, and weak optimization of skills in mastering new spaces. These are caused by the distortion of the analysis and synthesis of incoming information and disruption of control in the emotional-volitional sphere.

Keywords: EEG · brain · children · learning difficulties · analysis

1 Introduction

Coherence is a measure of synchronization between two different EEG leads. From a functional point of view, it reflects the degree of spatial-temporal relationships or the state of functional integration of neurons in anatomically connected structures, generating discharges synchronously [1]. Coherent analysis as a mathematical method for

Y. Dekhtyar and I. Saknite (Eds.): NBC 2023, IFMBE Proceedings 89, pp. 336–342, 2023.
https://doi.org/10.1007/978-3-031-37132-5_42

processing EEG is relatively new, and it is established that the information obtained has a fundamentally different qualitative level compared to power spectral analysis [2]. In clinical practice, it was concluded that coherent analysis has a greater diagnostic value than spectral analysis [3]. There is limited data on coherent EEG analysis in young children and even fewer comparative or correlational EEG analyses - this subtle indicator of the functional state of the brain and cognitive brain function indicators.

The age of children 6–8 years old chosen for the study is determined by the fact that it is a period when the acquisition of new skills, concepts, accumulation, and exchange of information in the brain is interrupted by a sharply demanded assessment of the quality of the perceived information, stored in the brain systems, and capable of reproduction in the form of an involuntary skill or formed elements of voluntariness: to endure a lesson, control one's behavior: raise a hand, attentively listen to the teacher, etc. Both of these qualities still have the character of immaturity, but this age period is crucial for assessing deviations in EEG in terms of analysis and synthesis of perceived information. It is important to determine the relationship between behavioral indicators and changes in EEG during various functional deviations from normal maturation of regulatory brain structures. Understanding the mechanisms of information processing by the brain is essential at all stages of ontogenesis, especially in preschool and early school age when children need to perceive large volumes of information.

In recent years, the number of students experiencing learning difficulties has been steadily increasing in all countries around the world, and these are children from all social strata of society. School failure in different age groups (6–9 years old) has been identified in 15–40% of children, according to various sources [4]. This is not only an emerging social problem but also a relevant medical, psychological, and pedagogical issue.

1.1 Research Goal

Comparative analysis of the features of neural connections formation in the brain cortex based on the dynamics of EEG coherence and indicators of behavior and cognitive functions in children aged 6–8 years with learning and behavior difficulties.

2 Research Methods

A functional EEG analysis was performed on children participating in state correctional programs in special education classes at various schools. These children underwent a medical-pedagogical commission at the Department of Education, where they were referred by kindergarten teachers and primary school teachers due to incomplete readiness for systematic education in school. School specialists referred children because of difficulties in mastering school subjects. The package of documents for the commission includes a clinical encephalogram and/or sleep encephalogram. Children are monitored by doctors and psychologists who provide ongoing correction and control of the functional state of the body, with the aim of accelerating the child's return to groups of their age norm.

Parents, psychologists, or doctors are always initiators of additional scientific EEG analysis. The functional analysis of quantitative deviations in EEG according to the "neurometric" methodology was named after two scientists - Roy John from New York University and Yu.D. Kropotov from the Institute of the Human Brain of the Russian Academy of Sciences [5]. This analysis provides additional knowledge about the functional features of EEG rhythms, the functions of brain regions and hemispheres, their connections, and uses additional parameters for assessing EEG (coherent analysis, mapping, etc.), which are still rarely used in clinical practice. Such analyses are usually carried out in scientific centers [5].

The age groups consisted of 80 children in a preparatory group aged 6 to 7 years (46 boys and 34 girls) and first-grade students aged 7 to 8 years (52 boys and 28 girls). Each age group consisted of two subgroups: the first - age norms, included children who, according to the teacher and psychologist's assessment, willingly and easily mastered the preparatory program in kindergarten, and the others - the school program - 30 children of normal age and 50 children with complaints of reduced attention activity and difficulties in learning.

The work was carried out in compliance with the conditions of the articles of the "Helsinki Declaration on Bioethics and Human Rights," category Medical Diagnostics and its subcategories 22:5 Diagnostic Methods, 12:1 Medical Semiotics, and 26:1 Medical Measuring Devices. Individual informed consent cards and personal parental initiatives for EEG registration and mathematical data analysis were completed.

The registration of brain electrical activity (EEG) was performed using an electroencephalograph - EEG analyzer - 21/26, "Encephalan 131–03", Medicom MTD, Russia. Electrodes were applied according to the "10–20" system. Computerized EEG recording was carried out in a monopolar ipsilateral ear lead, adhering to standard conditions for conducting this type of neurophysiological diagnostics [5]. To assess the level of intrahemispheric integration along cortico-cortical pathways, we conveniently used a limited number of lead pairs with short distances symmetrically in the left and right hemispheres (FP1-F3, F3-C3, C3-P3, P3-O1, FP1-F7, F7-T3, T3-T5, T5-O1), allowing for the analysis of short connection chains. Intrahemispheric connections Fp1 – Fp2, F7 – F8, T3 – T4, T5 – T6, O1 – O2, short and long chains. The results of all studies were statistically processed. A comparative assessment of quantitative features was performed using parametric statistical criteria (Student's t-test). Intergroup comparison of significance in the nonparametric distribution of related parameters was performed using the Wilcoxon paired test and in unrelated samples – the Mann-Whitney test. The critical significance level of the tests was determined at $p \leq 0.050$.

3 Research Results

Analysis of the dynamics of coherence values within the reference corridor in children with learning difficulties in both age groups revealed pronounced individual variability and instability, especially under cognitive load (mental arithmetic). In both groups, a symmetrical increase in the average coherence value (AC) up to significant values was observed: in preschoolers and high levels in first-graders in the frontal and anterior temporal (Fp1-F7 (F3); Fp2-F8-(F4) leads (Fig. 1).

As can be seen from Fig. 1, children with learning difficulties have more connections along the lateral chains of both hemispheres, while in groups of their age norm, they have more internal connections. The highest coherence value (AC) is registered in pairs: frontopolar-frontal and frontopolar-anterior temporal leads.

This is consistent with the concept of neurophysiologists about the leading role of the frontal regions of the cerebral cortex, which provide programming, regulation, and control of human behavior. Further, as you move away from the frontal regions, coherence values usually decrease (6). Such a pattern was found specifically in the age norm groups: coherence values gradually decreased as they moved away from the frontal regions. In children with learning difficulties in both groups, on the contrary, high coherence values were found not only in the frontal but also in the central-parietal-occipital and temporal leads (C3, P3, O1, T3-T5), more often in the left hemisphere. In 6–7-year-old children with learning difficulties, high coherence values could be registered in both hemispheres in the range of delta1, delta2, and theta waves. A comparative analysis with the spectral power of EEG rhythms revealed a significant decrease in alpha wave power compared to the age norm and a significant increase in EEG power in the range of slow rhythms delta1 and delta2, $p < 0.05$. Coherence indicators and spectral power of EEG in children with learning and behavioral difficulties correlated with characteristics of orientation in a closed maze: They were found to have communication and emotional skill disorders, fear of novelty, unstable attention, and low levels of working memory [4, 7].

In both groups of children with learning difficulties, compared to the age norm, a characteristic excessive number of predominantly intrahemispheric connections with a low and partially moderate level of coherence (0.3 - 0.33) was found in the slow wave band (delta-1,2, theta waves) in frontal and temporal, less often central leads of both brain hemispheres. These data correlated with the results of the EEG spectral analysis: the predominance of slow wave activity in the anterior leads, especially in 6–7-year-old children, and the slowed formation of almost all indicators of alpha activity (4). The analysis of psychological test data (R. Tammel, M. Lorki, V. Amen) revealed a high level of anxiety in both groups of children with learning difficulties, more pronounced in preschoolers, and girls had a greater fear than boys (7). Therefore, in both groups of children with learning difficulties, compared to the age norm, the excessive level of connections with a low and partially moderate level of coherence indicated their functional inappropriateness and was partly due to the high level of activation of behavioral systems and low activity of inhibitory systems.

As presented in Fig. 2, in 32–37% of children with learning difficulties of both age groups, excessive coherence was registered in the high-frequency wave band - gamma 35.0–70.0 Hz, in one or both hemispheres of the brain, exceeding the normative indicators at $p < 0.05$. Considering the pronounced individual variability and instability within, and less often between, central EEG connections in Fig. 2, typical neural connection schemes are presented.

As shown in Fig. 2, the spatial organization and quantitatively excessive expression of neural connections by coherence in the gamma range of the EEG is expressed in intra-hemispheric connections in one brain hemisphere, less often in both simultaneously. The acceleration of high-frequency rhythm formation processes can lead to the disruption

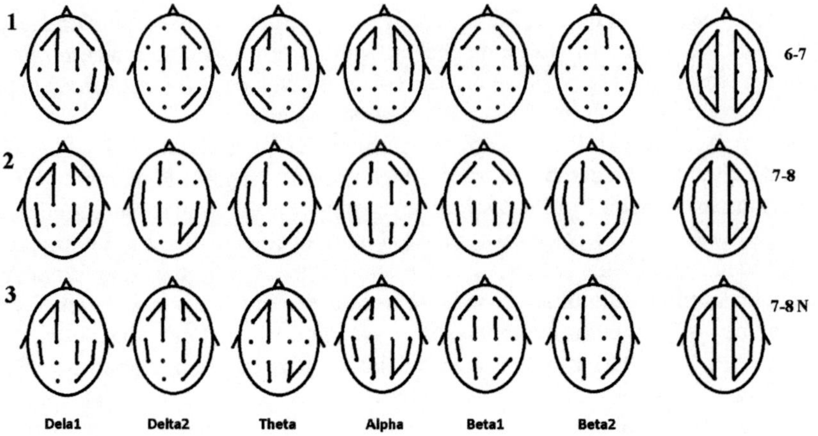

Fig. 1. The spatial organization of coherent EEG connections - oscillations of various ranges with coherence values within the reference corridor (0.51–1.00) when recording brain currents in a resting state in 6–7-year-old children - the upper row (1), in 7–8-year-old children with learning difficulties - the middle row (2), and in 7–8-year-old children, the lower row (3) is considered normal.

of spatial synchronization of potentials and, as a consequence, to the disruption of integrative brain functions (8,9,10). Notably, parents of these children complained about periods of unmotivated aggression and, less often, tearfulness in their behavior.

The neurophysiological features of the EEG correlated with both behavioral data and cognitive brain functions during orientation in the maze, as well as with psychological testing results. The dynamics of neural connections in children with learning difficulties indicated a pronounced age-related functional immaturity of the cortical regions of the brain and disturbances in regulatory structures - the balance of inhibitory and activating influences of the thalamocortical system. The data from both types of EEG analysis clearly predicted children's behavior in the maze: a high level of motor activity (often chaotic), fragmented perception, unstable attention, low working memory capacity, and weak optimization of skills in mastering new spaces, caused by distortion in the analysis and synthesis of incoming information.

Therefore, the coherence analysis of the EEG reflects the level of disturbances in the interconnectedness of the cortical areas of the cerebral hemispheres and the regulatory structures of the brain, which in turn cause difficulties in learning and behavior in children. The nature of neurodynamic and regulatory disturbances will enable the development of new methods to correct developmental deficiencies in higher cortical functions in the presence of these disorders.

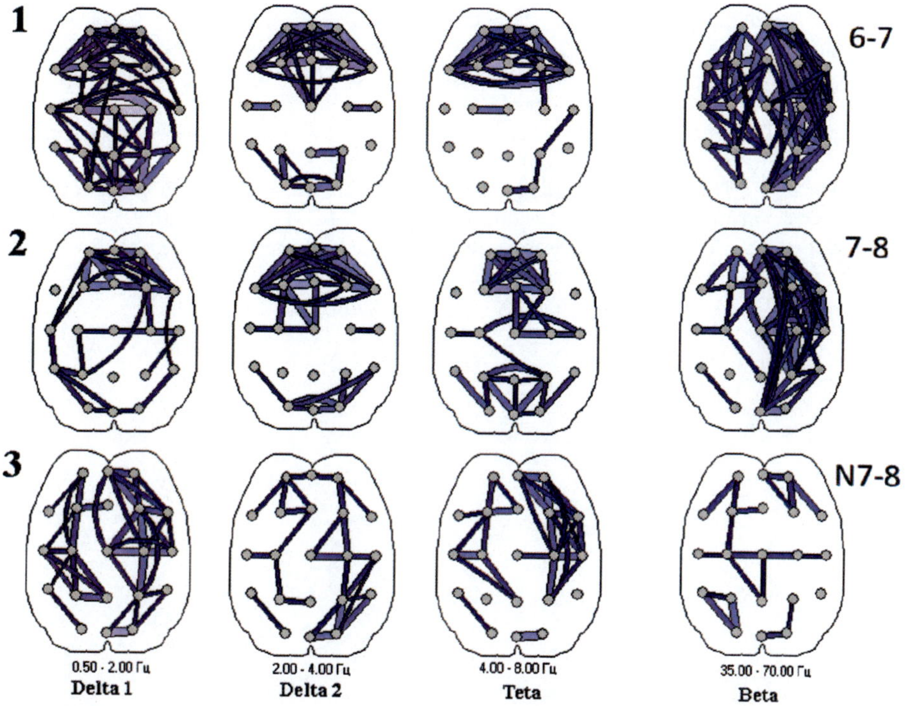

Fig. 2. Schemes of individual intra- and inter-central EEG connections by coherence in delta1, delta2, theta, and gamma ranges of the EEG. Brain bio-currents were recorded at rest in 6–7-year-old children - upper row (1), in 7–8-year-old children with learning difficulties - middle row (2), and in 7–8-year-old children - the norm - lower row (3).

References

1. Иванов Л.Б. Прикладная компьютерная электроэнцефалография. – М: АОЗТ Антидор», ISBN 5 -_93751–005–4 (2000)
2. Гриндель О.М.,Коптелов Ю.М., Машеров и др Очаги патологической активности в головном мозге человека и их влияние на пространственно-временные отношения ЭЭГ. Журн. в.с. нервн. Деят., 671–686с, Т48, №4 (1998)
3. Norman, R.M, Malla, A.K., Villiamson, P.C., et al.: EEG koherence and syndromes in schizophrenija Brit. J. Psychiatrry, **170**, 411–415 (1997)
4. Tolmacha, N., Pudvosky, K., Vandans, J., Bondarenko, A.: Brain Electrical Bioactivity Informational Specifics in Relation to Behavioural and Cognitive Functions in Children with Learning Difficulties https://www.biorxiv.org/content/https://doi.org/10.1101/2022.12.30.522305v1.full.pdf. Accessed 10 Apr 2023
5. Зенков Л.Р. Клиническая Электроэнцефалография с элементами эпилептологии.-М;МЕДпресс-информ, 368с (2017)
6. Астапов, В.М.: Тревожные расстройства у детей. Прикладная психология, No **2**, 32–44с (2002)

7. Рощина И.Ф.,.Зверева Н.В. Научное наследие А.Р.Лурия и современная клиническая психология развития. Медицинская психология в России Электронный научный журнал, №1 (30), 1c, (2015). http://mprj.ru. Accessed 10 Feb 2023
8. Mundy, P., Fox, N., Card, J: EEG coherence, joint attention and language development in the second year (2003). https://doi.org/10.1111/1467-7687.00253Haulon
9. Hanlon, H.W., Thatcher R.W., Rline M.J.: Gender Differences in the Development of EEG Coherence in Normal Children. Develop. Neuropsychol. **16**(3), 479–506 (2009)
10. Kulaichev, A.P.: The informativeness of Coherence Analysis in EEG Studies; Neuroscience and Behavioral Physiology Kluwer Academic Plenum Publishers (United States), t41, N3, 321–328 (2009)

Interpretable Model to Support Differential Diagnosis Between Ischemic Heart Disease, Dilated Cardiomyopathy and Healthy Subjects

Katerina Iscra[1](✉), Milos Ajcevic[1], Aleksandar Miladinovic[2], Laura Munaretto[3], Jacopo G. Rizzi[3], Marco Merlo[3], and Agostino Accardo[1]

[1] Department of Engineering and Architecture, University of Trieste, Trieste, Italy
katerina.iscra@phd.units.it
[2] Institute for Maternal and Child Health 'Burlo Garofolo', Trieste, Italy
[3] Cardiovascular Department, ASUGI and University of Trieste, Trieste, Italy

Abstract. The differential diagnosis between Ischemic Heart Disease (IHD) and Dilated Cardiomyopathy (DCM) can often be challenging, because only invasive, and not largely available exams can provide a definite diagnosis. The echocardiographic left ventricular ejection fraction (LVEF) and global longitudinal strain (GLS) as well as ECG heart rate variability (HRV) analysis are shown to be helpful tools for diagnosing several cardiac diseases. There is also a growing interest in application of interpretable machine learning techniques to guide the diagnosis. We aimed to produce an interpretable model applied for differential diagnosis between DCM, IHD and healthy subjects (HC) based on LVEF, GLS and HRV features. The study encompassed three groups: 130 DCM, 164 IHD, and 152 HC subjects. The novel GLS, LVEF, and linear and non-linear HRV features were extracted for each subject. Then, the interpretable models were produced by a logistic regression algorithm considering a set of features chosen with the ReliefF method. The results showed that the most informative features for classification between IHD, DCM e HC were: GLS, LVEF, age, FD, SD1/SD2 and sex, listed in order of importance. The obtained classification accuracy was 70% and the area under the ROC curve was 83.4%. The study demonstrates that a logistic regression model and its nomograms allow detailed clinical interpretation of the model and may be a powerful tools support differential diagnosis between IHD, DCM and HC.

Keywords: Echocardiographic imaging · HRV analysis · Interpretable model · IHD · DCM

1 Introduction

Chronic ischemic heart disease (IHD) is a condition that affects the coronary pathways and progresses slowly over time. IHD occurs when there is an imbalance between oxygen demand and supply, often due to atherosclerosis. When presenting acutely, IHD manifests with unstable angina, myocardial infarction, or sudden cardiac death, making

Y. Dekhtyar and I. Saknite (Eds.): NBC 2023, IFMBE Proceedings 89, pp. 343–349, 2023.
https://doi.org/10.1007/978-3-031-37132-5_43

early diagnosis crucial for better prognostic outcomes. The clinical diagnosis is often based on patient reported symptoms, including chest pain, as well as electrocardiography and echocardiography assessment. However, a definitive diagnosis can only be obtained through invasive coronary angiography, which can lead risks and complications. Dilated cardiomyopathy (DCM), on the other hand, is a non-ischemic and non-valvular heart disease characterized by an enlargement of the left ventricle or both ventricles and their progressive dysfunction. It is a leading cause of cardiac arrest and heart transplantation worldwide [1]. Since the pathological causes of DCM are varied, diagnosis may differ depending on the individual, but family history, physical examination, electrocardiogram, and transthoracic echocardiography are typically conducted. Furthermore, it is important to note that distinguishing between IHD and DCM can be difficult due to the fact that a definitive diagnosis can only be obtained through invasive or not widely and timely available diagnostic exams. However, certain non-invasive diagnostic tools, such as echocardiographic left ventricular ejection fraction (LVEF), global longitudinal strain (GLS), and electrocardiogram heart rate variability (HRV) analysis, have been shown to be useful in diagnosing various cardiac disease [2–5].

LVEF is presently the most frequently used cardiac imaging parameter in patients with heart failure (HF). Nonetheless, LVEF on its own may not be sufficient to categorize all patients and should be incorporated as part of a more comprehensive approach [6]. GLS, which is a parameter used to analyze myocardial deformation, has been shown to play a significant role in predicting cardiovascular outcomes [7]. In comparison to measuring LVEF alone, GLS has demonstrated greater effectiveness in detecting overall degeneration of the left ventricle [2]. On the other hand, HRV measurement is used to quantify cardiac autonomic activity, resulting from the interplay between sympathetic and parasympathetic activity. In addition to GLS and LVEF, pathophysiologic changes associated with heart disorders and accompanying changes in HRV can provide prognostic information that may not be captured by GLS and LVEF alone [8].

There is also a growing interest in application of machine learning techniques, which utilize HRV features in conjunction with other clinical data, to guide the diagnosis [9]. Nonetheless, it is important to acknowledge that utilizing black-box models in sensitive areas such as healthcare can be problematic. Medical personnel or patients themselves must have some understanding of why these models made certain decisions in order to accept or reject the outcome [9]. Therefore, interpretable models that offer output information pertaining to a particular disease and assist in assessing the credibility of the model are more appealing in the field of medicine [10].

The literature contains several medical studies based on interpretable models, such as logistic regression [11–13]. Logistic regression is a statistical model that illustrates the relationship between input parameters and can be interpreted using nomograms to determine the probability of an event occurring. The logistic regression nomogram not only facilitates predictions but also visually represents the model's architecture and the individual contributions of features to the class probability.

Therefore, the study aimed to produce an interpretable model applied for differential diagnosis between DCM, IHD and healthy subjects (HC) based on LVEF, GLS and HRV features.

2 Methods

2.1 Study Population and Protocol

A total of 488 subjects participated in this study, divided into three groups: 146 DCM (60F/86M, aged 61 ± 14y), 182 IHD (44F/138M, aged 73 ± 11y), and 160 HC (80F/80M, aged 61 ± 17y). Patients with DCM were enrolled after clinical evaluation, and those older than 35 years of age, with cardiovascular risk factors and/or without a family history of DCM underwent coronary angiography. The assessment of IHD was based on clinical, laboratory, and invasive findings [14], and patients with known risk factors such as alcohol or drug abuse, and tachyarrhythmias were excluded from the study. The LVEF was determined using the Simpson biplane method [15], and the GLS was determined using speckle tracking echocardiography. The GLS measurement was performed offline using dedicated software (TomTec Arena v2.0, TomTec Imaging Systems, Unterschleißheim, Germany), with visual assessment of the endocardial border and manual adjustment of the tracing, if necessary, to ensure accurate contour tracing. This study was conducted in accordance with the principles of the Declaration of Helsinki, and all participants provided written informed consent.

2.2 Heart Rate Variability Acquisition and Processing

All participants in the study underwent a 24-h Holter electrocardiogram (ECG) recording using the SpiderView ambulatory electrocardiographic recorder (Sorin Group, Italy) with a sampling rate of 200 Hz. The RR intervals were extracted and labeled using SyneScope analysis software (Sorin Group, Italy). The RR interval records were divided into 5-min segments without overlap, and each segment was included in the analysis only if the longest subsequence of ectopic beats (labeled as "ectopic" by the ECG Holter) or the longest artifact subsequence did not exceed 10 s, known as Heart Rate Total Variability [3]. The segments were then interpolated with cubic spline and resampled at 2 Hz. Linear and non-linear HRV features were extracted from each segment. Linear parameters, including MeanRR, SDNN, RMSSD, NN50, and pNN50, were calculated directly from the RR sequence. In the frequency domain, the absolute powers in the low (LF = 0.04–0.15 Hz) and high (HF = 0.15–0.40 Hz) frequency bands, related to vagal and sympathetic nerve control on heart rhythm, were estimated from the interpolated HRV signal. Normalized low and high-frequency powers (LFn, HFn) and their ratio (LF/HF) were also calculated from the latter parameters. Non-linear analysis was carried out by calculating Poincaré plot parameters (SD1, SD2), reflecting short- and long-term variability [16]. Fractal Dimension (FD) and beta exponent (betaExp) [17] were also extracted to quantify the complexity of the system generating the signal.

2.3 Outliers, Features Selection and Classification

Firstly, the dataset was preprocessed to identify any possible outliers using the local outlier factor detection method. Then, feature selection was performed using the information gain method, and subsequently, a logistic regression model was created. The local outlier factor detection method used a hamming distance measure with a contamination rate of 0.1 and a number of neighbors equal to 20 to obtain local density from the k-nearest neighbors ($k = 20$). A total of 42 subjects were excluded (16 DCM, 18 IHD, and 8 HC), leaving 130 DCM (57F/73M, aged 62 ± 13y), 164 IHD (41F/123M, aged 73 ± 10y), and 152 HC (77F/75M, aged 61 ± 17y) in the study. For feature selection, the Information Gain method [18] was used, with a cut-off of 0.086 for informative attribute estimates. The logistic regression algorithm was then used to create the model, which considered the selected HRV features, age, sex, LVEF, and GLS. To estimate the classification accuracy (CA), area under the curve (AUC), F1, precision, and recall of the dataset, 10-fold cross-validation was used. Finally, nomograms were created for each group to interpret and validate the obtained model.

3 Results

The results showed that the most informative features for classification between IHD, DCM e HC, according to obtained Information Gain coefficient were GLS, LVEF, age, FD, SD1/SD2 and sex, listed in order of importance.

The classification performance of produced logistic regression model based on selected features is reported in Table 1.

Table 1. Performance measures of the logistic regression model.

Model	AUC	CA	F1	Precision	Recall
Logistic Regression	0.83	0.70	0.70	0.71	0.70

The confusion matrix obtained for the logistic regression model is reported in Fig. 1. The confusion matrix shows that DCM patients were classified with a precision of 78.9%, IHD patients at 65.0%, and HC subjects at 69.4%.

		Predicted			
		DCM	HC	IHD	Σ
Actual	DCM	78.9%	11.9%	14.1%	130
	HC	3.7%	69.4%	20.9%	152
	IHD	17.4%	18.8%	65.0%	164
	Σ	109	160	177	446

Fig. 1. Confusion matrix obtained by logistic regression model.

The produced nomograms for logistic regression models are reported in Fig. 2. The features are listed in order of importance allowing us to select the subset of the most informative features.

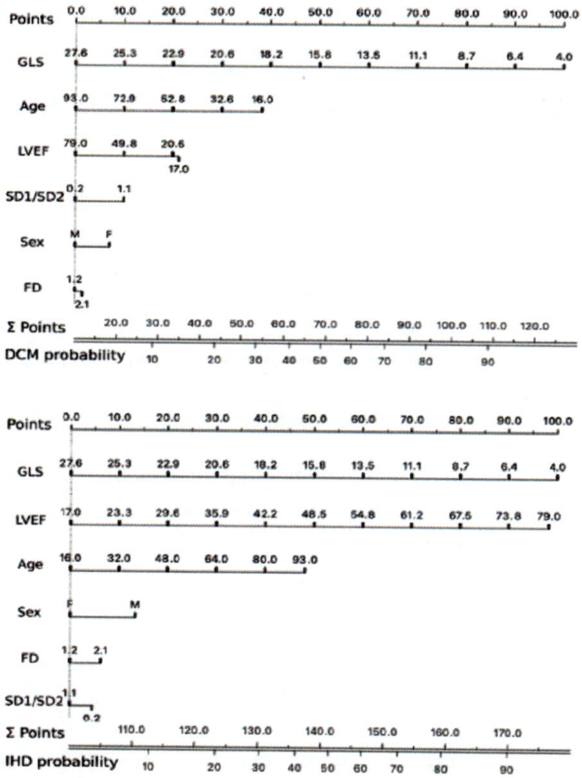

Fig. 2. Nomograms of the logistic regression model for DCM (top panel) and IHD (bottom panel) groups.

4 Discussion

Accurately distinguishing between DCM, IHD, and HC can be challenging, as a definitive diagnosis often requires invasive or unavailable diagnostic tests. However, diagnostic methods such as LVEF, GLS and HRV analysis are useful in detecting various cardiac conditions [2–6], but these parameters have not yet been used together to support the differential diagnosis. In recent years, there has been a growing interest in using interpretable machine learning approaches to aid in diagnosis. Therefore, the aim of this study was to develop an interpretable model that utilizes LVEF, GLS, and HRV features to aid in the differential diagnosis of DCM, IHD, and HC.

Our study found that GLS, LVEF, age, FD, SD1/SD2, and sex are the most informative features for accurately classifying DCM, IHD, and HC subjects. The model achieved an accuracy of 70%, and the confusion matrix for the test data showed that the DCM group had the highest precision (78.9%), while the IHD group had the lowest (65.0%). This difference in precision can be explained by the fact that the model's features characterized DCM patients better. This finding was partially confirmed by the created nomograms for the DCM and IHD groups.

In the IHD nomogram, GLS and LVEF appeared to be of equal importance. However, in the DCM nomogram, GLS was more crucial in the classification of these patients, as evidenced by its more extended line. Moreover, the model empirically determined the thresholds of 16 for GLS and 50 for LVEF for the diagnosis of DCM and IHD, which are consistent with recent research [19, 20]. It is noteworthy to underline that if GLS is less than 16, the probability of developing one of the two disorders increases. Conversely, an LVEF of less than 50 increases the chance of DCM while decreasing the likelihood of IHD. Age feature also played an important role, with the probability of developing DCM decreasing with age, consistent with the literature [21]. However, sex only partially influenced the outcome of the prediction: when examining the epidemiology of the IHD disease, it was found that women have the same incidence rates as men ten years younger [22]. FD, which quantifies the complexity of the system generating the signal, was more likely to be altered in patients with IHD and DCM [23]. Finally, SD1/SD2, which describes diastolic blood pressure, was found to be more variable in DCM patients [24].

In conclusion, our preliminary study produced a machine learning model with a reasonably high accuracy for the differential diagnosis of HC, DCM, and IHD. This level of accuracy can aid clinicians in deciding whether further invasive diagnostic tests are necessary. Our research also highlighted the importance of model interpretability, as evidenced by the usefulness of nomograms in evaluating the clinical plausibility of individual features and identifying thresholds for disease-related parameters. Overall, our study highlighted the utility of nomograms as a probabilistic classification tool to support clinical decision-making in the differential diagnosis of cardiac diseases.

Acknowledgement. Work partially supported by Master in Clinical Engineering, University of Trieste.

References

1. Lakdawala, N.K., Winterfield, J.R., Funke, B.H.: Dilated cardiomyopathy. Circ. Arrhythm. Electrophysiol. **6**, 228–237 (2013)
2. Ferrari, F., Menegazzo, W.R.: Global longitudinal strain or measurement of ejection fraction: Which method is better in stratifying patients with heart failure? Arq. Bras. Cardiol. **113**, 195–196 (2019)
3. Accardo, A., et al.: Toward a diagnostic CART model for Ischemic heart disease and idiopathic dilated cardiomyopathy based on heart rate total variability. Med. Biol. Eng. Comput. **60**, 2655–2663 (2022)
4. Agliari, E., Barra, A., Barra, O.A., Fachechi, A., Franceschi Vento, L., Moretti, L.: Detecting cardiac pathologies via machine learning on heart-rate variability time series and related markers. Sci. Rep. **10**, 8845 (2020)
5. Pastore, M.C., et al.: Speckle tracking echocardiography: Early predictor of diagnosis and prognosis in coronary artery disease. Biomed. Res. Int. **2021**, 6685378 (2021)
6. Mele, D., Nardozza, M., Ferrari, R.: Left ventricular ejection fraction and heart failure: An indissoluble marriage? Eur. J. Heart Fail. **20**, 427–430 (2018)
7. Ashish, K., Faisaluddin, M., Bandyopadhyay, D., Hajra, A., Herzog, E.: Prognostic value of global longitudinal strain in heart failure subjects: A recent prototype. Int. J. Cardiol. Heart Vasc. **22**, 48–49 (2018)

8. Kleiger, R.E., Miller, J.P., Bigger, J.T., Moss, A.J.: Decreased heart rate variability and its association with increased mortality after acute myocardial infarction. Am. J. Cardiol. **59**, 256–262 (1987)
9. Ahmad, M.A., Eckert, C., Teredesai, A.: Interpretable machine learning in healthcare. In: Proceedings of the 2018 ACM International Conference on Bioinformatics, Computational Biology, and Health Informatics, pp 559–560. Association for Computing Machinery, New York (2018)
10. Stiglic, G., Kocbek, P., Fijacko, N., Zitnik, M., Verbert, K., Cilar, L.: Interpretability of machine learning-based prediction models in healthcare. WIREs Data Min. Knowl. Discov. **10**, e1379 (2020)
11. Berkhemer, O.A., et al.: A randomized trial of intraarterial treatment for acute ischemic stroke. N. Engl. J. Med. **372**, 11–20 (2015)
12. Krahn, A.D., et al.: Prevention of arrhythmia device infection trial: The PADIT trial. J. Am. Coll. Cardiol. **72**, 3098–3109 (2018). https://doi.org/10.1016/j.jacc.2018.09.068
13. Cheng, X., et al.: Risk prediction of coronary artery stenosis in patients with coronary heart disease based on logistic regression and artificial neural network. Comput. Math. Methods Med. **2022**, e3684700 (2022)
14. Heart rate variability: standards of measurement, physiological interpretation and clinical use. Task Force of the European Society of Cardiology and the North American Society of Pacing and Electrophysiology. Circulation **93**, 1043–1065 (1996)
15. Lang, R.M., et al.: Recommendations for cardiac chamber quantification by echocardiography in adults: An update from the American Society of Echocardiography and the European Association of Cardiovascular Imaging. J. Am. Soc. Echocardiogr. **28**, 1-39.e14 (2015)
16. Woo, M.A., Stevenson, W.G., Moser, D.K., Trelease, R.B., Harper, R.M.: Patterns of beat-to-beat heart rate variability in advanced heart failure. Am. Heart J. **123**, 704–710 (1992)
17. Higuchi, T.: Approach to an irregular time series on the basis of the fractal theory. Phys. D Nonl. Phenom. (1988)
18. Lei, S.: A feature selection method based on information gain and genetic algorithm. In: 2012 International Conference on Computer Science and Electronics Engineering, pp. 355–358 (2012)
19. Chang, W.-T., et al.: The predictive value of global longitudinal strain in patients with heart failure mid-range ejection fraction. J. Cardiol. **77**, 509–516 (2021)
20. D'Elia, N., et al.: Normal global longitudinal strain: An individual patient meta-analysis. JACC Cardiovasc. Imaging **13**, 167–169 (2020)
21. Halliday, B.P., et al.: Sex- and age-based differences in the natural history and outcome of dilated cardiomyopathy. Eur. J. Heart Fail. **20**, 1392–1400 (2018)
22. Lloyd-Jones, D., et al.: Executive summary: heart disease and stroke statistics–2010 update: A report from the American Heart Association. Circulation **121**, 948–954 (2010)
23. Captur, G., Karperien, A.L., Hughes, A.D., Francis, D.P., Moon, J.C.: The fractal heart — embracing mathematics in the cardiology clinic. Nat. Rev. Cardiol. **14**, 56–64 (2017)
24. Voss, A., Schroeder, R., Truebner, S., Goernig, M., Figulla, H.R., Schirdewan, A.: Comparison of nonlinear methods symbolic dynamics, detrended fluctuation, and Poincare plot analysis in risk stratification in patients with dilated cardiomyopathy. Chaos **17**, 015120 (2007)

Assessment of Peripheral Perfusion Using Remote Photoplethysmography and Automated Capillary Refill Time Techniques in Severe COVID-19 Patients

Mara Klibus[1](✉), Veronika Eunapu[1], Zbignevs Marcinkevics[4], Uldis Rubins[2], Andris Grabovskis[2], Indulis Vanags[3], and Olegs Sabelnikovs[1]

[1] Department of Clinical Skills and Medical Technology, Rīga Stradiņš University, Riga, Latvia
mara.klibus@gmail.com
[2] Institute of Atomic Physics and Spectroscopy, University of Latvia, Riga, Latvia
[3] Department of Anaesthesiology and Reanimothology, Rīga Stradiņš University, Riga, Latvia
[4] Department of Human and Animal Physiology, Faculty of Biology, University of Latvia, Riga, Latvia

Abstract. The COVID-19 pandemic has posed significant challenges to healthcare systems worldwide, with critically ill patients requiring intensive care and fluid management to maintain systemic hemodynamics. This study aimed to investigate the potential of remote photoplethysmography (rPPG) and automated capillary refill time (aCRT) techniques for assessing alterations in peripheral perfusion among patients with severe COVID-19 infection during fluid resuscitation. Eight patients with severe acute respiratory distress syndrome (ARDS) due to COVID-19 were enrolled in a single-center prospective pilot study. Fluid responsiveness was evaluated using the passive leg raising test (PLRT) and a fluid challenge. Hemodynamic variables, perfusion index (PI) detected by rPPG, and manual and automated CRT parameters were collected at four different time intervals during the test. The results showed significant changes in hemodynamic variables, rPPG perfusion index, and manual and automated CRT parameters during fluid resuscitation, suggesting that these techniques could accurately evaluate peripheral perfusion changes in critically ill COVID-19 patients. The automated CRT technique, which provided objective assessments of capillary refill time, was found to be a potential alternative to the manual CRT technique. The present pilot study supports the hypothesis that rPPG and aCRT techniques, in conjunction with hemodynamic parameters, can reflect changes in peripheral perfusion during fluid resuscitation in patients with severe COVID-19. However, further studies with larger patient populations are required to validate these findings and develop guidelines for the application of these techniques in clinical practice.

Keywords: Remote photoplethysmography · automated capillary refill time · Peripheral perfusion

Y. Dekhtyar and I. Saknite (Eds.): NBC 2023, IFMBE Proceedings 89, pp. 350–356, 2023.
https://doi.org/10.1007/978-3-031-37132-5_44

1 Introduction

Although the COVID-19 pandemic, caused by the novel coronavirus SARS-CoV-2, has substantially subsided globally, new virus variants continue to emerge, potentially posing risks for intensive care overload in the near future. Since its outbreak in late 2019, the pandemic has presented unprecedented challenges to healthcare systems worldwide.

The clinical manifestations of COVID-19 encompass a wide range, from asymptomatic cases to severe respiratory symptoms necessitating mechanical ventilation [1–6]. Approximately 5% of cases progress to critical conditions, including severe lung damage, respiratory distress, hypoxia, respiratory failure, septic shock, and multi-organ dysfunction (MODS) [3, 4]. These critically ill patients often require admission to the intensive care unit (ICU) and aggressive fluid and vasopressor administration to maintain systemic hemodynamics.

Optimizing fluid management in patients with severe COVID-19 pneumonia or acute respiratory distress syndrome (ARDS) is crucial to balance the life-threatening risk of fluid overload and the consequences of hypovolemia, such as reduced tissue perfusion and multi-organ dysfunction (MOD) [3]. Current guidelines [6] recommend the use of dynamic parameters, including manual capillary refill time (mCRT), lactate level, and passive leg raising test (PLRT), to assess fluid responsiveness. However, lactate levels can be a non-specific indicator, as hyperlactatemia may arise from increased lactate production, decreased lactate clearance, or a combination of both factors [7].

Non-invasive assessment of peripheral perfusion remains challenging due to the lack of simple and objective diagnostic techniques. The development of innovative optical methods for peripheral perfusion evaluation, such as remote photoplethysmography and automated capillary refill time, is therefore of paramount importance.

Remote photoplethysmography (rPPG) is an emerging optical technique that measures blood volume changes in the microvascular bed of tissue by utilizing a remote camera to capture subtle variations in reflected light [8]. In contrast, automated capillary refill time (aCRT) is a technique that quantifies the time taken for capillaries to refill with blood following the application and release of pressure on the skin, providing an objective assessment of peripheral perfusion [9].

The primary aim of this study is to investigate alterations in peripheral perfusion among patients with severe COVID-19 infection using remote photoplethysmography and automated capillary refill time determination techniques.

Hypothesis: Remote photoplethysmography (rPPG) and automated capillary refill time techniques (aCRT), in conjunction with hemodynamic parameters, could reflect changes in peripheral perfusion during fluid resuscitation in patients with severe COVID-19.

2 Materials and Methods

The measurements were conducted in well ventilated and temperature controlled room (22 °C) at Pauls Stradins Clinical University Hospital (Riga, Latvia) in a general Intensive Care Unit with the approval of the Ethics Committee.

2.1 Measurement Protocol

Eight patients with severe ARDS due to COVID-19 were enrolled in a single-center prospective pilot study. To provide optimal resuscitation before further volume therapy, fluid responsiveness was evaluated using the passive leg raising test (PLRT) followed by a fluid challenge. During PLRT, cardiac output increases with volume expansion by transferring approximately 300 ml of venous blood from the lower body towards the right heart, achieved by tilting the patient's head down and elevating their feet at a 45-degree angle. A fluid challenge was performed using crystalloids at 10 ml/kg over 60 min. The test protocol was divided into four different stages: T1 - baseline conditions, T2 - during leg rising, T3 - after leg rising, and T4 - after fluid expansion, see Fig. 1.

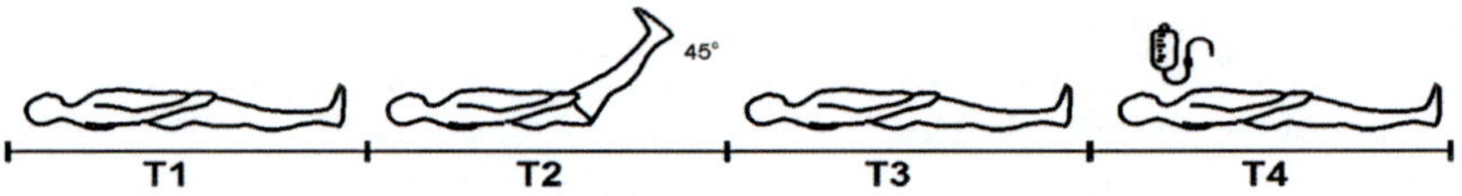

Fig. 1. Passive leg raising and fluid expansion test protocol. Four stages are denoted as follow: T1- baseline, T2-During leg rising, T3-after leg rising and T4- after fluid expansion.

Hemodynamic variables, such as cardiac output (CO l/min), measured by the MostcareUp Haemodynamic monitor (Vygon©), mean arterial blood pressure (MAP), perfusion index (PI) detected using rPPG, and manual CRT (mCRT) and automated CRT parameters T90 and Tst, were collected during all stages of protocol (T1-T4).While lactate levels were collected during T1 and T4 stages.

2.2 Optical Measurements

The components of the remote photoplethysmography system include a camera (Ximea-xiQ, CMOSIS CMV300 monochrome, 10-bit dynamic range, resolution 640 × 480), white light emitter (ALM Prismalix PRX800) for skin illumination with linearly polarized light, a narrow-band 540nm (bandwidth 10nm) optical filter (540 nm CWL 10nm FWHM bandpass filter Edmund Optics), an orthogonally oriented polarizer attached to the camera lens (low-distortion Edmund Optics, C- mount f = 6 mm lens) to decrease skin surface reflection, and a computational module for PPG signal processing and real-time hemodynamic parameter calculation, including mean perfusion index (PI) and spatial distribution of perfusion (perfusion map) [10, 11]. Prior to the passive leg raising test, the lamp was adjusted to achieve uniform illumination of the patient's palm. The distance between the palm and lamp was approximately 1 m.

The automated capillary refill time (aCRT) technique was utilized to improve the dynamic assessment of peripheral perfusion. The capillary refill time measurement device prototype is based on an embedded optoelectronic system. The major components of this system include an adjustable finger support frame (1), a moving opto-mechanical probe (2), a contactless temperature sensor (3), a button panel (4), and an LCD screen (5) for displaying measurement results as depicted in Fig. 2A. The aCRT measurement was performed by applying and removing an opto-mechanical probe on the skin surface of the middle finger while recording the signal at 525nm illumination. The prototype provides two parameters (T90 and Tst) characterizing capillary refill, along with a skin temperature reading. T90 represents the time taken for the light intensity to reach 90% from the moment pressure is released until Tst. Tst, on the other hand, signifies the stabilization of the signal or the maximum capillary refill following pressure release as shown in Fig. 2B. The detailed description of the aCRT technique is provided elsewhere [12].

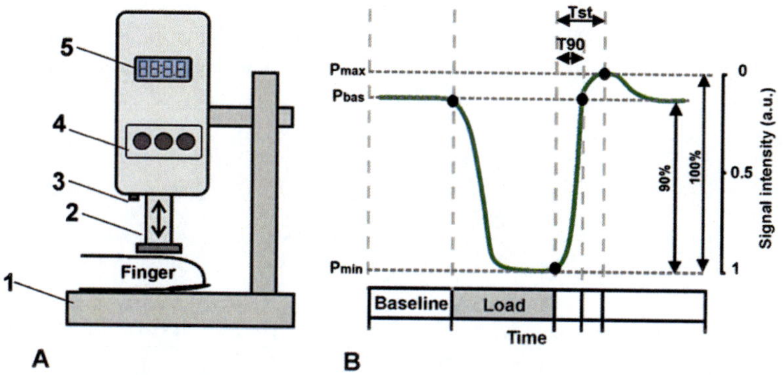

Fig. 2. Automated capillary refill time measurement prototype (A); Signal intensity during capillary refill measurement, Pbas (baseline perfusion), Pmin (minimal perfusion), Pmax (maximal perfusion). The time of the load is indicated by a gray square.

2.3 Statistics

Descriptive statistics, represented by mean ± standard deviation (SD) values, were calculated for continuous variables. The Friedman Repeated Measures Analysis of Variance on Ranks was employed to compare parameters from all protocol stages (T2-T4) against the baseline stage (T1). A statistically significant difference was considered at $p < 0.05$. All statistical analyses were conducted using SPSS for Windows 10.

3 Results

The titrated dosage of norepinephrine in the patient group ranged between 0.1–0.2 μg/kg/min. Hemodynamic variables were collected and analyzed at each stage of the protocol (T1–T4), with all changes compared to the baseline (T1).

During the passive leg raising test (T2), compared to the baseline, the mean arterial blood pressure (MAP) increased by 15%, the remote photoplethysmography (rPPG) perfusion index rose by 13% (see Fig. 3 T2), cardiac output (CO) showed an increase of 21%, the mean manual capillary refill time (mCRT) exhibited a decrease of 25.5%, the mean automated capillary refill time (aCRT) T90 was reduced by 30%, and Tst displayed a decrease of 22%. After fluid expansion (T4), in comparison to the baseline, the mean arterial blood pressure (MAP) increased by 17%, the remote photoplethysmography (rPPG) perfusion index rose by 30% (see Fig. 3. T4), cardiac output (CO) showed a 10% increase, the mean manual capillary refill time (mCRT) exhibited a 38% decrease, the mean automated capillary refill time (aCRT) T90 was reduced by 30%, Tst displayed a 14% reduction, and the mean lactate level decreased by 17%. These findings are detailed in Table 1.

Table 1. Parameters obtained during the passive leg raising and fluid expansion test protocol. The table displays the mean values ± standard deviation for the group (n = 8). Statistically significant differences ($p < 0.005$) between the baseline conditions (stage T1) and the subsequent stages: T2 - during leg raising, T3 - after leg raising, and T4 - after fluid expansion) are indicated by an asterisk (*).

Parameters	Baseline conditions T1	During Leg rising T2	After Leg rising T3	After fluid expansion T4
Mean arterial pressure (MAP), mmHg	77.9 ± 9.3	89.4 ± 12.8*	77.8 ± 10.1	91.7 ± 6.3
Cardiac output (CO), l/min	4.7 ± 0.8	5.7 ± 0.9*	4.8 ± 0.8	5.2 ± 1.0*
Lactate level, mmol/l	2.27 ± 1.0	n/a	n/a	1.9 ± 0.75*
Peripheral perfusion index, (PPI) a.u	30.5 ± 16.1	4.2 ± 16.4*	30.9 ± 15.8	39.1 ± 19.8*
Manual capillary refill time (mCRT), s	2.15 ± 0.59	1.6 ± 0.9*	1.9 ± 0.6	1.33 ± 0.5
Automated capillary refill time (aCRT) T90, s	1.5 ± 0.12	1.04 ± 0.12*	1.4 ± 0.6	1.05 ± 0.1
Automated capillary refill time aCRT. Tst, s	3.04 ± 0.35	2.59 ± 0.06	4.12 ± 2.25	1.9 ± 0.24
Temperature (T), °C	33.7 ± 1.9	33.4 ± 2.06	33.3 ± 2.4	33.1 ± 2.4

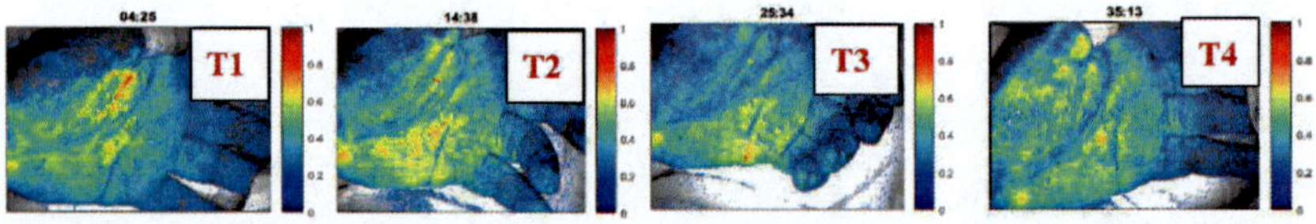

Fig. 3. Cutaneous rPPG perfusion index maps at different protocol stages: T1 (baseline), T2 (during leg raising), T3 (after leg raising), and T4 (after fluid expansion).After leg rising (T2) palm perfusion increased, as shown in higher density perfusion cluster.

4 Discussion

Our study results demonstrate, for the first time, that the perfusion index detected by remote photoplethysmography, and automated capillary refill time parameters can be used to evaluate peripheral perfusion changes during fluid resuscitation. To assess fluid responsiveness, the Surviving Sepsis Campaign guidelines suggest using dynamic parameters such as passive leg raising test, capillary refill time, and parameters estimating preload [6]. Predicting preload responsiveness is crucial for identifying patients who may benefit from fluid administration. Recent studies have shown that changes in the peripheral perfusion index (PI) correlate with changes in cardiac output. Therefore, when cardiac output is not monitored, changes in PI may help predict preload responsiveness and guide fluid therapy in ARDS patients [8]. In our study, we utilized remote photoplethysmography (rPPG) to detect the perfusion index and analyze its spatial distribution (perfusion maps) to characterize changes in peripheral perfusion during fluid resuscitation. Perfusion maps reveal the heterogeneity of cutaneous perfusion in peripheral tissues; therefore, we investigated PI not only at the finger level but also in the palm. Evidence from the Andromeda Shock Trial [9] suggests that a microcirculation-guided strategy based on CRT might reduce organ failure and patient mortality compared to a lactate-targeted approach. However, the manual CRT detection technique is subjective. To improve the dynamic assessment of peripheral perfusion, we used a custom-made automated capillary refill time (aCRT) technique. The detected parameters characterizing capillary refill (T90: 90% of capillary refill is achieved and Tst: capillary refill is entirely achieved) showed decreased time during fluid resuscitation, similar to the manual CRT method. Thus, the automated CRT technique may serve as an alternative to the manual CRT technique.

Our findings support the hypothesis that remote photoplethysmography (rPPG) and automated capillary refill time techniques (aCRT), in conjunction with hemodynamic parameters, can indeed reflect changes in peripheral perfusion during fluid resuscitation in patients with severe COVID-19. The observed changes in perfusion index, as detected by rPPG, and the automated capillary refill time parameters during fluid resuscitation were consistent with the expected physiological response, further validating the potential utility of these techniques in clinical practice. However, our study has several limitations.

First, we included only severe COVID-19 patients, while the results regarding PI might differ in other contexts, especially because PI is influenced by skin perfusion, which might be different in the perioperative period. Second, due to the small number of patients, further studies are required to clarify the potential applications of both methods.

5 Conclusions

This pilot study suggests that rPPG and aCRT techniques show promise as tools for accurately evaluating peripheral perfusion changes during fluid resuscitation in critically ill COVID-19 patients. However, further studies are required to reveal the applications of both methods in larger patient populations and to develop guidelines for their utilization.

Acknowledgements. We gratefully acknowledge the financial support provided by the Latvian Council of Science under the project FLPP-0326, "Development and Application of a Multi-parametric Optical Technique for Guiding Fluid Resuscitation and Vasopressor Therapy in Critically Ill COVID-19 Patients" (Project No. lzp-2022/1-0326). Our sincere appreciation goes to Edgars Laksa and Blazar Ltd. for their valuable contribution in supplying the capillary refill time measurement device prototype for this study.

References

1. World Health Organization (WHO). WHO Coronavirus (COVID-19) Dashboard. https://covid19.who.int
2. Koratala, A., Ronco, C., Kazory, A.: Need for objective assessment of volume status in critically ill patients with COVID-19: The tri-POCUS approach. Cardiorenal. Med. **10**(4), 209–216 (2020). https://doi.org/10.1159/000508544
3. Kazory, A., Ronco, C., McCullough, P.A.: SARS-CoV-2 (COVID-19) and intravascular volume management strategies in the critically ill. Proc. (Bayl. Univ. Med. Cent.) 1–6 (2020). https://doi.org/10.1080/08998280.2020.1754700
4. Colantuoni, A., Martini, R., Caprari, P., et al.: COVID-19 sepsis and microcirculation dysfunction. Front. Physiol. **11**, 747 (2020). https://doi.org/10.3389/fphys.2020.00747
5. Østergaard, L.: SARS CoV-2 related microvascular damage and symptoms during and after COVID-19: Consequences of capillary transit-time changes, tissue hypoxia and inflammation. Physiol Rep. **9**(3), e14726 (2021). https://doi.org/10.14814/phy2.14726
6. COVID-19 Treatment Guidelines Panel. Coronavirus Disease 2019 (COVID-19) Treatment Guidelines. National Institutes of Health. https://www.covid19treatmentguidelines.nih.gov/
7. Andersen, L.W., Mackenhauer, J., Roberts, J.C., Berg, K.M., Cocchi, M.N., Donnino, M.W.: Etiology and therapeutic approach to elevated lactate levels. Mayo Clin. Proc. **88**(10), 1127–1140 (2013). https://doi.org/10.1016/j.mayocp.2013.06.012. PMID: 24079682; PMCID: PMC3975915
8. Beurton, A., Teboul, J.L., Gavelli, F., et al.: The effects of passive leg raising may be detected by the plethysmographic oxygen saturation signal in critically ill patients. Crit. Care **23**, 19 (2019). https://doi.org/10.1186/s13054-019-2306-z
9. Hernandez, G., Ospina-Tascon, G., Petri Damiani, L., Estenssoro, E., Dubin, A., Hurtado, J., et al.: Effect of a resuscitation strategy targeting peripheral perfusion status vs serum lactate levels on 28-day mortality among patients with septic shock The ANDROMEDA-SHOCK Randomized Clinical Trial. JAMA **321**, 654–664 (2019)
10. Rubins, U., et al.: Simple and convenient remote photoplethysmography system for monitoring regional anesthesia effectiveness. IFMBE Proc. 378–381 (2017). https://doi.org/10.1007/978-981-10-5122-7_95
11. Rubins, U., et al.: Remote photoplethysmography for assessment of oral mucosa. Clin. Preclin. Opt. Diagnost. II. (2019). https://doi.org/10.1117/12.2526979
12. Marcinkevics, Z., Laksa, E., Rubenis, O., Blumfelde, M.: Reliability of automated optical determination of capillary refill time. Biophoton. Congr. JW1A.13 (2021). https://doi.org/10.1364/BODA.2021.JW1A.13

Influence of Metallized Coils on Human Leg Blood Circulation

Yuri Dekhtyar[1](✉), Ksenija Jašina[1], Viesturs Larins[2], Alexander Oks[1], Konstantins Pudovskis[3], Nelli Tolmača[3], and Vijay Vyas Vadhiraj[1]

[1] Riga Technical University, Kipsala str. 6A, Riga LV1048, Latvia
jurijs.dehtjars@rtu.lv
[2] Latvian Academy of Sport Education, Brivibas str. 333, Riga LV1006, Latvia
[3] Psychoneurophysiology and Bioregulation Research Center, Valdemara str. 17A, Riga LV1010, Latvia

Abstract. Chronic venous insufficiency occurs in 84% of the population. Methods, which employ external magnetic fields to enhance blood circulation, are used widely. However, the wired coils surrounding the blood flow could harvest the generated magnetic fields to use them for treatment. Such an approach has not been verified yet. The research aims to explore the possibility of the harvested field influence on blood circulation.

For this wired coils surrounded an extremity and the blood circulation in the extremity was identified using the rheography technique. The achieved results show that the rheography signal changed when the coil was used.

1 Introduction

Chronic venous insufficiency (CVI) occurs when the venous wall and/or valves in the veins of the extremities are not working effectively, causing difficulties to return blood to the heart from the extremities [1]. 84% of the population suffers from CVI [1].

External magnetic field enhances blood circulation in human extremities [2]. The applied fields have a low frequency (1–120 Hz) and induction (10^{-7}–10^{-5} T) [2].

Blood circulating in the vessels of a human produces a pulsed electrical current with the same frequency as the heart rate (~60 Hz). The current induces a weak magnetic field ~2 10^{-6} T [3]. These data are in accordance with the above values of frequency and induction of the fields applied from external sources. Therefore, there is a possibility to harvest the human magnetic field and redirect it back to the blood flow. In this case, blood circulation could be influenced without any external magnetic field source.

Such an approach has not been verified yet. The research for the first time aims to explore the possibility of harvesting human magnetic fields which could impact blood circulation.

Y. Dekhtyar and I. Saknite (Eds.): NBC 2023, IFMBE Proceedings 89, pp. 357–361, 2023.
https://doi.org/10.1007/978-3-031-37132-5_45

2 Methods

2.1 Harvesting of the Magnetic Field

A metallized coil was used to harvest the magnetic field. For this, the coil like a sock was put on the leg of the subject. Electrically conductive outlets of the coil were arranged in a perpendicular direction against the longitudinal axis of the leg.

The coil was made of a textile sleeve with woven electrically conductive threads (conductive coil). The distance between the outlets was equal to 1–2 mm. The length of the sleeve was ~30 cm. The sleeves without the electrically conductive threads (non-conductive coil) for placebo experiments were also fabricated. The sleeves with and without the electrically conductive threads were structurally and visually alike.

Twelve volunteers aged 18 to 25 (7 male and 5 female) participated in the study. All of them had similar heights (around 180 cm) and the length of their legs.

To intensify blood circulation in the legs, the volunteers were seating and were requested to move ten times their legs up and down stretching the calf and hamstring muscles. After that, the volunteers were seated for two minutes without any movement to relax (bringing the blood circulation to normal condition).

The conductive and non-conductive coils were put on the legs alternatively. However, the tested volunteer has not been informed about the specific coil that was put on the leg. The coils were replaced in two minutes.

Measurements of blood circulation during movement of the leg and rest time were done over three minutes.

Table 1 presents the consequence of the experiment stage to identify blood circulation change under coil influence.

Table 1. Experiment stages.

#	Right and left leg
1	Coil free/Rest
2	Conductive coil/Rest
3	Non-conductive coil/Rest
4	Conductive coil/Movements
5	Non-conductive coil/Movements

2.2 Blood Circulation Identification

Rheography poly-analysis (RGPA) was employed to assess changes of blood flow [4]. RGPA (also known as impedance plethysmography) measures the changes in the volume of blood vessels through an alteration in their electric impedance caused by the blood flow.

Measurements were done using the tool RGPA-6/12, REAN-POLY (Medicom, Russia). Ribbon (band) elastic electrodes connected corresponding places both at the right and left thigh and leg (Fig. 1). To bind the electrode to the skin the electro-conductive gel (Spectra 360 electrode gel, Parker Laboratories, USA) adhered to the moist skin (possibly sweated because of physical exercising) was used.

The measurements were done immediately after the exercises. To ensure that the electrodes were in connection with the skin, the volunteers were required not to make any movements during the recording. The volunteers located in a dimly lit room on a comfortable couch, and the experimental area was adjusted to reduce the subject's stress and anxiety.

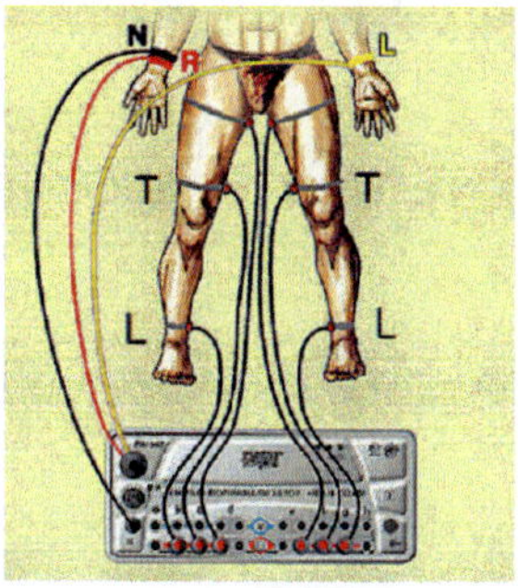

Fig. 1. Scheme of the electrodes standard binding: T and L –electrodes used for the research.

2.3 Processing of Acquired Data

Figure 2 presents an example of in-time (t) measured voltage (V) at some kind of electrode.

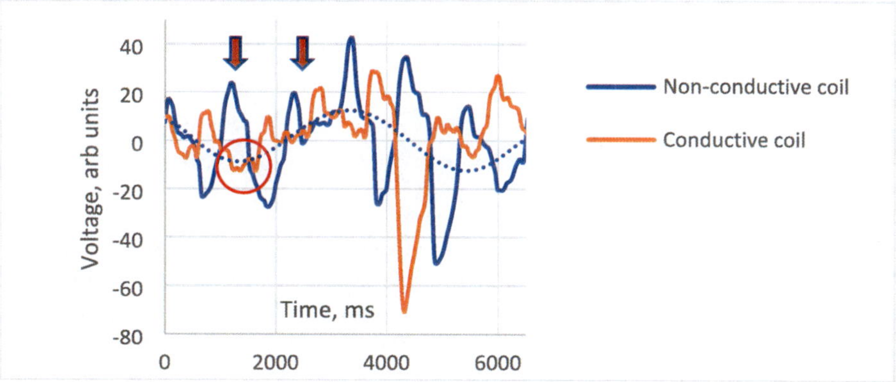

Fig. 2. Example of the V(t) diagrams.

In spite of the frequencies of the signals being similar in general, the voltage detected for the case of the conductive coil demonstrates the specific thin structure circled in Fig. 2.

The amplitudes of both the main harmonics and thick structures varied stochastically. The amplitudes of the V(t) diagram measured for the non-conductive coil fluctuated too. Therefore, a statistical contrast of the V(t) diagrams recorded for the cases of the conductive and non-conductive coils was employed. The data were collected with 4 ms measurement intervals for in Table 1 three-minute length of V(t) recording. As a result, the amount of the acquired numbers for the single experiment stage was equal to 45000 for each leg. Because of this, a comparison of statistics acquired for both conductive and non-conducted coils was achieved with a significance level ≤ 0.05.

To determine the impact of the conductive coil on blood circulation the statistical distributions of V values correspondingly collected for the conducted (SDC) and non-conducted (SDNC) coil cases were compared. A discrepancy between SDC and SDNC was classified as the influence of the conductive coil. The χ^2 test was applied to collate SDC with SDNC.

The values of V in the diagrams Fig. 2 are modulated with a wave-like process (dotted line as the example for the V(t) graph related to the non-conductive coil) which is connected with breathing. The dotted line was assumed as the baseline. To build it for the V(t) graphs, the Origin 2018 software was used. The baseline was subtracted from V(t).

3 Results

The achieved results for the stages in Table 1 are presented in Table 2.

Table 2. The number of cases (%) indicated the influence of the conductive coil on blood circulation

Compared experiment stage (Table 1)	Compared actions	Right leg		Left leg	
		Electrodes for rheography			
		T	L	T	L
1–2	Rest-Rest	100	100	100	100
1–3		0	0	0	0
2–3		100	100	100	100
4–5	Movement-Movement	100	100	100	100

The non-conductive coil does not have an influence on blood circulation in the rest of the leg (stages compared: 1–3). In contrast, the conductive coil changes circulation (stages 1–2 and 2–3). In the case of leg loading the conductive coil influences too. However, the non-conducted one does not (stages 4–5) impact the circulation.

4 Conclusion

The conductive coil put on a human leg has an influence on rheography electrical potential statistics. This is the evidence that metallized coils put on the human leg influences blood circulation.

Acknowledgment. The research was supported by Riga Technical University, Latvian Academy of Sport Education, and the Psychoneurophysiology and Bioregulation Research center, Latvia.

References

1. Rabe, E., Guex, J.J., Puskas, A., et al.: (2012) Epidemiology of chronic venous disorders in geographically diverse populations: results from the vein consult program. Int Angiol. **31**(2), 105–115 (2012)
2. Szopinski, J.Z.: The Biological Action of Physical Medicine: Controlling the Human Body's Information System, vol. 244 (2014)
3. Gross, S., Barmet, C., Dietrich, B., et al.: Dynamic nuclear magnetic resonance field sensing with part-per-trillion resolution. Nat. Commun. **7**, 13702 (2016)
4. Nikolaeva, T.M., Golubeva, E.K.: Feature of blood circulation in the upper limbs with dosed bicycle ergometry. Siberian J. Life Sci. Agricult. **12**(2), 84–88 (2020)

Hand Tracking for XR-Based Apraxia Assessment: A Preliminary Study

Giulia Pellegrino[1(✉)], Giovanni d'Errico[2], Valerio De Luca[1], Maria Cristina Barba[1], and Lucio Tommaso De Paolis[1]

[1] Department of Engineering for Innovation, University of Salento, Lecce, Italy
{giulia.pellegrino1,valerio.deluca,cristina.barba, lucio.depaolis}@unisalento.it

[2] Department of Applied Science and Technology, Polytechnic University of Turin, Turin, Italy
giovanni.derrico@polito.it

Abstract. The current assessment of upper limb apraxia consists in administering tests to patient, using analog instruments. The use of immersive Extended Reality (XR) environments allows the evaluation of such tests in a quantitative manner and not relying solely on the medical observations. The objective of the study is to provide a tool to support the assessment of hand apraxia using an immersive simulation scenario. The focus of the study is, therefore, to evaluate a hand tracking system, for analyzing the patient's interactions with the virtual environment, while performing apraxia assessment tests. It was necessary to preliminarily test the system on a pathology-free subject to verify its suitability for clinical objectives. Thanks to the trajectories stored by the hand tracking system, it is possible to estimate a minimum threshold value from which the detection made by the camera can be studied. Since this is a preliminary study, it is not possible to return results on the main objective. It can be seen from the results obtained that the selected hand tracking technology proves capable of providing the information required by the apraxia assessment. Further testing campaigns on specific patients and further interventions to improve the virtual environment are needed.

Keywords: Extended Reality · Hand Tracking · Apraxia · Virtual Reality

1 Introduction

Limb apraxia is a disorder of skilled movement that is characterized by difficulty performing purposeful movements with one or both hands in the absence of elemental motor deficits such as weakness or abnormal tone [1]. Limb apraxia has been primarily studied in patients with dementia or left hemisphere stroke (LHS). However, limb apraxia has also been reported in patients with right hemisphere stroke (RHS), multiple sclerosis (MS) or traumatic brain injury (TBI) [2]. Currently, the assessment of apraxia, in particular constructive apraxia, involves the administration of tests to the patient, such as imitating gestures, and evaluating their performance. These tests are typically performed using analog instruments [3]. Constructive apraxia presents as an

Y. Dekhtyar and I. Saknite (Eds.): NBC 2023, IFMBE Proceedings 89, pp. 362–369, 2023.
https://doi.org/10.1007/978-3-031-37132-5_46

isolated disorder of performing free or copied drawings or constructive, two- or three-dimensional tasks. The disorder may also be observed when performing puzzles or when copying figures, especially geometric ones. This ability may be impaired as a result of brain injury [4]. For many years, information and communication technologies (ICTs) have been assisting healthcare professionals in the process of diagnosis, treatment, monitoring, drug prescription, referral, information retrieval and communication, documentation, and transactions [5]. Emerging technologies, such as eXtended Reality (XR), also have numerous applications in biomedicine that are driving innovation in both basic research in the field and in healthcare. Extended reality refers to a set of methods that allow users to be immersed in a 3D or 4D (spatial and temporal) virtual environment at various levels, including virtual reality (VR), augmented reality (AR) and mixed reality (MR). While VR allows the user to be fully immersed in a virtual environment, AR and MR superimpose virtual objects on top of the real physical world. Although XR has broad applications in fields such as entertainment and education [6, 7], it also presents numerous opportunities in biomedicine that drive innovation in both basic research and healthcare. Current trends have demonstrated its widespread use in several areas, including biomedical data visualization and analysis, medical training and education, surgical procedures, digital therapy, rehabilitation, and telemedicine. Unlike conventional methods, XR adds an additional dimension that enables user-directed operations in an immersive and interactive context. For instance, in clinical settings, XR can facilitate faster education, training, and planning due to its immersive and interactive environment, simulating situations in which users can realistically practice for high-risk procedures such as surgery. The advantages of XR also enable the creation of digital therapies, for example, for anxiety and pain management, as the immersive interaction distracts users [8]. By providing a realistic and controllable simulation scenario with a high sense of presence, the use of immersive extended reality environments also allows tests for the diagnosis of constructive apraxia to be evaluated quantitatively, rather than solely relying on medical observations. The objective of the study is to provide a tool to support the assessment of hand apraxia using an immersive simulation scenario. Immersive scenarios and interfaces offer powerful experiences based on visualization and interaction design transforming the sense of space [9]. The need for testing with a high level of realism and intuitiveness in interaction with XR environments led to the use of a Natural User Interface (NUI). NUI was created to improve users' interaction with the system by using natural body movements to perform actions. The "natural" property does not refer to the interface, but to how users interact with it and what they experience using it [10]. Specifically, interaction techniques based on hand gesture recognition were applied. They are an important part of Human Computer Interaction (HCI), which gives computers the ability of capturing and interpreting hand gestures and executing commands afterwards [11]. This can be done through analysis of images captured by cameras or electromyographic signals detected by specific wearable devices [12, 13]. Furthermore, to analyze in detail the patient's interactions with the virtual environment, a hand-tracking system was necessary when performing apraxia assessment tests. This system enables users to see a virtual representation of their hands in the digital environment that is updated in real time [14]. The focus of the study is to evaluate a hand tracking and gesture recognition system. This is done in preparation for the possibility of submitting the prototype to an ethics committee of a healthcare facility for review. The acceptance of the prototype by the committee is a mandatory step to test and validate the tool on patients with apraxia. This introduction will be followed by a section on a brief analysis of the current state of the art, a description of the technologies employed and the system developed, a discussion of the results obtained, and, finally, the conclusions drawn and potential future avenues of research.

2 Related Work

Upon conducting a search for the keywords "Extended Reality" and "Apraxia" in major databases, few or no relevant results were found. Expanding the search to include all neuromotor disorders or using more outdated terms such as "virtual" and "augmented reality" yields more results, but they do not always provide useful background information or align with the focus of the present study. The literature examined reveals a strong focus on rehabilitation following neurological disorders, particularly stroke. This may be because grants are awarded to researchers who seek to address these universal problems. Upper limb research is more widespread than lower limb research and more diverse, routinely focusing on individual joints and muscles, which can be explained at least in part by the importance of hands for daily activities and independence [15]. XR has been shown to increase therapy time for post-stroke care without additional workload for therapists [16]. Immersive virtual environments have also been used to study limb apraxia via EEG [17]. AR has been employed for studying apraxia in tool usage, with visual stimuli shown to improve performance by facilitating access to the appropriate motor program in patients [18]. Some pilot studies suggest that virtual reality can aid in the rehabilitation of patients with buccal-facial [19], ideomotor [20], and constructive apraxia [21], but further validation and testing are required. From the research conducted, it appears that no previous studies have focused on the study and evaluation of trajectories detected through the Vive Hand Tracking system in the context of apraxia. The hand tracking capabilities of the same company's visors are typically achieved through proprietary controllers and trackers [22] or by integrating the device with LeapMotion [23]. The only existing scenario in which the system was used on patients with apraxia focused solely on rehabilitation rather than diagnostic assessment and only involved a single patient [20].

3 Materials and Methods

3.1 Hardware Platform

The platform used for the preliminary system study is based on a high-performance graphics workstation capable of supporting the HTC Vive [24], a wired VR device produced by Valve in collaboration with HTC. The commercially available kit includes a visor, two controllers, and two base stations that use "Lighthouse" technology to track the user's position and movement in physical space and replicate them in the virtual world. The base stations generate an infrared grid in physical space, which interacts with the 70 sensors on the HMD and controllers, allowing the user to move freely and naturally in a 5-m × 5-m area. The HMD connects to the PC via a USB 3.0 port and an HDMI port and requires a power source via the supplied cable. The base stations must also be connected to a power source. The HMD [25] includes two 1080 × 1200-pixel AMOLED (Active Matrix Organic Light Emitting Diode) screens and has a refresh rate of 90 Hz, a field of view of 110°, and a pair of non-circular Fresnel lenses that adjust the user's interpupillary distance (IPD). The HTC Vive has the advantage of performing all processing on the machine to which it is connected, without imposing any limitations on the app to be developed. Additionally, it allows on-screen monitoring of everything that the subject sees and does in the virtual environment, which is one of the main requirements of the developed prototype. The device also allows for intervention in the game with on-screen mouse clicks and using graphical elements not visible in VR. However, this issue is not relevant in the designed prototype as it only targets the upper limbs and not the motion of the lower limbs.

3.2 Software Development

The preliminary system study was conducted using the Unity platform [26]. Unity is a cross-platform graphics engine developed by Unity Technologies that allows real-time development of video games and interactive content. Compared to other graphics engines, Unity offers a wide range of plugins and SDKs that facilitate virtual reality creation and interaction as well as an extensive developer community, web-based tutorials, documentation, and examples [27]. Connecting Unity to the Vive headset and its accessories can be done quickly and easily with the OpenVR System Development Kit (SDK) and its SteamVR management interface [28]. To meet the system's requirement of free upper limbs movement in the virtual environment without additional devices (e.g., controllers or trackers), the only SDK available from the proprietary company for this purpose was the Vive Hand Tracking SDK [29]. A GUI has been developed within the VR environment. It includes a timer that starts as soon as the doctor starts the game and will stop only when the game is finished; a counter of the tiles that the user has put back into the correct position. The design in [30] for 2D games was used to divide the image into perfectly fitting puzzle tiles. It was then necessary to make the changes to the objects and scripts of the game to use it in a VR environment.

3.3 Experimental Setup

The system involves two actors: the medical staff and the user (or healthy subject).

The medical staff is responsible for properly setting up the HTC Vive in the room and guiding the user in positioning and lowering the VR helmet. He or she is also responsible for starting the application and, with the timing he or she deems appropriate, initiating the game and evaluating its possible interruption. The user's task is to reconstruct the puzzle. As soon as the clinician deems it appropriate, he/she will initiate the game session in which the artwork will crumble into 40 puzzle tiles that must be put back into their starting position. The testing phase was carried out by a healthy male individual between the ages of 35 to 40. When administering the test to users, special care was taken to ensure that they did not encounter any physical obstacles in their surroundings, thus ensuring total freedom of their upper limbs.

4 Results

The positions that both limbs took during the session were recorded in a csv file using the hand tracking and gesture recognition system. Specifically, the paths taken once a puzzle card was grasped until it was released were recorded. The grasping and release of a card occur with two distinct gestures that upon recognition activate the system to track a central point in each palm of the hands processing 90 fps [29]. The coordinates recorded are two-dimensional since in the virtual environment it is not allowed to move the tiles outside the game board. To validate the hand tracking system, it was assumed that in a healthy subject the motor planning of the trajectories to be performed would have to be very close to the ideal trajectory, that is, the length of the segment that is obtained by calculating the Euclidean distance between two pairs of points in space. The real trajectory was, on the other hand, calculated by applying the Euclidean distance formula to each pair of points traversed by the user after grasping a tile. We define the absolute error as the difference between the length of the path taken and the ideal path. To be able to compare errors, it is necessary to eliminate their dependence on the length of the path. Therefore, we use the relative error, i.e. the normalization of the absolute error on the basis of the length of the ideal trajectory. By illustration, we report graphs for the left hand but reflecting the behavior of the right hand. From the graph in Fig. 1 It is easy to see that it is very small relative errors that are most frequent. This confirms the initial thesis, according to which a healthy user should have made trajectories very similar to the ideal ones. In order to be able to estimate a minimum threshold value from which the camera detection can be studied, the path lengths corresponding to the most frequent relative errors, i.e. those with more than 80 occurrences and those with more than 60 occurrences (Fig. 2) within a game session, were identified. It is evident that in both ranges of relative errors, the displacements involved are close to half a centimeter in length and therefore insignificant for the purposes of any clinical evaluation. The small relative errors could be caused either by inaccuracies in the detection system or could be due to the human being.

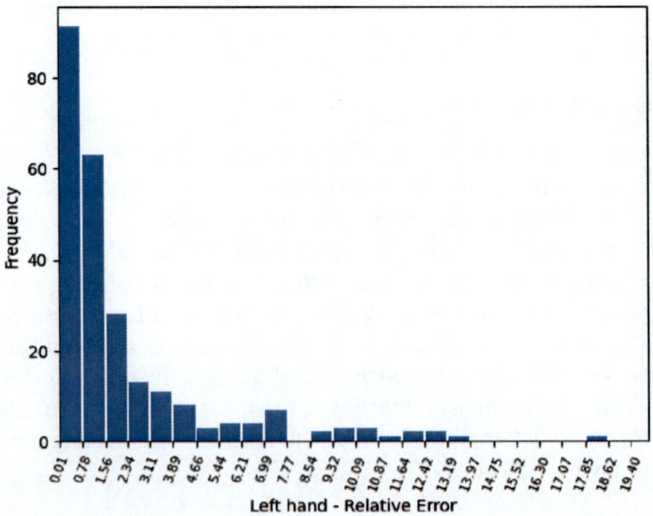

Fig. 1. Frequency of relative left-handed errors in a playing session

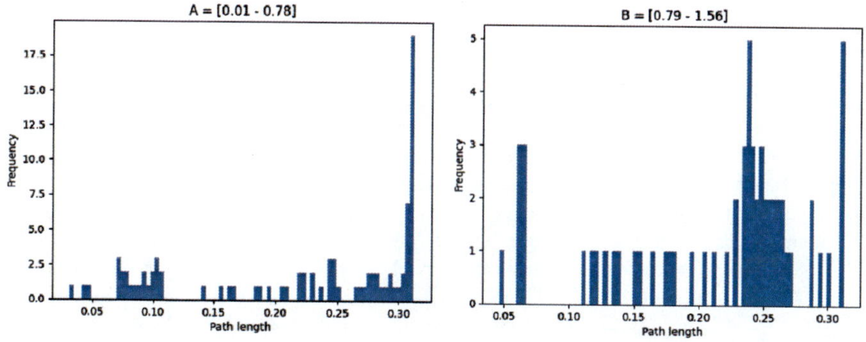

Fig. 2. Frequency of path lengths with relative error in intervals A and B

5 Conclusions and Limitations of the Work

Since this is a preliminary study, it is not possible to provide results on the main objective. The results may say that the chosen hand tracking technology may be able to provide the necessary information for the assessment of apraxia. To confirm this hypothesis, it is necessary to conduct a test campaign on a significant experimental sample of healthy subjects. A time-correlated analysis of the data and an expansion of the pool of outliers, so far limited to the detection of the two gestures in the same spatial coordinates, may also be useful.

This preliminary study is preparatory to the possibility of submitting the prototype for consideration by an ethics committee of a health care facility. Acceptance of the prototype by the committee is a mandatory step to be able to test and validate the tool on patients with apraxia, developing other specific scenarios with the hardware and software tools described here. Interaction with such new environments would also permit expansion of the gestures to be performed and recognized. The analysis could be useful for application of the tracking system in other fields as well. Early detection of the disease can help slow down its course and significantly improve patients' quality of life: in particular, mitigating the effects of the disease can prevent patients from losing basic limb and hand functions, enabling them to maintain a good level of independence in daily life and efficiency at work for longer. From a scientific point of view, the use of the application opens possibilities beyond supporting the diagnosis and rehabilitation of individual patients. Once many patients with apraxia are reached, the data collected by the platform could allow the discovery of specific features of the disease that are currently unknown. Inferential statistical analysis could allow the identification of new patterns and trends in the overall evolution of the disease. In addition, comparison with traditional neurological scales could lead more skeptical medical personnel to gradually gain more confidence in the system, on the one hand, and suggest the development of more robust integrated models, on the other.

References

1. Buxbaum, L.J., Randerath, J.: Limb apraxia and the left parietal lobe. Handb. Clin. Neurol. **151**, 349–363 (2018). https://doi.org/10.1016/B978-0-444-63622-5.00017-6
2. Buchmann, I., et al.: Limb apraxia profiles in different clinical samples. Clin. Neuropsychol. **34**, 217–242 (2020). https://doi.org/10.1080/13854046.2019.1585575
3. (PDF) Manual for Diagnostic Instrument for Limb Apraxia - Short Version (DILA-S) english version. https://www.researchgate.net/publication/322266689_Manual_for_Diagnostic_Instrument_for_Limb_Apraxia_-_Short_Version_DILA-S_english_version
4. Migliaccio, R., Bourgeois, A., Bartolomeo, P.: Aprassie. EMC -Neurologia **21**, 1–8 (2021). https://doi.org/10.1016/S1634-7072(21)44500-9
5. Dumay, A.C.M., Freriks, G.: Quality management issues for medical ICT. Stud Health Technol Inform. **103**, 93–100 (2004). https://doi.org/10.3233/978-1-60750-946-2-93
6. De Paolis, L.T., Chiarello, S., D'Errico, G., Gatto, C., Nuzzo, B.L., Sumerano, G.: Mobile extended reality for the enhancement of an underground oil mill: a preliminary discussion. In: De, L.T., Arpaia, P., Bourdot, P. (eds.) AVR 2021. LNCS, vol. 12980, pp. 326–335. Springer, Cham (2021). https://doi.org/10.1007/978-3-030-87595-4_24
7. De Paolis, L.T., De Luca, V., D'Errico, G.: augmented reality to understand the leonardo's machines. In: De, L.T., Bourdot, P. (eds.) AVR 2018. LNCS, vol. 10851, pp. 320–331. Springer, Cham (2018). https://doi.org/10.1007/978-3-319-95282-6_24
8. Yuan, J., et al.: Primer Extended reality for biomedicine. https://doi.org/10.1038/s43586-023-00198-y
9. Çöltekin, A., et al.: Geo-Information Extended Reality in Spatial Sciences: A Review of Research Challenges and Future Directions. https://doi.org/10.3390/ijgi9070439
10. Guerino, G.C., Valentim, N.M.C.: Usability and user experience evaluation of natural user interfaces: a systematic mapping study. IET Softw. **14**, 451–467 (2020). https://doi.org/10.1049/IET-SEN.2020.0051
11. Yasen, M., Jusoh, S.: A systematic review on hand gesture recognition techniques, challenges and applications. PeerJ Comput. Sci. 5 (2019). https://doi.org/10.7717/PEERJ-CS.218
12. De Paolis, L.T., De Luca, V.: The impact of the input interface in a virtual environment: the Vive controller and the Myo armband. Virtual Real. **24**(3), 483–502 (2019). https://doi.org/10.1007/s10055-019-00409-6
13. De Paolis, L.T., De Luca, V.: The effects of touchless interaction on usability and sense of presence in a virtual environment. Virtual Real. **26**, 1551–1571 (2022). https://doi.org/10.1007/S10055-022-00647-1
14. Palacios-Ibá, A., Alonso-García, M., Contero, M., Camba, J.D.: The Influence of Hand Tracking and Haptic Feedback for Virtual Prototype Evaluation in the Product Design Process (2022). https://doi.org/10.1115/1.4055952
15. Butz, B., Jussen, A., Rafi, A., Lux, G., Gerken, J.: A taxonomy for augmented and mixed reality applications to support physical exercises in medical rehabilitation & mdash; a literature review. Healthcare **10**, 646 (2022). https://doi.org/10.3390/HEALTHCARE10040646
16. Das, A., Day, T.W., Kulkarni, V., Buchanan, A., Cottrell, K., John, N.W., Chatterjee, K.: Towards intelligent extended reality in stroke rehabilitation: application of machine learning and artificial intelligence in rehabilitation. Augmenting Neurol, Disorder Pred. Rehabilit. Using Artifi. Intell., 309–329 (2022). https://doi.org/10.1016/B978-0-323-90037-9.00006-0
17. Spinelli, G., Pezzetta, R., Canzano, L., Tidoni, E., Aglioti, S.M.: Brain dynamics of action monitoring in higher-order motor control disorders: the case of apraxia. eNeuro. **9**, (2022). https://doi.org/10.1523/ENEURO.0334-20.2021
18. Rohrbach, N., et al.: Improvement of apraxia with augmented reality: influencing pantomime of tool use via holographic cues. Front Neurol. **12**, 1491 (2021). https://doi.org/10.3389/FNEUR.2021.711900/BIBTEX

19. Emedoli, D., Arosio, M., Tettamanti, A., Iannaccone, S.: Virtual reality augmented feedback rehabilitation associated to action observation therapy in buccofacial apraxia: Case Report. Clin. Med. Insights Case Rep. **14** (2021). https://doi.org/10.1177/1179547621994579
20. Park, W., Kim, J., Kim, M.Y.: Efficacy of virtual reality therapy in ideomotor apraxia rehabilitation: a case report. Medicine **100**, E26657 (2021). https://doi.org/10.1097/MD.0000000000026657
21. Maggio, M.G., et al.: Limb apraxia in individuals with multiple sclerosis: Is there a role of semi-immersive virtual reality in treating the Cinderella of neuropsychology? Mult. Scler. Relat. Disord. **69**, 104405 (2023). https://doi.org/10.1016/J.MSARD.2022.104405
22. Gonçalves, G., Melo, M., Barbosa, L., Vasconcelos-Raposo, J., Bessa, M.: Evaluation of the impact of different levels of self-representation and body tracking on the sense of presence and embodiment in immersive VR. Virtual Real. **26**, 1–14 (2021). https://doi.org/10.1007/s10055-021-00530-5
23. Cardoso, S., et al.: Personalized virtual reality environments for intervention with people with disability. Electronics **11**, 1586 (2022). https://doi.org/10.3390/ELECTRONICS11101586
24. VIVE - VR Headsets, Games, and Metaverse Life | United States. https://www.vive.com/us/
25. VIVE Specs & User Guide - Developer Resources. https://developer.vive.com/resources/hardware-guides/vive-specs-user-guide/
26. Unity Real-Time Development Platform | 3D, 2D, VR & AR Engine. https://unity.com/
27. Unity - Manual: Unity User Manual 2021.3 (LTS). https://docs.unity3d.com/Manual/UnityManual.html
28. SteamVR Plugin | Integration | Unity Asset Store. https://assetstore.unity.com/packages/tools/integration/steamvr-plugin-32647
29. Overview - Developer Resources. https://developer.vive.com/resources/vive-sense/hand-tracking-sdk/overview/
30. GitHub - shamim-akhtar/jigsaw-puzzle: This project is a Jigsaw Puzzle game. The tiles are dynamically created at runtime based on Bezier Curve. You can play the WebGL version of the game on Faramira., https://github.com/shamim-akhtar/jigsaw-puzzle

Training Ophthalmoscopic Skills in Extended Reality: Assessment of User Experience

Albina Abdullayeva(✉), Karola Panke, and Tatjana Pladere

Department of Optometry and Vision Science, Faculty of Physics, Mathematics and Optometry, University of Latvia, Riga, Latvia
albina.abdullayeva@lu.lv

Abstract. In medical education and training, extended reality-based simulators have grown in popularity. In many countries, trainee ophthalmologists and optometrists develop their ophthalmoscopic skills using interactive scenarios with different eye fundus conditions shown via extended reality headsets. This positively affects learning outcomes and students' confidence levels. Less is known about user experience. Therefore, we aimed to assess the objective and subjective parameters of user experience to elucidate the effect of using an extended reality-based ophthalmoscope simulator on human vision and user comfort. To the best of our knowledge, this is the first assessment of visual aftereffects and asthenopia complaints following the use of an extended reality-based ophthalmoscope simulator. On average, the near point of convergence, amplitude of accommodation, and fusional reserves did not change considerably in participants with normal or corrected-to-normal visual acuity following the training in extended reality. Nevertheless, some changes were observed in accommodation lag and microfluctuations at near. Moreover, the discomfort increased for some participants more than for others. Overall, the most widespread complaints were headaches, dry eyes, and eye strain. To reduce the potential discomfort, users can be recommended to follow the changes in their comfort and take breaks regularly.

Keywords: extended reality · ophthalmoscope simulator · human vision · accommodation · user comfort

1 Introduction

Extended reality (XR) technology, including virtual reality (VR) and augmented reality (AR), has been increasingly used in various fields such as education and healthcare. Particularly, its ability to generate three-dimensional environments and ensure immersive experience can foster medical training, allowing students and medical professionals to learn and practice in a safe and controlled environment.

Eyesi indirect ophthalmoscope simulator is the latest tool which is based on video see-through AR. The implementation of ophthalmoscope simulators in ophthalmology and optometry training has been shown to improve clinical skills and diagnostic accuracy when combined with traditional learning methods [1–4]. Moreover, it has been reported to positively affect students' attitude towards studies and their level of confidence [3, 4].

Y. Dekhtyar and I. Saknite (Eds.): NBC 2023, IFMBE Proceedings 89, pp. 370–376, 2023.
https://doi.org/10.1007/978-3-031-37132-5_47

Nevertheless, there has been limited focus on the fact that this simulator relies on technology that may pose challenges to the human visual system. Specifically, prolonged viewing of stereo digital content through near-eye displays has been shown to negatively affect ocular accommodation and vergence [5–7], and user comfort [6, 7]. While the adverse effects are generally short-term in nature [5, 6], there is a lack of long-term studies that would allow to elucidate the potential impact of repeated exposure to near-eye displays on visual health.

Understanding the impact of using stereo XR technology on humans is crucial for optimizing its usage and minimizing potential adverse effects [8]. Therefore, we aimed to assess the objective and subjective parameters of user experience to elucidate the effect of training ophthalmoscopic skills in XR on human vision and comfort.

2 Method

2.1 Participants

Sixteen ophthalmology and optometry students with a mean age of 25 ± 5 years were included in the study. None of them had previous experience of using the XR-based ophthalmoscope simulator. All participants underwent an eye examination to confirm that the following inclusion criteria are met:

- visual acuity at far and near is at least 1.0 (in decimal units), measured using a Snellen eye chart;
- no binocular vision and accommodation disorders.

The study was approved by the Ethics Committee of the University of Latvia and conducted according to the principles of the Declaration of Helsinki. Participants provided written informed consent before participating in the study. Confidentiality and anonymity were ensured throughout the study by assigning a unique identification code to each participant.

2.2 Experimental Setup

The training of ophthalmoscopic skills was performed using the Eyesi indirect ophthalmoscope simulator (VRmagic Holding AG, Mannheim, Germany). It utilizes near-eye displays mounted in the headset that allows the front-located head model to be seen as a three-dimensional simulation of a patient's face. The participants wore the headset and had access to two magnifying lenses (16× and 20×) that were required to be correctly placed in front of the head model to examine the eye fundus. Additionally, there was a touch screen that displayed the user interface and a live view of the examination.

During the 40-min training, the participants' task was to identify location of all the colored geometric shapes arranged in the eye fundus images (see Fig. 1). The abstract cases built in the simulator program were used for this purpose.

Fig. 1. A person using the Eyesi indirect ophthalmoscope simulator.

Before and immediately after the training in XR, the set of measurements was performed in the following order:

1. objective measures of visual functions,
2. subjective measures of visual functions,
3. subjective assessment of discomfort.

The eye refraction at far (6 m), accommodation response at near (33 cm), and microfluctuations at both near and far distances were assessed as objective measures using the PowerRef3 device. The PowerRef3 eccentric photo refractometer (Plusoptix GmbH, Nuremberg, Germany) provided real-time measurements of eye refraction at a sampling rate of 50 Hz. The spherical measurements were conducted in 0.01 D steps, and to ensure precise measurements, the light condition was regulated individually to maintain a pupil size between 4 mm and 8 mm. Detailed information about the calculations can be found in our previous studies [9, 10].

The subjective measures of visual functions including accommodation amplitude, near point of convergence, and break points of near fusional reserves (base in and base out) were assessed using standard clinical tests as described in [11] by using RAF ruler and prism bar. All visual functions were analyzed using a non-parametric test, the Wilcoxon signed-rank test [12], to identify any significant changes in response to training in XR. The statistical significance was set to $p < 0.05$ (95% confidence interval).

The subjective assessment of discomfort was performed using the symptom questionnaire. It included eleven questions. Seven symptoms were selected from the Sickness Simulator Questionnaire developed by [13] and broadly adopted for the assessment of user experience in the digital immersive environments [14, 15]. Those were fatigue, headache, eye strain, nausea, difficulty concentrating, blurred vision, and dizziness (eyes open). Four more symptoms (eye pain, dry eyes, double vision, and drowsiness) were included inspired by the research done by [7]. For each symptom, the participants assessed its relevance and severity by choosing one of the responses: none (0), slight (1), moderate (2), severe (3), and very severe (4). For the further analysis of changes in user comfort, difference in symptoms (after – before) was calculated and described as slight improvement (-1), no changes (0), slight deterioration (+1), moderate deterioration (+2), and severe deterioration (at least $+ 3$).

3 Results

The visual functions of participants were assessed using objective and subjective methods before and after 40 min of training in XR. The corresponding results are shown in Table 1.

Table 1. Summary of the changes in visual functions before and after using the Eyesi indirect ophthalmoscope simulator for 40 min.

Visual functions	Before	After	p-value
Eye refraction at far, diopter	−0.43 (0.62)	−0.39 (0.81)	0.28
Accommodation lag, diopter	1.8 (0.7)	1.7 (0.7)	0.04
Accommodation microfluctuations at near, diopter	0.11 (0.11)	0.13 (0.07)	<0.001
Accommodation microfluctuations at far, diopter	0.05 (0.02)	0.06 (0.05)	0.25
Accommodation amplitude, diopter	9.0 (3.5)	10.0 (4.5)	0.19
Near point of convergence, cm	5.5 (3.0)	5.8 (3.8)	0.54
Near fusional reserves (base in), prism diopter	16 (5)	16 (5)	0.79
Near fusional reserves (base out), prism diopter	18 (20)	23 (22)	0.21

Data presented as median (interquartile range). P-values calculated using Wilcoxon signed-rank test.

Overall, the use of the Eyesi indirect ophthalmoscope simulator led to a change in accommodation lag, suggesting that the accommodation response was slightly stronger after training in XR compared to the pre-XR situation. While considerable differences in accommodation microfluctuations at near were noted, the accommodation microfluctuations at far remained more stable on average. Moreover, similar to eye refraction at far, no significant differences were detected in subjective measures of visual functions.

In addition to the objective and subjective measures of visual functions, we explored the relevance and severity of different symptoms before and after the use of the simulator. As seen in Fig. 2, most participants showed slight to no changes in the symptoms, while moderate deterioration in user comfort was primarily associated with headache, dry eyes, and eye strain. Across all participants, there were several cases of severe deterioration in user comfort, specifically related to nausea, difficulty concentrating, dry eyes, and double vision. Nevertheless, some participants initially reported minor discomfort, which later decreased. For instance, there were four cases of slightly reduced drowsiness. As there was one case of reported visual disturbance (slightly blurred and double vision) before the training in XR, it should be emphasized that all participants underwent an eye examination and met the inclusion criteria, but they were also free to report any symptoms they experienced just prior to the start of the training.

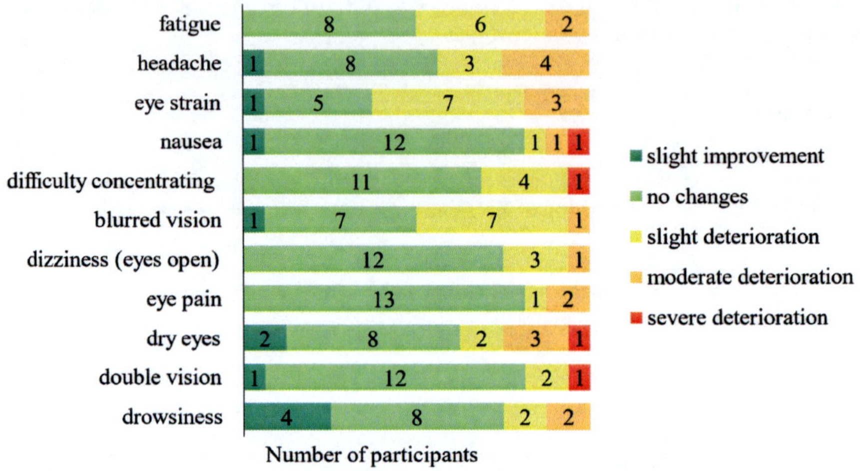

Fig. 2. The changes in user comfort following the 40-min use of the Eyesi indirect ophthalmoscope simulator are shown for each of the eleven questionnaire items.

4 Discussion

The XR-based simulations are considered valuable for training clinical skills, and their popularity grows. As more people start using XR technology for different purposes, it becomes increasingly important to understand how to use it properly to avoid potential harm to visual health.

We conducted a study to assess the user experience of ophthalmology and optometry students training clinical skills in XR. Our main focus was to elucidate how the use of the Eyesi indirect ophthalmoscope simulator for 40 min affected visual functions and user comfort in individuals with normal or corrected-to-normal visual acuity. As a result, statistically significant ($p < 0.05$) changes have been found in several visual functions. Specifically, the accommodation lag decreased, and accommodation microfluctuations at near increased after the training in XR compared to the pre-XR baseline. Moreover, user comfort moderately to severely worsened for some participants.

Over the years, extensive research has been conducted to investigate the impact of using near-eye displays on the functionality of the human visual system. However, a consensus has not been reached due to the variability in the methods and results of these studies, e.g. [5–7, 11]. Our study indicates that the objective measures of ocular accommodation may be more useful for the assessment of user experience and visual ergonomics of the digital immersive environments compared to subjective measures.

Additionally, our study highlighted that changes in user comfort when using near-eye displays vary among individuals, reflecting the variability of their reactions. While most participants had moderate to no changes in symptoms, some experienced notable discomfort after training ophthalmoscopic skills in XR for 40 min. This finding has important implications. Firstly, before incorporating XR technology with near-eye displays into ophthalmology and optometry program curricula, it is necessary to verify if students' visual functions and sensitivity to visual stress allow for their use. Secondly, students should be advised to actively monitor their comfort and take regular breaks during training. To further advance our understanding of user experience in XR, future research should investigate the association between measures of visual functions and changes in user comfort.

Acknowledgement. The research was funded by the Latvian Council of Science (project No. lzp-2021/1-0399). We thank Toms Grinbergs (University of Latvia) for taking the photo used as Fig. 1.

References

1. Chou, J., Kosowsky, T., Payal, A.R., Gonzalez Gonzalez, L.A., Daly, M.K.: Construct and face validity of the Eyesi indirect ophthalmoscope simulator. Retina **37**(10), 1967–1976 (2017)
2. Chu, R., Lu, E., Lee, J.: Assessment of competency following use of Eyesi indirect ophthalmoscope simulators within a first-year optometric curriculum. Optom. Educ. **45**(2), 1–6 (2020)
3. Ricci, L., Ferraz, C.: Ophthalmoscopy simulation: advances in training and practice for medical students and young ophthalmologists. Adv. Med. Educ. Pract. **8**, 435–439 (2017)
4. Tso, H.L., Young, J., Yung, C.W.: Comparing Eyesi virtual reality simulator and traditional teaching methods for direct ophthalmoscopy: Students' perspectives at Indiana University School of Medicine. J. Acad. Ophthalmol. **13**(01), e66–e72 (2021)
5. Banstola, S., Hanna, K., O'Connor, A.: Changes to visual parameters following virtual reality gameplay. Br. Irish Orthop. J. **18**(1), 57–64 (2022)
6. Szpak, A., Michalski, S.C., Loetscher, T.: Exergaming with Beat Saber: An investigation of virtual reality aftereffects. J. Med. Internet Res. **22**(10), e19840 (2020)
7. Yoon, H.J., Kim, J., Park, S.W., Heo, H.: Influence of virtual reality on visual parameters: Immersive versus non-immersive mode. BMC Ophthalmol. **20**(1), 200 (2020)
8. Pladere, T., Svarverud, E., Krumina, G., Gilson, S.J., Baraas, R.C.: Inclusivity in stereoscopic XR: Human vision first. Front. Virt. Real. **3**, 1006021 (2022)
9. Panke, K., Pladere, T., Velina, M., Svede, A., Ikaunieks, G., Krumina, G.: Ocular performance evaluation: how prolonged near work with virtual and real 3D image modifies our visual system. In: Proceedings of the 2nd International Conference on Applications of Intelligent Systems (APPIS 2019), vol. 14, pp. 1–5. Association for Computing Machinery, New York (2019)
10. Panke, K., Pladere, T., Velina, M., Svede, A., Krumina, G.: Objective user visual experience evaluation when working with virtual pixel-based 3D system and real voxel-based 3D system. Photonics **6**, 106 (2019)

11. Livitcuka, R., Alksnis, R., Pladere, T.: Impact of interpupillary distance mismatch on visual aftereffects of virtual reality gameplay. In: Proceedings of SPIE 12449, Optical Architectures for Displays and Sensing in Augmented, Virtual, and Mixed Reality (AR, VR, MR) IV, 1244925 (2023)
12. Rosner, B., Glynn, R.J., Lee, M.-L.T.: The Wilcoxon signed rank test for paired comparisons of clustered data. Biometrics **62**(1), 185–192 (2006)
13. Kennedy, R. S., Lane, N. E., Berbaum, K. S., Lilienthal, M. G.: Simulator sickness questionnaire: An enhanced method for quantifying simulator sickness. Int. J. Aviat. Psychol. **3**(3), 203–220 (1993)
14. Balk, S.A., Bertola, M.A., Inman, V.W.: Simulator sickness questionnaire: Twenty years later. In: Proceedings of the Seventh International Driving Symposium on Human Factors in Driver Assessment, Training and Vehicle Design, Bolton Landing, pp. 257–263. Public Policy Center, University of Iowa, New York, Iowa (2013)
15. Bimberg, P., Weissker, T., Kulik, A.: On the usage of the Simulator Sickness Questionnaire for virtual reality research. In: 2020 IEEE Conference on Virtual Reality and 3D User Interfaces Abstracts and Workshops (VRW), Atlanta, pp. 464–467 (2020)

Retrieving the Refractive Index of a Biological Material via Symbolic Regression

Julián Sierra-Vélez[1](✉), Demetrio Macias[1], Alexandre Vial[1], and Marco Antonio Giraldo[2]

[1] Laboratory Light, nanomaterials & nanotechnologies - L2n, University of Technology of Troyes & CNRS EMR 7004, 12 Rue Marie Curie, 10004 Troyes, France
julian.sierra@utt.fr

[2] Biophysics Group - Institute of Physics, University of Antioquia, Calle 70 N° 52-21 A.A. 1226, 050010 Medellin, Colombia

Abstract. Modeling the optical properties of biological structures allows the development of novel biocompatible materials. Using two different approaches, we were able to extract a closed-form mathematical expression that accurately models the refractive index of two distinct biological materials. Our approaches rely on Symbolic Regression (SR), a generic numerical implementation based on Genetic Programming (GP) that does not require any initial assumptions about the algebraic form of the model under search. We explore the possibilities and limitations of our SR approach, and the possible extension to solve more complex configurations.

Keywords: Genetic Programming · biomaterials · refractive index · reflectance · thin film

1 Introduction

The design of novel materials by mimicking biological structures, and by possibly controlling their optical response, has been quite a successful approach in the last two decades [1]. Optical biomimetics has yielded promising results in the development of bioinspired and highly engineered optical materials [2, 3]. Nevertheless, the optical properties of organic materials, as those that mimic human tissues have been poorly investigated. To better understand the mechanical and optical properties of those materials, which would lead to further improvement, a number of theoretical methods have been developed.

Recently, Machine Learning (ML) methods have been used in biomaterials research. Approaches based on Deep Learning (DL) have proven to be very effective in the development of innovative biomaterials [4]. However, DL does not provide a mathematical expression that relates the different variables implicated in the specific studied system, and it requires knowing the algebraic form of the model. We propose the alternative use of Symbolic Regression (SR) analysis. In this method, a closed-form expression (i.e. an equation) is obtained, and no assumptions about the system are required. Recently,

Y. Dekhtyar and I. Saknite (Eds.): NBC 2023, IFMBE Proceedings 89, pp. 377–384, 2023.
https://doi.org/10.1007/978-3-031-37132-5_48

SR was successfully used to model the optical properties of transparent and absorbing dielectrics [5]. Motivated by the potential of the method, we consider it a suitable approach for the study of organic biomaterials.

This work is organized as follows: Sect. 2 is devoted to the formulation of the problem and to briefly outline the methods we employ throughout our numerical simulations. In Sect. 3 we discuss typical results obtained when we retrieve the optical properties of the materials considered. We give our concluding remarks in Sect. 4.

2 Formulation of the Problem and Methodology

2.1 Statement of the Problem

Although there would not be evident limitations considering biological structures such as hair [6], nails, or skin [7], their complex morphology would increase the complexity of the problem, as specific methods are required to compute their spectral response. This would deviate the attention from the main objective of this contribution, which is to illustrate the potential of SR when use to find the closed-form expression that characterizes the optical properties of a given material. Let us then consider the geometry shown in Fig. 1.

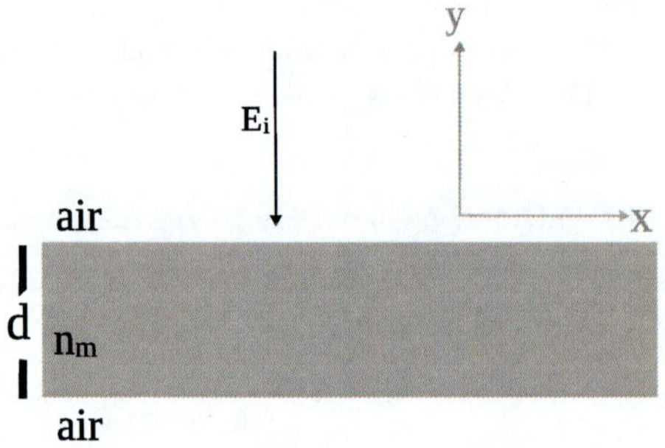

Fig. 1. Scheme of the structure studied in this project.

The unsupported thin film in Fig. 1 is made of a non-absorbing, isotropic, homogeneous, and dispersive material with a refractive index $n_m(\lambda)$. The film has a thickness d and it is surrounded by air ($n_{air} = 1.0$). Also, we consider the geometry invariant along the y direction. Our objective is then to retrieve a closed-form expression that models the dispersive refractive index of the film in Fig. 1.

2.2 Symbolic Regression

Symbolic Regression (SR) is a method that finds a mathematical expression that relates the data within a given set, without previous knowledge of the data's nature [8]. The choice of SR, as the most suitable method for our study, stems from the fact that it does not require any preliminary hypothesis about the algebraic form of the closed-form expression to be retrieved.

Although detailed descriptions of the operational principles of SR can be found elsewhere [8], for the sake of completeness we consider it convenient to briefly outline them in this section. SR is a technique that identifies a mathematical expression that fits interval-valued data [9]. The SR scheme to be used throughout this work is based on Genetic Programming (GP) [10] and we employ an open-source Python machine-learning library known as "gplearn" [11].

The flux diagram in Fig. 2 shows the main stages of the regression process.

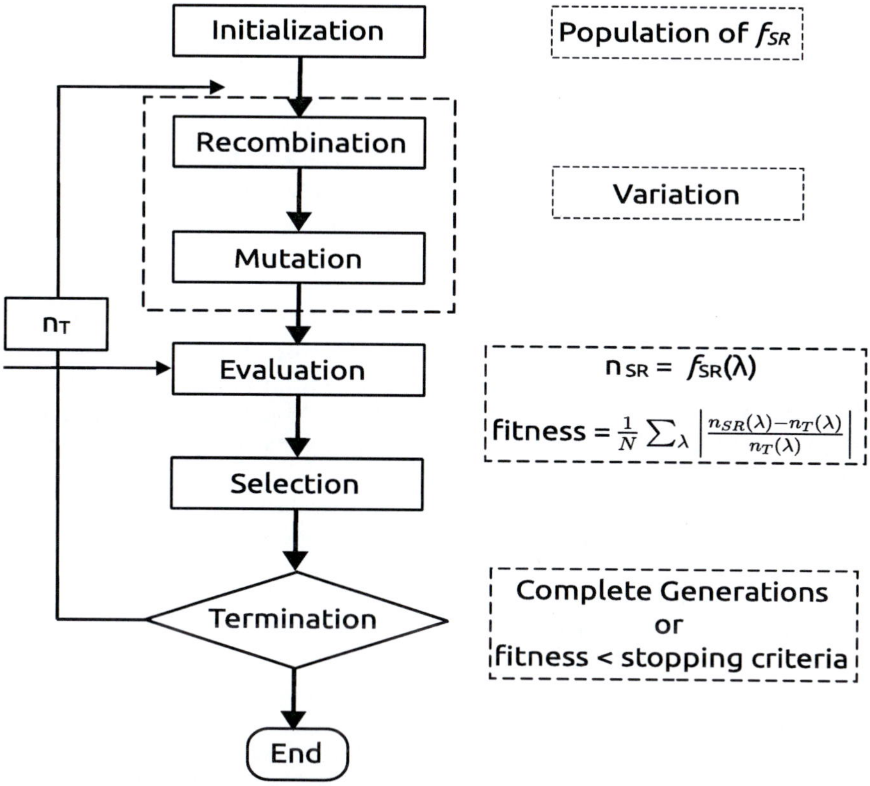

Fig. 2. Flux diagram of the SR optimization process based on Fig. 1 after [5].

The first step is to randomly generate an initial population of size μ composed by functions f_{SR} that, in the present case, represent the physical dispersion model we sought to retrieve. Each f_{SR} is defined by a combination of variables, constants, and allowed operations such as addition, subtraction, multiplication, or division. Some elements from this initial population are then randomly selected to undergo genetic operations of variation, such as recombination and mutation [9], to allow the evolution of the initially predicted functions.

Next, each modified function f_{SR} is evaluated on the problem's domain (λ in this problem), and the refractive indices n_{SR} predicted by the different f_{SR} are compared with the reference or target data n_T. The metric used to perform the comparison is the Mean Absolute Percentage Error (MAPE), which assigns a fitness value Eq. 1 to each f_{SR}.

$$fitness = \frac{1}{N}\sum_{\lambda}\left|\frac{n^{SR}(\lambda) - n^{T}(\lambda)}{n^{T}(\lambda)}\right| \tag{1}$$

where N are the number of wavelengths λ considered.

When the fitness value matches a stopping criterion or a maximum number of generations is reached, we considered that the closed-form expression that models the refractive index had been found.

3 Results

In this section we assess the performance of our SR scheme. Although for space reasons we only consider two different biological materials, the results of extensive numerical simulations suggest that the application of our SR approach could be extended to more complex cases.

In order to proceed in a gradual way, we divide this section into two case studies, where the aim of both is to find the closed-form expression that models the refractive index of the material under consideration. In the first case, that we denote as "direct approach", we search for the model directly from the refractive index models reported in the literature. In the second case, we search for the model from spectral data, e.g. reflectance, transmittance or absorption spectra, which often are the only experimental information that has been reported. We denote this second case as "indirect approach". This is due to the fact that the fitness function compares the target spectrum, generated considering the reported data, with the spectra generated with the refractive indices predicted by the SR.

We make use of Eq. 1 to measure the closeness between the target indices and those predicted by the SR. Also, to facilitate the visualization of our results, we establish the following notation: m denotes the materials, $n_m^T(\lambda)$ is the target index, $n_m^{SR\epsilon}(\lambda)$ and $n_m^{SRR}(\lambda)$ respectively represent the closed-form expressions found through the direct and the indirect approaches.

3.1 Target Data

We take as target the refractive index of two distinct biological materials: keratin and chitin. The refractive indices of these non absorbing materials are described by Cauchy's equation

$$n_m(\lambda) = A + \frac{B}{\lambda^2} \tag{2}$$

where A and B are distinct real coefficients. For keratin, we use the values $A = 1.532$ and $B = 5.89 \times 10^3 nm^2$ obtained from white goose feathers [12]. For chitin, the values $A = 1.517$ and $B = 8.80 \times 10^3 nm^2$ were used, which were obtained from glass scales of the butterfly [13].

3.2 Direct Approach

In this case we search for the mathematical expression for the refractive index provided by Eq. 2. The closed-form expressions, retrieved through the SR, for keratin and chitin are respectively given by Eqs. 3 and 4

$$n_{ke}^{SRR}(\lambda) = 1.522 + \frac{15.718}{\lambda} + \frac{162.333}{\lambda^2} \tag{3}$$

$$n_{ch}^{SRR}(\lambda) = 1.495 + \frac{28.171}{\lambda} \tag{4}$$

Although is clear that Eqs. 3 and 4 involve fractions with the wavelength as the denominator, their algebraic forms are not identical to that of Eq. 2. This could be interpreted as a failure of the SR scheme. However, the visual comparison in Fig. 3 shows good agreement between the target indices and those predicted by the direct approach. This means that SR provides a closed-form expression that accurately models the dispersive behavior of the material considered. This fact can be useful in situations where a well established model is not available or the existing models are not well suited for the problem studied.

3.3 Indirect Approach

As mentioned in previous paragraphs, in certain situations one does not have access to the ellipsometric data. It seems thus well worth exploring the possibilities of SR when searching for the refractive index using spectral information. In this case we take as target the reflectance spectra generated when the structure in Fig. 1 is composed by keratin and chitin.

In order to generate the spectra required for the indirect approach, we assume that the structure in Fig. 1 is illuminated with a plane wave at normal incidence. Also, we considered only visible light with wavelengths in the range of $\lambda = [400,750]$ nm. As stated previously, we consider non-absorbing materials. Therefore, we can compute the reflectance spectra using Fresnel equations [14].

Let us note $R_m^T(\lambda)$ is the target reflectance spectrum generated with the target data (Eq. 2), and $R_m^{SR}(\lambda)$ is the reflectance spectrum generated with each index $n_m^{SRR}(\lambda)$ predicted by the SR scheme. We can then rewrite the fitness function (Eq. 1) as

$$fitness = \frac{1}{N}\sum_{\lambda}\left|\frac{R_m^{SR}(n_m|\lambda) - R_m^T(n_m|\lambda)}{R_m^T(n_m|\lambda)}\right| \tag{5}$$

where the explicit dependence of $R_m(\lambda)$ on the refractive index n_m emphasizes the indirect nature of this approach. That is, Eq. 5 compares the closeness between the spectrum generated with the target index and the spectrum generated with the dispersion models predicted by the SR.

The closed-form expressions retrieved from spectral information are given by Eqs. 6 and 7 for keratin and chitin, respectively.

$$n_{ke}^{SRR}(\lambda) = 1.511 + \frac{21.519}{\lambda} + \frac{0.007}{\lambda^2} \tag{6}$$

$$n_{ch}^{SRR}(\lambda) = 1.501 + \frac{27.116}{\lambda} \tag{7}$$

As for the direct approach, the visual comparison in Fig. 3 shows good agreement between the target data, blue dotted line, and the predictions of the SR, green crosses, in spite of the difference in the algebraic form.

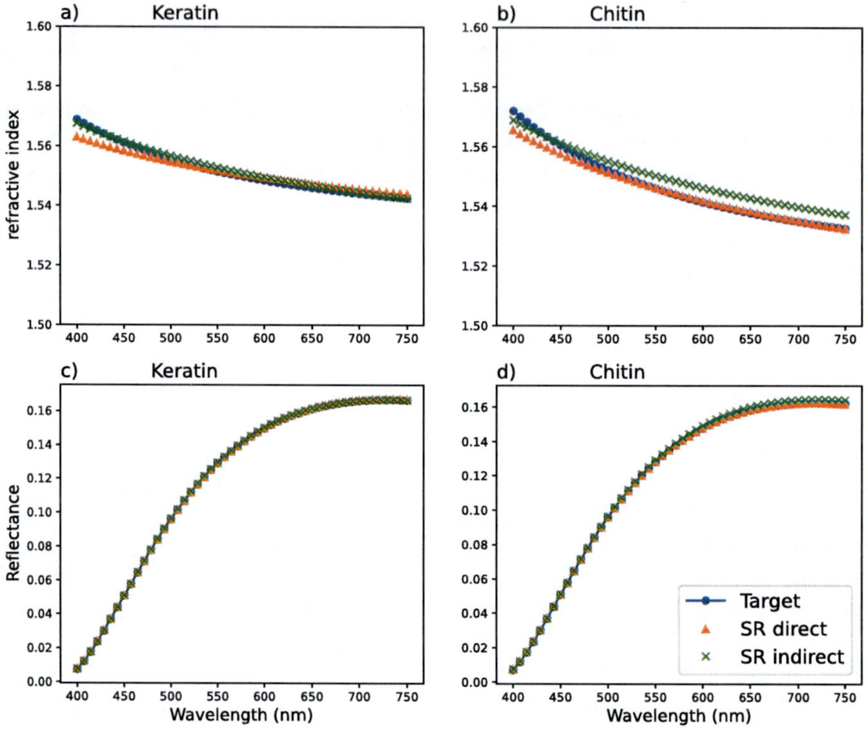

Fig. 3. Refractive indexes of a) keratin and b) chitin for the target data [12, 13] - blue dotted line, and the values obtained by direct and indirect approaches of SR, orange triangles and green crosses respectively. Reflectance spectrum of c) keratin and d) chitin obtained following Eq. 5, for the target and retrieved refractive indexes.

It is important to mention, that despite the good agreement shown in Fig. 3(a, b), none of the approaches studied provides a perfect fit between the target data and the indices predicted by the SR scheme. This somewhat expected result is as a direct consequence of SR's operational principles, rather than to the physical nature of the processed data. That is, throughout the regression process, the SR will search for the best combination of operators that fits the target data, independently of its kind. Nevertheless, as is the case for other metaheuristic methods, it is not possible to guarantee convergence to the global optimum but to its neighborhood.

4 Conclusions

In this study, we have successfully retrieved the closed-form mathematical expressions that model the refractive index of keratin and chitin by means of SR. There are no evident limitations to extend the potential of this method to more complex configurations, such as biological tissues that can be modeled as multilayers. Furthermore, mixed media, which more accurately represent actual tissue, could also be modeled as having an effective refractive index that depends on their relative concentrations.

Our findings have several implications for the modeling of biological structures. SR enables the exploration of a vast search space of possible models, which can lead to the discovery of previously unknown relationships. This has potential applications in prosthetics and the cosmetic industry, where more realistic and bio-compatible materials can be produced. However, it is important to note that the use of SR is not without limitations, as it may suffer from overfitting and requires careful tuning of its meta-parameters.

In conclusion, our study highlights the potential of SR as an effective tool for the modeling and understanding of biological structures, with applications in bio-optics and beyond. The ability to generate closed-form expressions that provide physical insights into the underlying properties of the biological system under study is a valuable asset in the field of optical modeling. Future work can build on our findings by exploring other materials and including other physical parameters in the SR scheme.

References

1. Sanchez, C., Arribart, H., Giraud Guille, M.: Biomimetism and bioinspiration as tools for the design of innovative materials and systems. Nat. Mater. **4**, 277–288 (2005)
2. Tadepalli, S., Slocik, J.M., Gupta, M.K., Naik, R.R., Singamaneni, S.: Bio-optics and bio-inspired optical materials. Chem. Rev. (2017)
3. Yu, K., et al.: Biomimetic optical materials: Integration of nature's design for manipulation of light. Prog. Mater. Sci. **58**(6), 825–873 (2013)
4. Suwardi, A., et al.: Machine learning-driven biomaterials evolution. Adv. Mater. **34**, 1 (2022)
5. Li, Q., Macias, D., Vial, A.: Modeling the optical properties of transparent and absorbing dielectrics by means of symbolic regression. Opt. Exp. **30**(23), 41862 (2022)
6. Jones, L.N.: Hair structure anatomy and comparative anatomy. Clin. Dermatol. **19**(2), 95–103 (2001)
7. Lai-Cheong, J.E., McGrath, J.A.: Structure and function of skin, hair and nails. Medicine (United Kingdom) **41**(6), 317–320 (2013)
8. Billard, L., Edwin, D.: Symbolic regression analysis. In: Classification, Clustering, and Data Analysis: Recent Advances and Applications. Springer, Heidelberg (2002)
9. Poli, R., et al.: A Field Guide to Genetic Programming (2008). ISBN:9781409200734
10. Koza, J.R.: Genetic programming as a means for programming computers by natural selection. Statist. Comput. **4**(2), 87–112 (1994)
11. Stephens, T.: Genetic Programming in Python, with a scikit-learn inspired API: gplearn (2016)
12. Stavenga, D., et al.: High refractive index of melanin in shiny occipital feathers of a bird of paradise. Light. Sci. Appl. **4**, e243 (2015)
13. Giraldo, M.A., Stavenga, D.G.: Brilliant iridescence of Morpho butterfly wing scales is due to both a thin film lower lamina and a multilayered upper lamina. J. Comp. Physiol. A **202**, 381–388 (2016)

14. Born, M., Wolf, E.: Principles of Optics: Electromagnetic Theory of Propagation, Interference and Diffraction of Light. Elsevier (2013)

Author Index

Y. Dekhtyar and I. Saknite (Eds.): NBC 2023, IFMBE Proceedings 89, pp. 385–387, 2023.
https://doi.org/10.1007/978-3-031-37132-5